SOLIDWORKS 2020
Tutorial

A Step-by-Step Project Based Approach
Utilizing 3D Solid Modeling

David C. Planchard
CSWP & SOLIDWORKS Accredited Educator

PUBLICATIONS

SDC Publications
P.O. Box 1334
Mission, KS 66222
913-262-2664
www.SDCpublications.com
Publisher: Stephen Schroff

The Y14 ASME Engineering Drawing and Related Documentation Publications utilized in this text are as follows: ASME Y14.1 1995, ASME Y14.2M-1992 (R1998), ASME Y14.3M-1994 (R1999), ASME Y14.41-2003, ASME Y14.5-1982, ASME Y14.5-1999, and ASME B4.2. Note: By permission of The American Society of Mechanical Engineers, Codes and Standards, New York, NY, USA. All rights reserved.

Download all needed model files (SW-TUTORIAL-2020 folder) from the SDC Publication website (www.SDCpublications.com/downloads/978-1-63057-317-1).

Additional information references the American Welding Society, AWS 2.4:1997 Standard Symbols for Welding, Braising, and Non-Destructive Examinations, Miami, Florida, USA.

ISBN-13: 978-1-63057-317-1
ISBN-10: 1-63057-317-5

Printed and bound in the United States of America.

INTRODUCTION

SOLIDWORKS® 2020 Tutorial is written to assist students, designers, engineers and professionals who are new to SOLIDWORKS. The text provides a step-by-step project based learning approach featuring machined components with additional semester design projects.

Desired outcomes and usage competencies are listed for each chapter. The book is divided into four sections with 11 Chapters.

Chapter 1 - Chapter 5: Explore the SOLIDWORKS User Interface and CommandManager, Document and System properties, simple and complex parts and assemblies, proper design intent, design tables, configurations, multi-sheet, multi-view drawings, BOMs, and Revision tables using basic and advanced features.

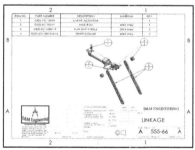

Follow the step-by-step instructions and develop multiple assemblies that combine over 100 extruded machined parts and components. Formulate the skills to create, modify and edit sketches and solid features.

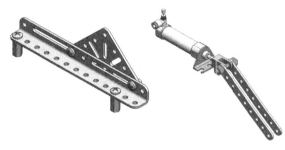

Learn the techniques to reuse features, parts and assemblies through symmetry, patterns, copied components, apply proper design intent, design tables, and configurations.

Chapter 6: Create the final ROBOT Assembly. The physical components and corresponding Science, Technology, Engineering and Math (STEM) curriculum are available from Gears Educational Systems. All assemblies and components for the final ROBOT assembly are provided.

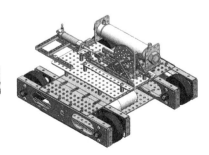

Chapter 7 - Chapter 10: Prepare for the Certified SOLIDWORKS Associate (CSWA) exam. The certification indicates a foundation in and apprentice knowledge of 3D CAD and engineering practices and principles.

Each chapter addresses one of the five categories in the CSWA exam: Drafting Competencies, Basic Part Creation and Modification, Intermediate Part Creation and Modification, Advanced Part Creation and Modification, Assembly Creation and Modification.

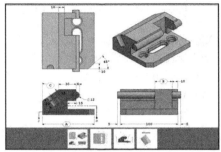

Chapter 11: Provide a basic understanding between Additive vs. Subtractive manufacturing. Discuss Fused Filament Fabrication (FFF), STereoLithography (SLA), and Selective Laser Sintering (SLS) printer technology. Select suitable filament material. Comprehend 3D printer terminology. Knowledge of preparing, saving, and printing a model on a Fused Filament Fabrication 3D printer. Information on the Certified SOLIDWORKS Additive Manufacturing (CSWA-AM) exam.

Download all needed model files from the SDC Publications website (www.sdcpublications.com) to a local hard drive.

About the Author

David Planchard is the founder of D&M Education LLC. Before starting D&M Education, he spent over 35 years in industry and academia holding various engineering, marketing, and teaching positions. He holds five U.S. patents. He has published and authored numerous papers on Machine Design, Product Design, Mechanics of Materials, and Solid Modeling. He is an active member of the SOLIDWORKS Users Group and the American Society of Engineering Education (ASEE). David holds a BSME, MSM with the following professional certifications: CCAI, CCNP, CSWA-SD, CSWSA-FEA, CSWA-AM, CSWP, CSWP-DRWT and SOLIDWORKS Accredited Educator. David is a SOLIDWORKS Solution Partner, an Adjunct Faculty member and the SAE advisor at Worcester Polytechnic Institute in the Mechanical Engineering department. In 2012, David's senior Major Qualifying Project team (senior capstone) won first place in the Mechanical Engineering department at WPI. In 2014, 2015 and 2016, David's senior Major Qualifying Project teams won the Provost award in the Mechanical Engineering department for design excellence. In 2018, David's senior Major Qualifying Project team (Co-advisor) won the Provost award in the Electrical and Computer Engineering department. Subject area: Electrical System Implementation of Formula SAE Racing Platform.

David Planchard is the author of the following books:

- **SOLIDWORKS® 2020 Reference Guide**, 2019, 2018, 2017, 2016, 2015, 2014, 2013, 2012, 2011, 2010, and 2009

- **Engineering Design with SOLIDWORKS® 2020**, 2019, 2018, 2017, 2016, 2015, 2014, 2013, 2012, 2011, 2010, 2009, 2008, 2007, 2006, 2005, 2004, and 2003

- **Engineering Graphics with SOLIDWORKS® 2020,** 2019, 2018, 2017, 2016, 2015, 2014, 2013, 2012, and 2011

- **SOLIDWORKS® 2020 Quick Start**, 2019, 2018

- **SOLIDWORKS® 2017 in 5 Hours with video instruction**, 2016, 2015, and 2014

- **SOLIDWORKS® 2020 Tutorial**, 2019, 2018, 2017, 2016, 2015, 2014, 2013, 2012, 2011, 2010, 2009, 2008, 2007, 2006, 2005, 2004, and 2003

- **Drawing and Detailing with SOLIDWORKS® 2014**, 2012, 2010, 2009, 2008, 2007, 2006, 2005, 2004, 2003, and 2002

- **Official Certified SOLIDWORKS® Professional (CSWP) Certification Guide Version 5: 2018 - 2020**, Version 4: 2015 - 2017, Version 3: 2012 - 2014, Version 2: 2012 - 2013, Version 1: 2010 - 2010

- **Official Guide to Certified SOLIDWORKS® Associate Exams: CSWA, CSWA-SD, CSWSA-FEA, CSWA-AM Version 4: 2017 - 2019**, Version 3: 2015 - 2017, Version 2: 2012 - 2015, Version 1: 2012 -2013

- **Assembly Modeling with SOLIDWORKS® 2012**, 2010, 2008, 2006, 2005-2004, 2003, and 2001Plus

- **Applications in Sheet Metal Using Pro/SHEETMETAL & Pro/ENGINEER**

Acknowledgements

Writing this book was a substantial effort that would not have been possible without the help and support of my loving family and of my professional colleagues. I would like to thank Professor John M. Sullivan Jr., Professor Jack Hall, Professor Mehul A. Bhatia and the community of scholars at Worcester Polytechnic Institute who have enhanced my life, my knowledge and helped to shape the approach and content to this text.

The author is greatly indebted to my colleagues from Dassault Systèmes SOLIDWORKS Corporation for their help and continuous support: Mike Puckett, Avelino Rochino, Yannick Chaigneau, Terry McCabe and the SOLIDWORKS Partner team.

Thanks also to Professor Richard L. Roberts of Wentworth Institute of Technology, Professor Dennis Hance of Wright State University, Professor Jason Durfess of Eastern Washington University and Professor Aaron Schellenberg of Brigham Young University - Idaho who provided vision and invaluable suggestions.

SOLIDWORKS certification has enhanced my skills and knowledge and that of my students. Thank you to Ian Matthew Jutras (CSWE), technical contributor, and Stephanie Planchard, technical procedure consultant.

Contact the Author

We realize that keeping software application books current is imperative to our customers. We value the hundreds of professors, students, designers, and engineers that have provided us input to enhance the book. Please contact me directly with any comments, questions or suggestions on this book or any of our other SOLIDWORKS books at dplanchard@msn.com or planchard@wpi.edu.

Note to Instructors

Please contact the publisher **www.SDCpublications.com** for classroom support materials (.ppt presentations, labs and more) and the Instructor's Guide with model solutions and tips that support the usage of this text in a classroom environment.

Trademarks, Disclaimer and Copyrighted Material

SOLIDWORKS®, eDrawings®, SOLIDWORKS Simulation®, SOLIDWORKS Flow Simulation, and SOLIDWORKS Sustainability are a registered trademark of Dassault Systèmes SOLIDWORKS Corporation in the United States and other countries; certain images of the models in this publication courtesy of Dassault Systèmes SOLIDWORKS Corporation.

Microsoft Windows®, Microsoft Office® and its family of products are registered trademarks of the Microsoft Corporation. Other software applications and parts described in this book are trademarks or registered trademarks of their respective owners.

The publisher and the author make no representations or warranties with respect to the accuracy or completeness of the contents of this work and specifically disclaim all warranties, including without limitation warranties of fitness for a particular purpose. No warranty may be created or extended by sales or promotional materials. Dimensions of parts are modified for illustration purposes. Every effort is made to provide an accurate text. The authors and the manufacturers shall not be held liable for any parts, components, assemblies or drawings developed or designed with this book or any responsibility for inaccuracies that appear in the book. Web and company information was valid at the time of this printing.

The Y14 ASME Engineering Drawing and Related Documentation Publications utilized in this text are as follows: ASME Y14.1 1995, ASME Y14.2M-1992 (R1998), ASME Y14.3M-1994 (R1999), ASME Y14.41-2003, ASME Y14.5-1982, ASME Y14.5-1999, and ASME B4.2. Note: By permission of The American Society of Mechanical Engineers, Codes and Standards, New York, NY, USA. All rights reserved.

Additional information references the American Welding Society, AWS 2.4:1997 Standard Symbols for Welding, Braising, and Non-Destructive Examinations, Miami, Florida, USA.

References

- SOLIDWORKS Help Topics and What's New, SOLIDWORKS Corporation, 2020.

- 80/20 Product Manual, 80/20, Inc., Columbia City, IN, 2012.

- Ticona Designing with Plastics - The Fundamentals, Summit, NJ, 2009.

- SMC Corporation of America, Product Manuals, Indiana, USA, 2012.

- Emerson-EPT Bearing Product Manuals and Gear Product Manuals, Emerson Power Transmission Corporation, Ithaca, NY, 2009.

- Emhart - A Black and Decker Company, On-line catalog, Hartford, CT, 2012.

During the initial SOLIDWORKS installation, you are requested to select either the ISO or ANSI drafting standard. ISO is typically a European drafting standard and uses First Angle Projection. The book is written using the ANSI (US) overall drafting standard and Third Angle Projection for drawings.

Screen shots in the book were made using SOLIDWORKS 2020 SP0 running Windows® 10.

Download all needed model files (SW-TUTORIAL-2020 folder) from the SDC Publication website (www.SDCpublications.com/downloads/978-1-63057-317-1). All templates, logos and model documents along with additional support materials for the book are available.

To obtain additional CSWA exam information, visit the SOLIDWORKS VirtualTester Certification site at https://SOLIDWORKS.virtualtester.com/.

Bracket
Chapter 2 Homework
Chapter 3 Homework
Chapter 4 Homework
Chapter 5 Homework
Chapter 6 Homework
Chapter 6 Models
Graph paper
LOGO
MY-TEMPLATES
Pneumatic Components
SOLIDWORKS CSWA Model Folder
Decimal - Millimeters - Points
simulation_theory_manual

Additional semester design projects are included in the exercise section of Chapter 6. Copy the components from the Chapter 6 Homework folder. View all components. Create an ANSI assembly document.

Chapter 6 Homework
Name
Bench Vice Assembly Project
Butterfly Valve Assembly Project
Drill Guide Assembly Project
Kant Twist Clamp Assembly Project
Pipe Vice Assembly Project
Pulley Assembly Project
Quick Acting Clamp Assembly Project
Radial Engine Assembly Project
Shaper Tool Head Assembly Project
Shock Assembly Project
Welder Arm Assembly Project

Insert and create all needed components and mates to assemble the assembly and to simulate proper movement per the provided avi file.

TABLE OF CONTENTS

The Instructor's information contains over 60 classroom presentations along with helpful hints, What's new, sample quizzes, avi files of assemblies, projects and all initial and final SOLIDWORKS model files.

Download all needed model files (SW-TUTORIAL-2020 folder) from the SDC Publication website (www.SDCpublications.com/downloads/978-1-63057-317-1). View the provided models to enhance the user experience.

The book provides information on creating and storing special Part, Assembly and Drawing templates in the MY-TEMPLATES folder. The MY-TEMPLATES folder is added to the New SOLIDWORKS Document dialog box. Talk to your IT department *before you set* any new locations on a network system. The procedure in the book is designed for your personal computer.

If you do not create the MY-TEMPLATE tab or the special part, drawing, or assembly templates, use the standard SOLIDWORKS default template and apply all of the needed document properties and custom properties.

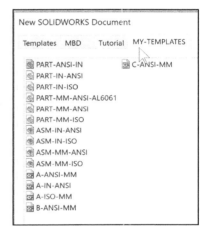

What is SOLIDWORKS?

SOLIDWORKS® is a mechanical design automation software package used to build parts, assemblies and drawings that takes advantage of the familiar Microsoft® Windows graphical user interface.

SOLIDWORKS is an easy to learn design and analysis tool (SOLIDWORKS Simulations, SOLIDWORKS Motion, SOLIDWORKS Flow Simulation etc.), which makes it possible for designers to quickly sketch 2D and 3D concepts, create 3D parts and assemblies and detail 2D drawings.

In SOLIDWORKS, you create 2D and 3D sketches, 3D parts, 3D assemblies and 2D drawings. The part, assembly and drawing documents are related. Additional information on SOLIDWORKS and its family of products can be obtained at their URL, www.SOLIDWORKS.com.

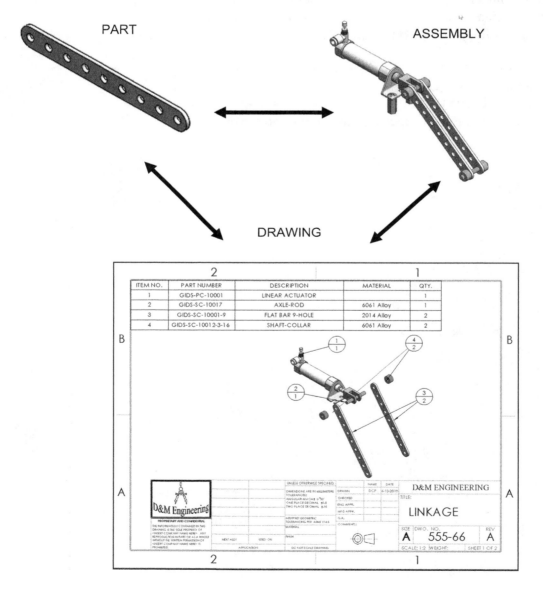

Features are the building blocks of parts. Use features to create parts, such as Extruded Boss/Base and Extruded Cut. Extruded features begin with a 2D sketch created on a Sketch plane.

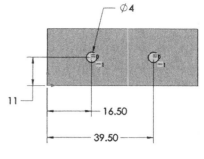

The 2D sketch is a profile or cross section. Sketch tools such as lines, arcs and circles are used to create the 2D sketch. Sketch the general shape of the profile. Add Geometric relationships and dimensions to control the exact size of the geometry.

Create features by selecting edges or faces of existing features, such as a Fillet. The Fillet feature rounds sharp corners.

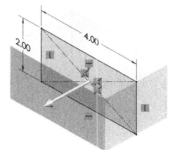

Dimensions drive features. Change a dimension, and you change the size of the part.

Apply Geometric relationships: Vertical, Horizontal, Parallel, etc. to maintain Design intent.

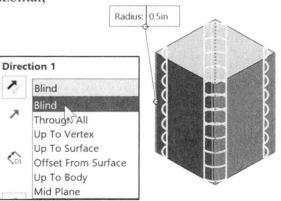

Create a hole that penetrates through a part (Through All). SOLIDWORKS maintains relationships through the change.

The step-by-step approach used in this text allows you to create parts, assemblies and drawings by doing, not just by reading.

The book provides the knowledge to modify all parts and components in a document.

Change is an integral part of design.

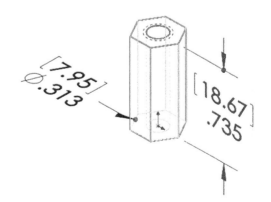

Design Intent

What is design intent? All designs are created for a purpose. Design intent is the intellectual arrangement of features and dimensions of a design. Design intent governs the relationship between sketches in a feature, features in a part and parts in an assembly.

The SOLIDWORKS definition of design intent is the process in which the model is developed to accept future modifications. Models behave differently when design changes occur.

Design for change. Utilize geometry for symmetry, reuse common features, and reuse common parts. Build change into the following areas that you create:

- **Sketch**

- **Feature**

- **Part**

- **Assembly**

- **Drawing**

When editing or repairing geometric relations, it is considered best practice to edit the relation vs. deleting it.

Design Intent in a sketch

Build design intent in a sketch as the profile is created. A profile is determined from the Sketch Entities. Example: Rectangle, Circle, Arc, Point, Slot etc. Apply symmetry into a profile through a sketch centerline, mirror entity and position about the reference planes and Origin. Always know the location of the Origin in the sketch.

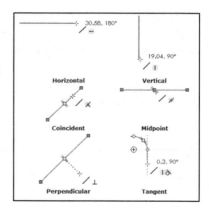

Build design intent as you sketch with automatic Geometric relations. Document the decisions made during the up-front design process. This is very valuable when you modify the design later.

A rectangle (Center Rectangle Sketch tool) contains Horizontal, Vertical and Perpendicular automatic Geometric relations.

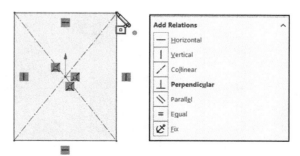

Apply design intent using added Geometric relations if needed. Example: Horizontal, Vertical, Collinear, Perpendicular, Parallel, Equal etc.

Example A: Apply design intent to create a square profile. Sketch a rectangle. Apply the Center Rectangle Sketch tool. Note: No construction reference centerline or Midpoint relation is required with the Center Rectangle tool. Insert dimensions to fully define the sketch.

Example B: If you have a hole in a part that must always be 16.5mm≤ from an edge, dimension to the edge rather than to another point on the sketch. As the part size is modified, the hole location remains 16.5mm≤ from the edge as illustrated.

Design intent in a feature

Build design intent into a feature by addressing End Conditions (Blind, Through All, UpToVertex, etc.), symmetry, feature selection, and the order of feature creation.

Example A: The Extruded Base feature remains symmetric about the Front Plane. Utilize the Mid Plane End Condition option in Direction 1. Modify the depth, and the feature remains symmetric about the Front Plane.

Example B: Create 34 teeth in the model. Do you create each tooth separately using the Extruded Cut feature? No.

Create a single tooth and then apply the Circular Pattern feature. Modify the Circular Pattern from 32 to 24 teeth.

Design intent in a part

Utilize symmetry, feature order and reuse common features to build design intent into a part. Example A: Feature order. Is the entire part symmetric? Feature order affects the part.

Apply the Shell feature before the Fillet feature and the inside corners remain perpendicular.

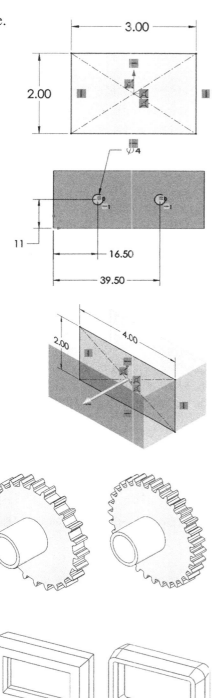

Design intent in an assembly

Utilizing symmetry, reusing common parts and using the Mate relation between parts builds the design intent into an assembly.

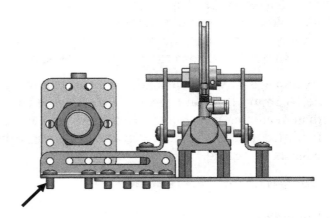

Example A: Reuse geometry in an assembly. The assembly contains a linear pattern of holes. Insert one screw into the first hole. Utilize the Component Pattern feature to copy the machine screw to the other holes.

Design intent in a drawing

Utilize dimensions, tolerance and notes in parts and assemblies to build the design intent into a drawing.

Example A: Tolerance and material in the drawing. Insert an outside diameter tolerance +.000/-.002 into the TUBE part. The tolerance propagates to the drawing.

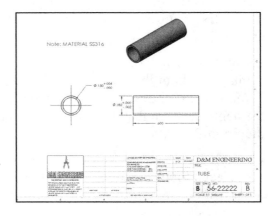

Define the Custom Property Material in the Part. The Material Custom Property propagates to your drawing.

🔍 Additional information on design process and design intent is available in SOLIDWORKS Help.

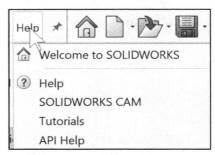

🔆 The book is designed to expose the new user to many tools, techniques and procedures. It may not always use the most direct tool or process.

Overview of Chapters

Chapter 1: **Overview of SOLIDWORKS and the User Interface**.

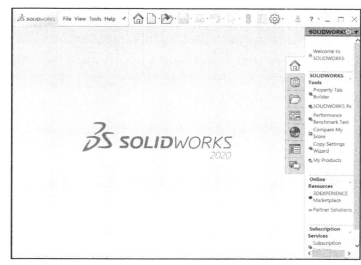

SOLIDWORKS is a design software application used to create 2D and 3D sketches, 3D parts and assemblies and 2D drawings.

Chapter 1 introduces the user to the SOLIDWORKS Welcome dialog box, User Interface (UI) and the CommandManager: Menu bar toolbar, Menu bar menu, Drop-down menus, Context toolbars, Consolidated drop-down toolbars, System feedback icons, Confirmation Corner, Heads-up View toolbar, Document Properties and more.

Start a new SOLIDWORKS session. Create a new part. Open an existing part and view the created features and sketches using the Rollback bar. Design the part using proper design intent.

Chapter 2: **Parts and Assembly Creation**.

Create three parts: AXLE, SHAFT-COLLAR and FLATBAR

Understand and apply the following Sketch tools: Circle, Smart dimension and Centerpoint Straight Slot.

Insert geometric relations and dimensions: Equal, Vertical, Horizontal, Parallel, Perpendicular, Coincident and MidPoint.

Utilize the following features: Extruded Boss/Base, Extruded Cut and Linear Pattern.

Create the LINKAGE assembly. The LINKAGE assembly utilizes the provided AirCylinder assembly. Insert the following Standard mates: Coincident, Concentric and Parallel.

Utilize the Standard and Quick mate procedure.

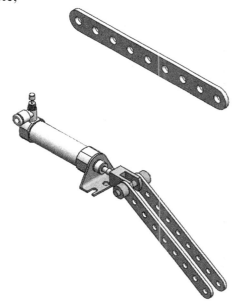

Chapter 3: **Front Support Assembly**.

Chapter 3 introduces various Sketch planes to create parts.
The Front, Top and Right Planes each contain the Extruded
Boss/Base feature for the TRIANGLE, HEX-STANDOFF
and ANGLE-13HOLE parts.

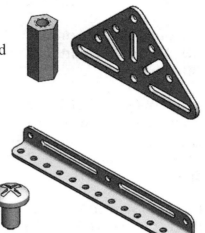

Utilize Geometric relationships in your sketch.

Create the SCREW part using the following features:
Revolved Base, Extruded Cut, Fillet and Circular
Pattern.

Create the FRONT-SUPPORT assembly.

Utilize additional parts from the Web or the provided
components to create the RESERVOIR SUPPORT
assembly in the chapter exercises.

Apply the following mate types: Concentric, Coincident,
Parallel, Distance, Gear and Cam.

Chapter 4: **Fundamentals of Drawing**.

Chapter 4 covers the development of a customized Sheet
format and Drawing templates.

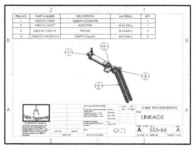

Review the differences between the Edit Sheet and the
Edit Sheet Format modes.

Develop a company logo from a bitmap or picture file.

Create a FLATBAR drawing.

Insert dimensions created from the part features.

Create a LINKAGE assembly drawing with multiple
views.

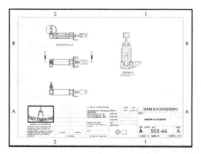

Develop and incorporate a Bill of Materials into the
drawing Custom Properties in the parts and assemblies.

Add information to the Bill of Materials
in the assembly drawing.

Insert a Design Table to create multiple
configurations of parts and assemblies.

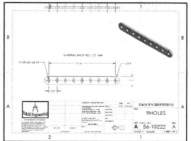

Chapter 5: Advanced Features.

Chapter 5 focuses on creating six parts for the PNEUMATIC-TEST-MODULE Assembly: WHEEL, HOOK, WHEEL, HEX-ADAPTER, AXLE-3000 and SHAFTCOLLAR-500.

Obtain the ability to reuse geometry by modifying existing parts and to create new parts.

Knowledge of the following SOLIDWORKS features: Plane, Lofted Base, Extruded Cut, Swept Base, Dome, Thread, Extruded Boss/Base, Revolved Cut, Extruded Cut, Circular Pattern, Axis, Instant3D, Hole Wizard, Advanced Hole, Split Line and Thread Wizard.

Chapter 6: PNEUMATIC-TEST-MODULE Assembly and Final ROBOT Assembly.

Chapter 6 focuses on the PNEUMATIC-TEST-MODULE Assembly and the final ROBOT Assembly.

Create the WHEEL-AND-AXLE assembly. First, create the 3HOLE-SHAFTCOLLAR assembly and the 5HOLE-SHAFTCOLLAR assembly.

Insert the WHEEL part, AXLE 3000 part, HEX-ADAPTER part and SHAFTCOLLAR-500 part.

Insert the FLAT-PLATE part that was created in the Chapter 3 exercises. Insert the LINKAGE assembly and add components: HEX-STANDOFF, AXLE and SHAFT-COLLAR.

Insert the AIR-RESERVOIR-SUPPORT assembly. Insert the SCREW part. Utilize the Pattern Driven Component Pattern tool and the Linear Component Pattern tool.

Insert the FRONT-SUPPORT assembly and apply the Mirror Components tool to complete the Pneumatic Test Module Assembly.

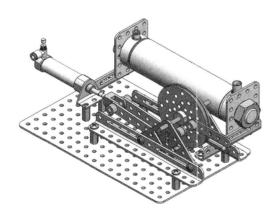

Create the final ROBOT Assembly as illustrated with the Robot-platform sub-assembly, PNEUMATIC-TEST-MODULE sub-assembly, basic_integration sub-assembly and the HEX-ADAPTER component. Add additional components in the chapter exercises.

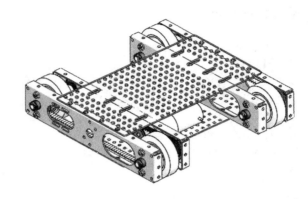

Learn the process to work with multiple documents between parts and assemblies and to apply the following Assembly tools: Insert Component, Standard Mates: Concentric, Coincident, and Parallel, Linear Component Pattern, Pattern Driven Component Pattern, Circular Component Pattern, Mirror Components and Replace Components.

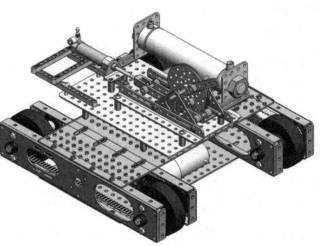

Download all needed model files (SW-TUTORIAL-2020 folder) from the SDC Publication website (www.SDCpublications.com/downloads/978-1-63057-317-1). All assemblies and components for the final ROBOT assembly are provided.

- Bracket
- Chapter 2 Homework
- Chapter 3 Homework
- Chapter 4 Homework
- Chapter 5 Homework
- Chapter 6 Homework
- Chapter 6 Models
- Graph paper
- LOGO
- MY-TEMPLATES
- Pneumatic Components
- SOLIDWORKS CSWA Model Folder
- Decimal - Millimeters - Points
- simulation_theory_manual

Additional projects are included in the exercise section of the chapter. Copy the components from the Chapter 6 Homework folders. View all components.

Create an ANSI assembly document. Insert and create all needed components and mates to assemble the assembly and to simulate proper movement per the provided avi file.

Chapter 6 Homework

Name

- Welder Arm Assembly Project
- Shock Assembly Project
- Quick Acting Clamp Assembly Project
- Radial Engine Assembly Project
- Shaper Tool Head Assembly Project
- Kant Twist Clamp Assembly Project
- Pipe Vice Assembly Project
- Pulley Assembly Project
- Bench Vice Assembly Project
- Butterfly Valve Assembly Project
- Drill Guide Assembly Project

Chapter 7 - 10: Introduction to the Certified SOLIDWORKS Associate (CSWA) Exam.

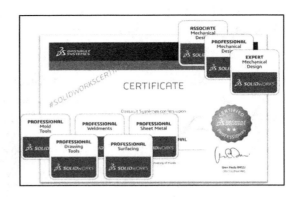

DS SOLIDWORKS Corp. offers various stages of certification representing increasing levels of expertise in 3D CAD design as it applies to engineering: Certified SOLIDWORKS Associate CSWA, Certified SOLIDWORKS Professional CSWP and Certified SOLIDWORKS Expert CSWE along with specialty fields in Drawing, Simulation, Sheet Metal, Surfacing and more.

The CSWA certification indicates a foundation in and apprentice knowledge of 3D CAD design and engineering practices and principles.

The CSWA Academic exam is provided either in a single 3 hour segment, or 2 - 90 minute segments.

Part 1 of the CSWA Academic exam is 90 minutes, minimum passing score is 80, with 6 questions. There are two questions in the Basic Part Creation and Modification category, two questions in the Intermediate Part Creation and Modification category and two questions in the Assembly Creation and Modification category.

Part 2 of the CSWA Academic exam is 90 minutes, minimum passing score is 80 with 8 questions. There are three questions on the CSWA Academic exam in the Drafting Competencies category, three questions in the Advanced Part Creation and Modification category and two questions in the Assembly Creation and Modification category.

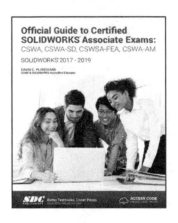

The CSWA exam for industry is only provided in a single 3 hour segment. The exam consists of 14 questions five categories.

All exams cover the same material.

To obtain additional CSWA exam information, visit the SOLIDWORKS VirtualTester Certification site at https://SOLIDWORKS.virtualtester.com/.

Chapter 11: Additive Manufacturing - 3D Printing.

Provide a basic understanding between Additive vs. Subtractive manufacturing. Discuss Fused Filament Fabrication (FFF), STereoLithography (SLA), and Selective Laser Sintering (SLS) printer technology. Select suitable filament material. Comprehend 3D printer terminology. Knowledge of preparing, saving, and printing a model on a Fused Filament Fabrication 3D printer. Information on the Certified SOLIDWORKS Associate Additive Manufacturing (CSWA-AM) exam.

On the completion of this project, you will be able to:

- Discuss Additive vs Subtractive manufacturing.

- Review 3D printer technology: Fused Filament Fabrication (FFF), STereoLithography (SLA), and Selective Laser Sintering (SLS).

- Select the correct filament material: PLA (Polylactic acid), FPLA (Flexible Polylactic acid), ABS (Acrylonitrile butadiene styrene), PVA (Polyvinyl alcohol), Nylon 618, and Nylon 645.

- Create an STL (*.stl) file, an Additive Manufacturing (*.amf) file and a 3D Manufacturing format (*.3mf) file.

- Prepare G-code.

- Comprehend general 3D printer terminology.

- Understand optimum build orientation.

- Enter slicer parameters:

  o Raft, brim, skirt, layer height, percent infill, infill pattern, wall thickness, fan speed, print speed, bed temperature, and extruder (hot end) temperature.

- Address fit tolerance for interlocking parts.

- Define general 3D Printing tips.

- Print directly from SOLIDWORKS.

- Knowledge of the Certified SOLIDWORKS Associate Additive Manufacturing exam.

About the Book

You will find a wealth of information in this book. The book is a project based step-by-step text written for new and intermediate users. The following conventions are used throughout this book:

- The term document refers to a SOLIDWORKS part, drawing or assembly file.

- The list of items across the top of the SOLIDWORKS interface is the Menu bar menu or the Menu bar toolbar. Each item in the Menu bar has a pull-down menu. When you need to select a series of commands from these menus, the following format is used: Click **Insert**, **Reference Geometry**, **Plane** from the Menu bar. The Plane PropertyManager is displayed.

- The book is organized into chapters. Each chapter is focused on a specific subject or feature.

- The ANSI overall drafting standard and Third Angle projection is used as the default setting in this text. IPS (inch, pound, second) and MMGS (millimeter, gram, second) unit systems are used.

- Download all needed model files (SW-TUTORIAL-2020 folder) from the SDC Publication website (www.SDCpublications.com/downloads/978-1-63057-317-1). All assemblies and components for the final ROBOT assembly are located in the Chapter 6 Models folder.

The following command syntax is used throughout the text. Commands that require you to perform an action are displayed in **Bold** text.

Format:	Convention:	Example:
Bold	• All commands actions. • Selected icon button. • Selected icon button. • Selected geometry: line, circle. • Value entries.	• Click **Options** ⚙ from the Menu bar toolbar. • Click **Corner Rectangle** ▭ from the Sketch toolbar. • Click **Sketch** ▦ from the Context toolbar. • Select the **centerpoint**. • Enter **3.0** for Radius.
Capitalized	• Filenames. • First letter in a feature name.	• Save the **FLATBAR** assembly. • Click the **Fillet** ◳ feature.

Windows Terminology in SOLIDWORKS

The mouse buttons provide an integral role in executing SOLIDWORKS commands. The mouse buttons execute commands, select geometry, display Shortcut menus and provide information feedback.

A summary of mouse button terminology is displayed below:

Item:	Description:
Click	Press and release the left mouse button.
Double-click	Double press and release the left mouse button.
Click inside	Press the left mouse button. Wait a second, and then press the left mouse button inside the text box.
	Use this technique to modify Feature names in the FeatureManager design tree.
Drag	Point to an object, press and hold the left mouse button down.
	Move the mouse pointer to a new location.
	Release the left mouse button.
Right-click	Press and release the right mouse button.
	A Shortcut menu is displayed. Use the left mouse button to select a menu command.
Tool Tip	Position the mouse pointer over an Icon (button). The tool name is displayed below the mouse pointer.
Large Tool Tip	Position the mouse pointer over an Icon (button). The tool name and a description of its functionality are displayed below the mouse pointer.
Mouse pointer feedback	Position the mouse pointer over various areas of the sketch, part, assembly or drawing.
	The cursor provides feedback depending on the geometry.

A mouse with a center wheel provides additional functionality in SOLIDWORKS. Roll the center wheel downward to enlarge the model in the Graphics window. Hold the center wheel down. Drag the mouse in the Graphics window to rotate the model.

Visit SOLIDWORKS website: http://www.SOLIDWORKS.com /sw/support/hardware.html to view their supported operating systems and hardware requirements.

Hardware & System Requirements
Research graphics cards hardware, system requirements, and other related topics.

SolidWorks System Requirements
Hardware and system requirements for SolidWorks 3D CAD products.

Graphics Card Drivers
Find graphics card drivers for your system to ensure system performance and stability.

Data Management System Requirements
Hardware and system requirements for SolidWorks Product Data Management (PDM) products.

Anti-Virus
The following Anti-Virus applications have been tested with SolidWorks 3D CAD products.

SolidWorks Composer System Requirements
Hardware and system requirements for SolidWorks Composer and other 3DVIA related products.

Hardware Benchmarks
Applications and references that can help determine hardware performance.

SolidWorks Electrical System Requirements
Hardware and system requirements for SolidWorks Electrical products.

The Instructor's information contains over 60 classroom presentations, along with helpful hints, What's new, sample quizzes, avi files of assemblies, projects, and all initial and final SOLIDWORKS model files.

The book does not cover starting a SOLIDWORKS session in detail for the first time. A default SOLIDWORKS installation presents you with several options. For additional information for an Education Edition, visit the following site: http://www.SOLIDWORKS.com/sw/engineering-education-software.htm

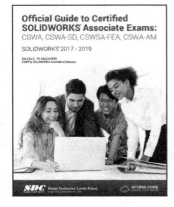

The CSWA certification indicates a foundation in and apprentice knowledge of 3D CAD design and engineering practices and principles.

The CSWA Academic exam is provided either in a single 3 hour segment, or 2 - 90 minute segments. The CSWA exam for industry is only provided in a single 3 hour segment. All exams cover the same material.

To obtain additional CSWA exam information, visit the SOLIDWORKS VirtualTester Certification site at https://SOLIDWORKS.virtualtester.com/.

Chapter 1

Overview of SOLIDWORKS® 2020 and the User Interface

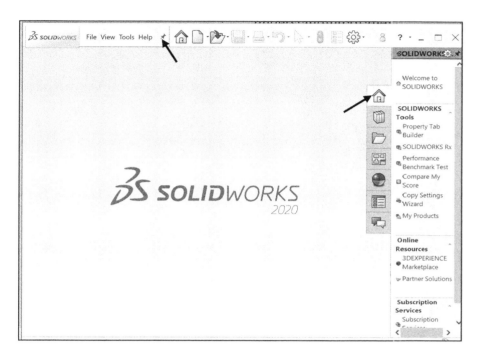

Below are the desired outcomes and usage competencies based on the completion of Chapter 1.

Desired Outcomes:	Usage Competencies:
• A comprehensive understanding of the SOLIDWORKS® 2020 User Interface (UI) and CommandManager.	• Ability to establish a SOLIDWORKS session. • Aptitude to utilize the following items: *Menu bar toolbar, Menu bar menu, Drop-down menus, Context toolbars, Consolidated drop-down toolbars, System feedback icons, Confirmation Corner, Heads-up View toolbar, Document Properties and more.* • Open a new and existing SOLIDWORKS part. • Knowledge to zoom, rotate and maneuver a three-button mouse in the SOLIDWORKS Graphics window.

Notes:

Chapter 1 - Overview of SOLIDWORKS® 2020 and the User Interface

Chapter Objective

Provide a comprehensive understanding of the SOLIDWORKS® default User Interface and CommandManager: Menu bar toolbar, Menu bar menu, Drop-down menu, Right-click Pop-up menus, Context toolbars/menus, Fly-out tool button, System feedback icons, Confirmation Corner, Heads-up View toolbar and more.

On the completion of this chapter, you will be able to:

- Utilize the SOLIDWORKS Welcome dialog box.

- Establish a SOLIDWORKS session.

- Comprehend the SOLIDWORKS 2020 User Interface.

- Recognize the default Reference Planes in the FeatureManager.

- Open a new and existing SOLIDWORKS part.

- Utilize SOLIDWORKS Help and SOLIDWORKS Tutorials.

- Zoom, rotate and maneuver a three-button mouse in the SOLIDWORKS Graphics window.

What is SOLIDWORKS®?

- SOLIDWORKS® is a mechanical design automation software package used to build parts, assemblies and drawings that takes advantage of the familiar Microsoft® Windows graphical user interface.

- SOLIDWORKS is an easy to learn design and analysis tool (SOLIDWORKS Simulation, SOLIDWORKS Motion, SOLIDWORKS Flow Simulation, Sustainability, etc.), which makes it possible for designers to quickly sketch 2D and 3D concepts, create 3D parts and assemblies and detail 2D drawings.

- Model dimensions in SOLIDWORKS are associative between parts, assemblies and drawings. Reference dimensions are one-way associative from the part to the drawing or from the part to the assembly.

- This book is written for the beginner to intermediate user.

Start a SOLIDWORKS 2020 Session

Start a SOLIDWORKS 2020 session and familiarize yourself with the SOLIDWORKS User Interface. As you read and perform the tasks in this chapter, you will obtain a sense of how to use the book and the structure. Actual input commands or required actions in the chapter are displayed in bold.

The book does not cover starting a SOLIDWORKS session in detail for the first time. A default SOLIDWORKS installation presents you with several options. For additional information, visit http://www.SOLIDWORKS.com.

Activity: Start a SOLIDWORKS 2020 Session.

Start a SOLIDWORKS 2020 session.

1)   Type **SOLIDWORKS 2020** in the Search window.

2)   Click the **SOLIDWORKS 2020** application (or if available, **double-click** the SOLIDWORKS icon on the desktop). The SOLIDWORKS Welcome dialog box is displayed by default.

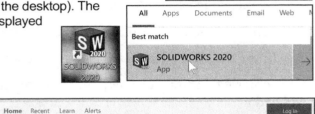

The Welcome dialog box provides a convenient way to open recent documents (Parts, Assemblies and Drawings), view recent folders, access SOLIDWORKS resources, and stay updated on SOLIDWORKS news.

3)   **View** your options. Do not open a document at this time.

4)   If the Welcome dialog box is not displayed, click the **Welcome to SOLIDWORKS** 🏠 icon from the Standard toolbar or click **Help** > **Welcome to SOLIDWORKS** from the Main menu.

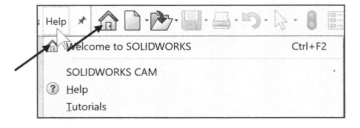

Home Tab

The Home tab lets you open new and existing documents, view recent documents and folders, and access SOLIDWORKS resources (*Part, Assembly, Drawing, Advanced mode, Open*).

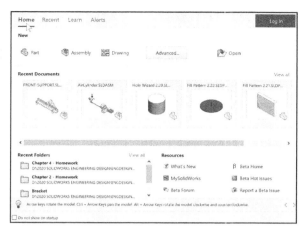

Recent Tab

The Recent tab lets you view a longer list of recent documents and folders. Sections in the Recent tab include *Documents* and *Folders*.

The Documents section includes thumbnails of documents that you have opened recently.

Click a thumbnail to open the document, or hover over a thumbnail to see the document location and access additional information about the document. When you hover over a thumbnail, the full path and last saved date of the document appears.

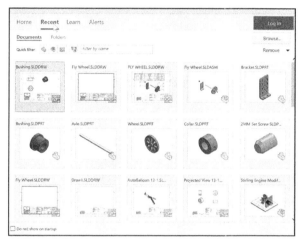

Learn Tab

The Learn tab lets you access instructional resources to help you learn more about the SOLIDWORKS software.

Sections in the Learn tab include:

- **Introducing SOLIDWORKS**. Open the Introducing SOLIDWORKS book.

- **Tutorials**. Open the step-by-step tutorials in the SOLIDWORKS software.

- **MySolidWorks Training**. Open the Training section at MySolidWorks.com.

- **Introducing SOLIDWORKS (Samples)**. Open local folders containing sample models.

- **Tutorials (Samples)**. Open the SOLIDWORKS Tutorials (videos) section at solidworks.com.

- **What's New (Samples)**. List of new changes.

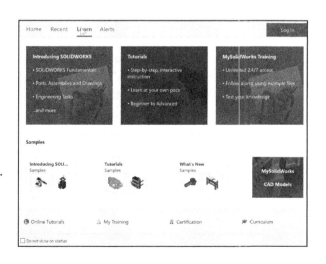

- **MySOLIDWORKS - CAD Models**. Open models in 3D ContentCentral and the Community Library.

- **My Training**. Open the My Training section at MySolidWorks.com.

- **Certification**. Open the SOLIDWORKS Certification Program section at solidworks.com.

- **Curriculum**. Open the Curriculum section at solidworks.com.

💡 When you install the software, if you do not install the Help Files or Example Files, the Tutorials and Samples links are unavailable.

Alerts Tab

The Alerts tab keeps you updated with SOLIDWORKS news.

Sections in the Alerts tab include Critical, Troubleshooting, and Technical.

💡 The Critical section does not appear if there are no critical alerts to display.

- **Troubleshooting**. Includes troubleshooting messages and recovered documents that used to be on the SOLIDWORKS Recovery tab in the Task Pane.

💡 If the software has a technical problem and an associated troubleshooting message exists, the Welcome dialog box opens to the Troubleshooting section automatically on startup, even if you selected **Do not show at startup** in the dialog box.

- **Technical Alerts**. Open the contents of the SOLIDWORKS Support Bulletins RSS feed at solidworks.com.

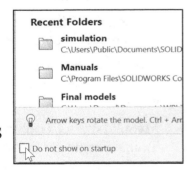

Close the Welcome dialog box.

5) Click **Close** ✕ from the Welcome dialog box. The SOLIDWORKS Graphics window is displayed. Note: You can also click outside the Welcome dialog box, in the Graphics window.

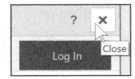

View the SOLIDWORKS Graphics window.

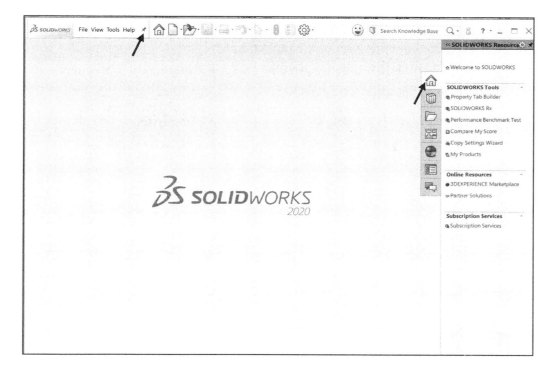

If you do not see this screen, click the SOLIDWORKS Resources 🏠 icon on the right side of the Graphics window located in the Task Pane.

6)   **Hover** the mouse pointer over the SOLIDWORKS icon.

7)   **Pin** ✚ the Menu Bar toolbar.

8)   **View** your options from the Menu bar menu: **File**, **View**, **Tools** and **Help**.

Menu Bar toolbar

The SOLIDWORKS (UI) is designed to make maximum use of the Graphics window. The Menu Bar toolbar contains a set of the most frequently used tool buttons from the Standard toolbar.

The following default tools are available:

- **Welcome to SOLIDWORKS** ⌂ - Open the Welcome dialog box, **New** ▯ - Create a new document; **Open** ▱ - Open an existing document; **Save** ▤ - Save an active document; **Print** ▤ - Print an active document; **Undo** ↺ - Reverse the last action; **Select** ▱ - Select Sketch entities, components and more; **Rebuild** ▮ - Rebuild the active part, assembly or drawing; **File Properties** ▤ - Show the summary information on the active document; and **Options** ⚙ ▾ - Change system options and Add-Ins for SOLIDWORKS.

Menu Bar menu

Click SOLIDWORKS in the Menu Bar toolbar to display the Menu Bar menu. SOLIDWORKS provides a context-sensitive menu structure. The menu titles remain the same for all three types of documents, but the menu items change depending on which type of document is active.

Example: The Insert menu includes features in part documents, mates in assembly documents, and drawing views in drawing documents. The display of the menu is also dependent on the workflow customization that you have selected. The default menu items for an active document are File, Edit, View, Insert, Tools, Window, Help and Pin.

The Pin ⚲ option displays the Menu bar toolbar and the Menu bar menu as illustrated. Throughout the book, the Menu bar menu and the Menu bar toolbar are referred to as the Menu bar.

Drop-down menu

SOLIDWORKS takes advantage of the familiar Microsoft® Windows user interface. Communicate with SOLIDWORKS through drop-down menus, Context sensitive toolbars, Consolidated toolbars or the CommandManager tabs.

🔅 A command is an instruction that informs SOLIDWORKS to perform a task.

To close a SOLIDWORKS drop-down menu, press the Esc key. You can also click any other part of the SOLIDWORKS Graphics window or click another drop-down menu.

Create a New Part Document

In the next section create a new part document.

Activity: Create a new Part Document.

A part is a 3D model which consists of features. What are features?

Features are geometry building blocks.

Most features either add or remove material.

Some features do not affect material (Cosmetic Thread).

Features are created either from 2D or 3D sketched profiles or from edges and faces of existing geometry.

Features are individual shapes that combined with other features make up a part or assembly. Some features, such as bosses and cuts, originate as sketches. Other features, such as shells and fillets, modify a feature's geometry.

Features are displayed in the FeatureManager as illustrated (Boss-Extrude1, Cut-Extrude1, Cut-Extrude2, Mirror1, Cut-Extrude3 and CirPattern1).

🔅 The first sketch of a part is called the Base Sketch. The Base sketch is the foundation for the 3D model. The book focuses on 2D sketches and 3D features.

🔅 FeatureManager tabs and tree folders will vary depending on system setup and Add-ins.

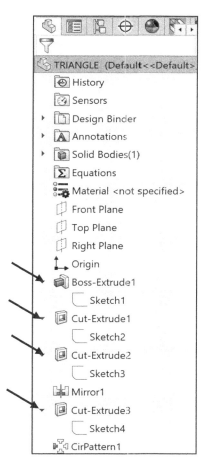

There are two modes in the New SOLIDWORKS Document dialog box: Novice and Advanced. The Novice option is the default option with three templates. The Advanced mode contains access to additional templates and tabs that you create in system options. Use the Advanced mode in this book.

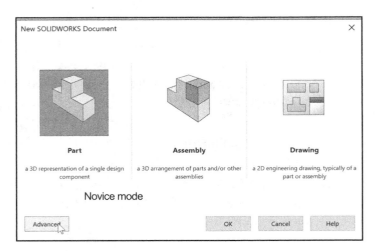

Create a new part.

9) Click **New** from the Menu bar. The New SOLIDWORKS Document dialog box is displayed.

Select the Advanced mode.

10) Click the **Advanced** button as illustrated. The Advanced mode is set.

11) Click the **Templates** tab.

12) Click **Part**. Part is the default template from the New SOLIDWORKS Document dialog box.

13) Click **OK** from the New SOLIDWORKS Document dialog box.

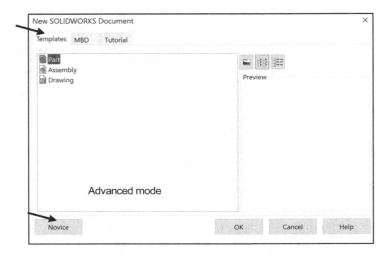

 Illustrations may vary depending on your SOLIDWORKS version and operating system.

The Advanced mode remains selected for all new documents in the current SOLIDWORKS session. When you exit SOLIDWORKS, the Advanced mode setting is saved.

The default SOLIDWORKS installation contains three tabs in the New SOLIDWORKS Document dialog box: *Templates, MBD,* and *Tutorial*. The *Templates* tab corresponds to the default SOLIDWORKS templates. The *MBD* tab corresponds to the templates utilized in the SOLIDWORKS (Model Based Definition). The *Tutorial* tab corresponds to the templates utilized in the SOLIDWORKS Tutorials.

Part1 is displayed in the FeatureManager and is the name of the document. Part1 is the default part window name.

The Part Origin ↳ is displayed in blue in the center of the Graphics window. The Origin represents the intersection of the three default reference planes: *Front Plane*, *Top Plane* and *Right Plane*. The positive X-axis is horizontal and points to the right of the Origin in the Front view. The positive Y-axis is vertical and points upward in the Front view. The FeatureManager contains a list of features, reference geometry, and settings utilized in the part.

Edit the document units directly from the Graphics window as illustrated.

CommandManager and FeatureManager tabs will vary depending on system setup and Add-ins.

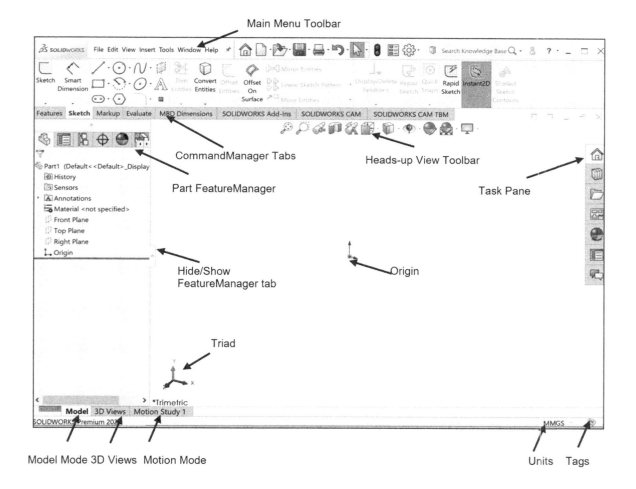

View the Default Sketch Planes.

14)   Click the **Front Plane** from the FeatureManager.

15)   Click the **Top Plane** from the FeatureManager.

16)   Click the **Right Plane** from the FeatureManager.

17)   Click the **Origin** from the FeatureManager. The Origin is the intersection of the Front, Top, and Right Planes.

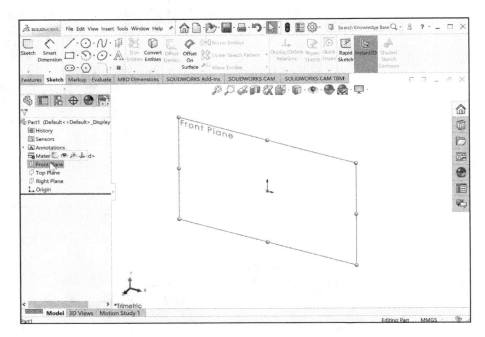

Download all needed model files (SW-TUTORIAL-2020 folder) from the SDC Publication website (www.SDCpublications.com/downloads/978-1-63057-317-1). Open the Bracket part. Review the features and sketches in the Bracket FeatureManager. Work directly from a local hard drive.

Activity: Download the SOLIDWORKS folder. Open the Bracket Part.

Download the SOLIDWORKS folder. Open an existing SOLIDWORKS part.

18)   **Download** the SW-TUTORIAL-2020 folder.

19)   Click **Open** from the Menu bar menu.

20)   Browse to the **SW-TUTORIAL-2020\Bracket** folder.

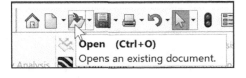

21)   Double-click the **Bracket** part. The Bracket part is displayed in the Graphics window.

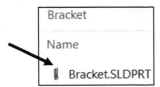

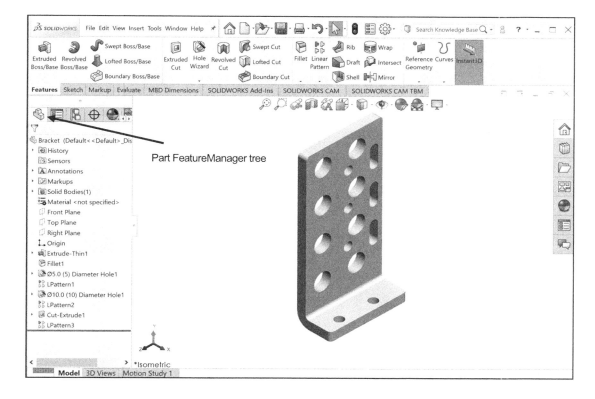

Part FeatureManager tree

The FeatureManager design tree is located on the left side of the SOLIDWORKS Graphics window. The FeatureManager provides a summarized view of the active part, assembly, or drawing document. The tree displays the details on how the part, assembly or drawing document was created.

Use the FeatureManager rollback bar to temporarily roll back to an earlier state, to absorbed features, roll forward, roll to previous, or roll to the end of the FeatureManager design tree. You can add new features or edit existing features while the model is in the rolled-back state. You can save models with the rollback bar placed anywhere.

In the next section, review the features in the Bracket FeatureManager using the Rollback bar.

Activity: Use the FeatureManager Rollback Bar option.

Apply the FeatureManager Rollback Bar. Revert to an earlier state in the model.

22) Place the **mouse pointer** over the rollback bar in the FeatureManager design tree as illustrated. The pointer changes to a hand 🖐. Note the provided information on the feature. This is called Dynamic Reference Visualization.

23) Drag the **rollback bar** up the FeatureManager design tree until it is above the features you want rolled back, in this case 10.0 (10) Diameter Hole1.

24) **Release** the mouse button.

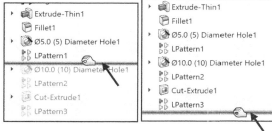

View the first feature in the Bracket Part.

25) Drag the **rollback bar** up the FeatureManager above Fillet1. View the results in the Graphics window.

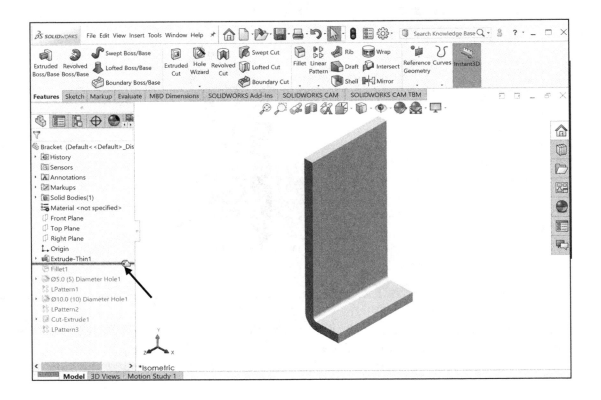

Return to the original Bracket Part FeatureManager.

26) Right-click **Extrude-Thin1** in the FeatureManager. The Pop-up Context toolbar is displayed.

27) Click **Roll to End**. View the results in the Graphics window.

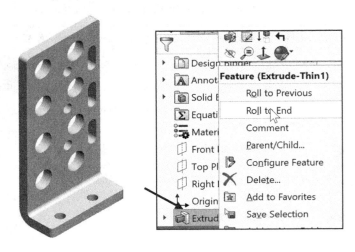

Heads-up View toolbar

SOLIDWORKS provides the user with numerous view options. One of the most useful tools is the Heads-up View toolbar displayed in the Graphics window when a document is active.

💡 *Dynamic Annotation Views* 🖾 : Only available with SOLIDWORKS MBD (Model Based Definition). Provides the ability to control how annotations are displayed when you rotate models.

In the next section, apply the following tools: Zoom to Fit, Zoom to Area, Zoom out, Rotate and select various view orientations from the Heads-up View toolbar.

Activity: Utilize the Heads-up View toolbar.

Zoom to Fit the model in the Graphics window.

28)   Click the **Zoom to Fit** 🔎 icon. The tool fits the model to the Graphics window.

Zoom to Area on the model in the Graphics window.

29)   Click the **Zoom to Area** 🔍 icon. The Zoom to Area 🔍 icon is displayed.

Zoom in on the top left hole.

30)   **Window-select** the top left corner as illustrated. View the results.

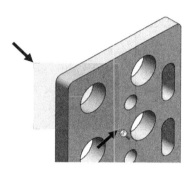

De-select the Zoom to Area tool.

31)   Click the **Zoom to Area** 🔍 icon.

Fit the model to the Graphics window.

32)   Press the **f** key.

Rotate the model.

33)   Hold the **middle mouse button** down. Drag **upward** ↻, **downward** ↻, to the **left** ↻ and to the **right** ↻ to rotate the model in the Graphics window.

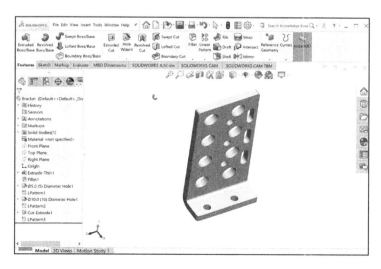

Display a few Standard Views.

34) Click **inside** the Graphics window.

35) Click **Front** 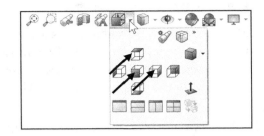 from the drop-down Heads-up view toolbar. The model is displayed in the Front view.

36) Click **Right** from the drop-down Heads-up view toolbar. The model is displayed in the Right view.

37) Click **Top** from the drop-down Heads-up view toolbar. The model is displayed in the Top view.

Display a Trimetric view of the Bracket model.

38) Click **Trimetric** from the drop-down Heads-up view toolbar as illustrated. Note your options. View the results in the Graphics window.

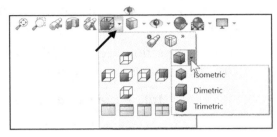

SOLIDWORKS Help

Help in SOLIDWORKS is context-sensitive and in HTML format. Help is accessed in many ways, including Help buttons in all dialog boxes and PropertyManager (or press F1) and Help ⑦ tool on the Standard toolbar for SOLIDWORKS Help.

🔆 CommandManager and FeatureManagers tabs will vary depending on system setup and SOLIDWORKS Add-ins.

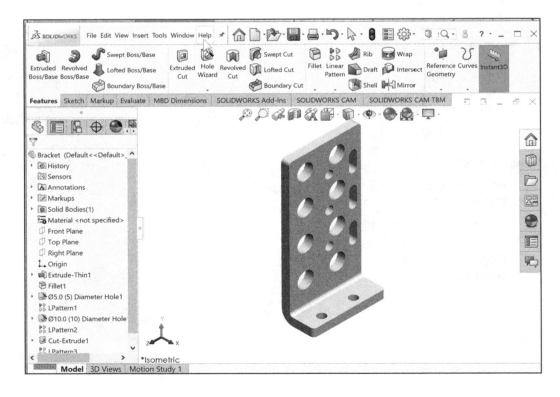

39)   Click **Help** from the Menu bar.

40)   Click **Help** ⑦ from the drop-down menu. The
         SOLIDWORKS Home Page is displayed by default.
         View your options.

🔅 SOLIDWORKS Web Help is active by default
under Help in the Main menu.

Close Help. Return to the SOLIDWORKS Graphics window.

41)   **Close** ✖ SOLIDWORKS Home.

SOLIDWORKS Tutorials

Display and explore the SOLIDWORKS tutorials.

42)   Click **Help** from the Menu bar.

43)   Click **Tutorials**. The SOLIDWORKS Tutorials are
         displayed. The SOLIDWORKS Tutorials are
         presented by category.

44)   Click the **Getting Started** category. The Getting
         Started category provides lessons on parts,
         assemblies, and drawings.

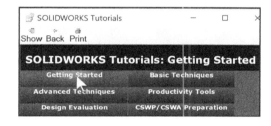

In the next section, close all models, tutorials and
view the additional User Interface tools.

Activity: Close all Tutorials and Models.

Close SOLIDWORKS Tutorials and models.

45)   **Close** ✖ SOLIDWORKS Tutorials.

46)   Click **Window, Close All** from the Menu bar menu.

User Interface Tools

The book utilizes additional areas of the
SOLIDWORKS User Interface. Explore an overview
of these tools in the next section.

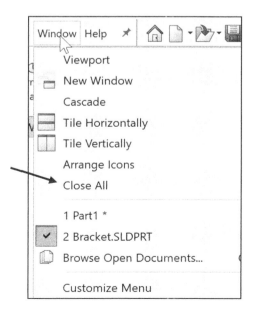

Right-click

Right-click in the Graphics window on a model, or in the FeatureManager on a feature or sketch to display the Context-sensitive toolbar. If you are in the middle of a command, this toolbar displays a list of options specifically related to that command.

Right-click an empty space in the Graphics window of a part or assembly, and a selection context toolbar above the shortcut menu is displayed. This provides easy access to the most commonly used selection tools.

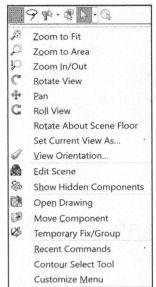

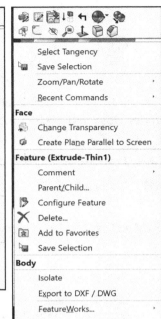

Consolidated toolbar

Similar commands are grouped together in the CommandManager. For example, variations of the Rectangle sketch tool are grouped in a single fly-out button as illustrated.

If you select the Consolidated toolbar button without expanding:

For some commands such as Sketch, the most commonly used command is performed. This command is the first listed and the command shown on the button.

For commands such as rectangle, where you may want to repeatedly create the same variant of the rectangle, the last used command is performed. This is the highlighted command when the Consolidated toolbar is expanded.

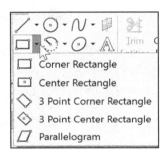

System feedback icon

SOLIDWORKS provides system feedback by attaching a symbol to the mouse pointer cursor.

The system feedback symbol indicates what you are selecting or what the system is expecting you to select.

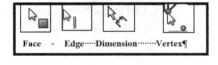

As you move the mouse pointer across your model, system feedback is displayed in the form of a symbol, riding next to the cursor as illustrated. This is a valuable feature in SOLIDWORKS.

Confirmation Corner

When numerous SOLIDWORKS commands are active, a symbol or a set of symbols is displayed in the upper right-hand corner of the Graphics window. This area is called the Confirmation Corner.

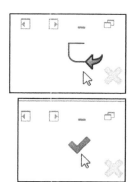

When a sketch is active, the confirmation corner box displays two symbols. The first symbol is the sketch tool icon. The second symbol is a large red X. These two symbols supply a visual reminder that you are in an active sketch. Click the sketch symbol icon to exit the sketch and to save any changes that you made.

When other commands are active, the confirmation corner box provides a green check mark and a large red X. Use the green check mark to execute the current command. Use the large red X to cancel the command.

Confirm changes you make in sketches and tools by using the D keyboard shortcut to move the OK and Cancel buttons to the pointer location in the Graphics window.

Heads-up View toolbar

SOLIDWORKS provides the user with numerous view options from the Standard Views, View and Heads-up View toolbar.

The Heads-up View toolbar is a transparent toolbar that is displayed in the Graphics window when a document is active.

You can hide, move or modify the Heads-up View toolbar. To modify the Heads-up View toolbar, right-click on a tool and select or deselect the tools that you want to display.

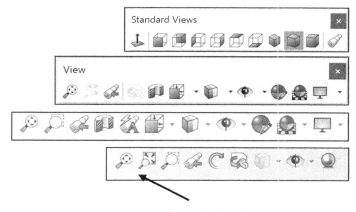

For a drawing document

The following views are available.
Note: available views are document dependent.

- *Zoom to Fit* : Fit the model to the Graphics window.

- *Zoom to Area* : Zoom to the areas you select with a bounding box.

- *Previous View* : Display the previous view.

- *Section View* 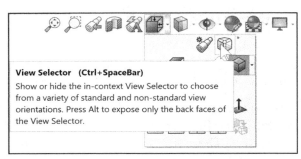: Display a cutaway of a part or assembly, using one or more cross section planes.

- *Dynamic Annotation Views* : Only available with SOLIDWORKS MBD. Control how annotations are displayed when you rotate a model.

The Orientation dialog has an option to display a view cube (in-context View Selector) with a live model preview. This helps the user to understand how each standard view orientates the model. With the view cube, you can access additional standard views. The views are easy to understand and they can be accessed simply by selecting a face on the cube.

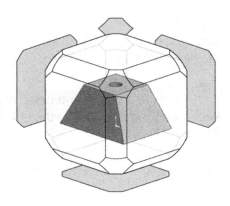

To activate the Orientation dialog box, press (Ctrl + spacebar) or click the View Orientation icon from the Heads up View toolbar. The active model is displayed in the View Selector in an Isometric orientation (default view).

Click the View Selector icon in the Orientation dialog box to show or hide the in-context View Selector.

Press **Ctrl + spacebar** to activate the View Selector.

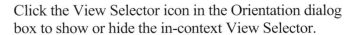

Press the **spacebar** to activate the Orientation dialog box.

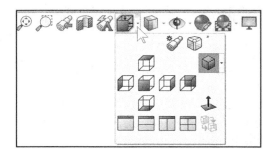

- *View Orientation box* : Select a view orientation or the number of viewports. The options are *Top, Left, Front, Right, Back, Bottom, Single view, Two view - Horizontal, Two view - Vertical, Four view*. Click the drop-down arrow to access Axonometric views: *Isometric, Dimetric* and *Trimetric*.

- *Display Style* : Display the style for the active view. The options are *Wireframe, Hidden Lines Visible, Hidden Lines Removed, Shaded, Shaded With Edges*.

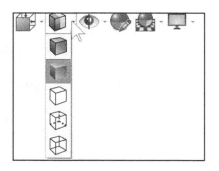

- *Hide/Show Items* 👁 : Select items to hide or show in the Graphics window. The available items are document dependent. Note the View Center of Mass ✚ icon.

- *Edit Appearance* 🔵 : Edit the appearance of entities of the model.

- *Apply Scene* 🎨 : Apply a scene to an active part or assembly document. View the available options.

- *View Setting* 🖥 : Select the following settings: *RealView Graphics, Shadows In Shaded Mode, Ambient Occlusion, Perspective* and *Cartoon.*

- *Rotate view* ↻ : Rotate a drawing view. Input Drawing view angle and select the ability to update and rotate center marks with view.

- *3D Drawing View* 🔲 : Dynamically manipulate the drawing view in 3D to make a selection.

To display a grid for a part, click Options ⚙ ▾, Document Properties tab. Click Grid/Snaps, check the Display grid box.

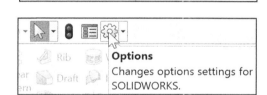

🔅 Add a custom view to the Heads-up View toolbar. Press the space key. The Orientation dialog box is displayed. Click the New View 🔧 tool. The Name View dialog box is displayed. Enter a new named view. Click OK.

Use commands to display information about the triad or to change the position and orientation of the triad. Available commands depend on the triad's context.

🔅 Save space in the CommandManager, limit your CommandManager tabs. Right-click on a CommandManager tab. Click Tabs. View your options to display CommandManager tabs.

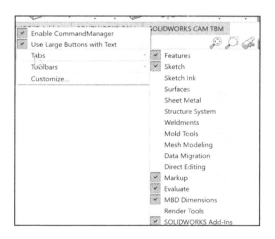

SOLIDWORKS CommandManager

The SOLIDWORKS CommandManager is a Context-sensitive toolbar. By default, it has toolbars embedded in it based on your active document type. When you click a tab below the CommandManager, it updates to display that toolbar. For example, if you click the Sketch tab, the Sketch toolbar is displayed.

For commercial users, SOLIDWORKS Model Based Definition (MBD) and SOLIDWORKS CAM is a separate application. For education users, SOLIDWORKS MBD and SOLIDWORKS CAM is included in the SOLIDWORKS Education Edition.

Below is an illustrated CommandManager for a **Part** document. Tabs will vary depending on system setup and Add-ins.

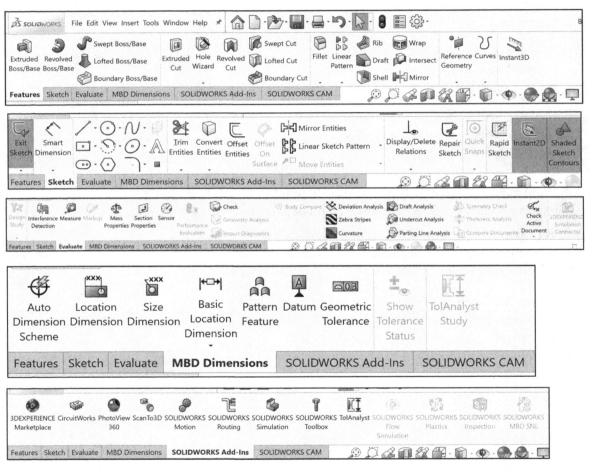

Set button size from the Toolbars tab of the Customize dialog box. To facilitate element selection on touch interfaces such as tablets, you can set up the larger Size buttons and text from the Options menu (Standard toolbar).

The SOLIDWORKS CommandManager is a Context-sensitive toolbar that automatically updates based on the toolbar you want to access. By default, it has toolbars embedded in it based on your active document type. The available tools are feature and document dependent.

Below is an illustrated CommandManager for a **Drawing** document. Tabs will vary depending on system setup and Add-ins.

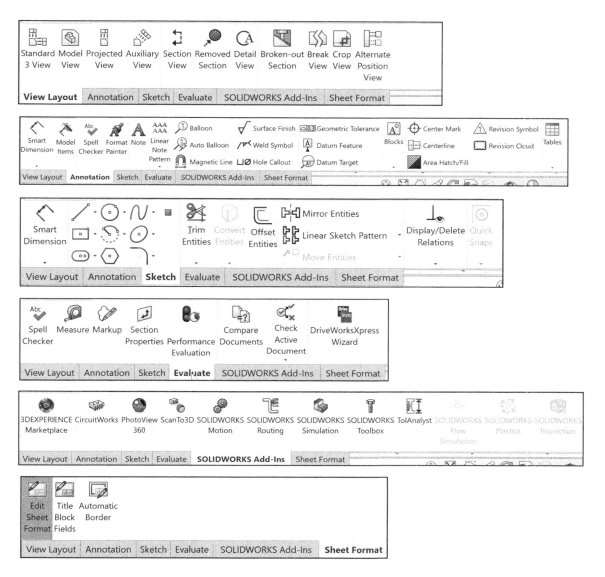

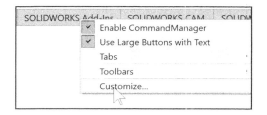

💡 To add a custom tab, right-click on a tab and click Customize CommandManager. You can also select to add a blank tab and populate it with custom tools from the Customize dialog box.

The SOLIDWORKS CommandManager is a Context-sensitive toolbar that automatically updates based on the toolbar you want to access. By default, it has toolbars embedded in it based on your active document type. The available tools are feature and document dependent.

Below is an illustrated CommandManager for an **Assembly** document. Tabs will vary depending on system setup and Add-ins.

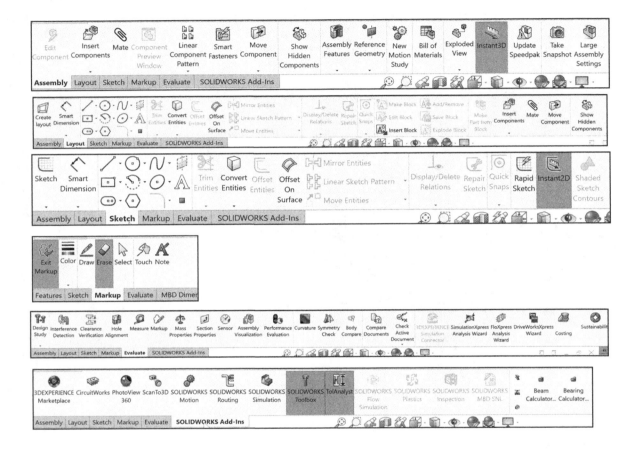

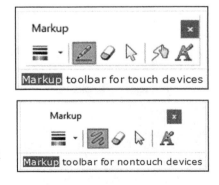

The Ink Markup toolbar is renamed to the Markup toolbar. The Markup tab is displayed by default in the CommandManager on some systems. You can draw markups with a mouse on non-touch devices, display bounding boxes for markups, create markups in drawings, and use the context toolbar to access markup options.

The Markup toolbar displays different options depending on the device. Draw ✏ and Touch ✋ are **not** available for non-touch devices.

Float the CommandManager. Drag the Features, Sketch or any CommandManager tab. Drag the CommandManager anywhere on or outside the SOLIDWORKS window.

To dock the CommandManager, perform one of the following:

While dragging the CommandManager in the SOLIDWORKS window, move the pointer over a docking icon -

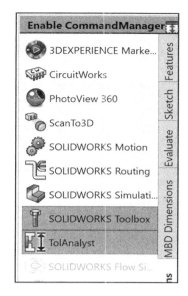

Dock above, Dock left, Dock right and click the needed command.

Double-click the floating CommandManager to revert the CommandManager to the last docking position.

Screen shots in the book were made using SOLIDWORKS 2020 SP0 running Windows® 10.

Selection Enhancements

Right-click an empty space in the Graphics window of a part or assembly; a selection context toolbar above the shortcut menu provides easy access to the most commonly used selection tools.

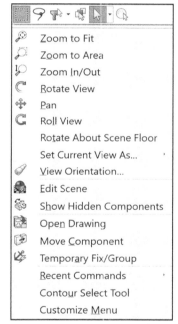

- **Box Selection** ⬚. Select entities in parts, assemblies, and drawings by dragging a selection box with the pointer.

- **Lasso Selection** ⌇. Select entities by drawing a lasso around the entities.

- **Selection Filters** ⛛. List of selection filter commands.

- **Select Other** ⬚. Display the Select Other dialog box.

- **Select** ⬚. Display a list of selection commands.

- **Magnified Selection** ⬚. Display the magnifying glass, which gives you a magnified view of a section of a model.

☼ Save space in the CommandManager, right-click in the CommandManager and un-check the Use Large Buttons with Text box. This eliminates the text associated with the tool.

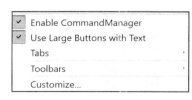

FeatureManager Design Tree

The FeatureManager consists of various tabs:

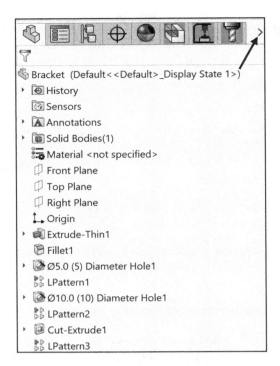

- *FeatureManager design tree* ▤ tab.

- *PropertyManager* ▤ tab.

- *ConfigurationManager* ▤ tab.

- *DimXpertManager* ⊕ tab.

- *DisplayManager* ● tab.

- *CAM FeatureManager tree* ▤ tab.

- *CAM Operation tree* ▤ tab.

- *CAM Tools tree* ▤ tab.

Click the ▤ direction arrows to expand or collapse the FeatureManager design tree.

CommandManager and FeatureManager tabs and folder files will vary depending on system setup and Add-ins.

Select the Hide/Show FeatureManager Area tab

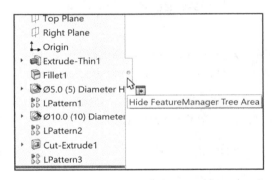

▤ as illustrated to enlarge the Graphics window for modeling.

🔆 The Sensors tool ▤ located in the FeatureManager monitors selected properties in a part or assembly and alerts you when values deviate from the specified limits. There are five sensor types: Simulation Data, Mass properties, Dimensions, Measurement and Costing Data.

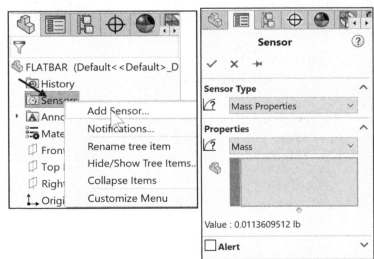

Various commands provide the ability to control what is displayed in the FeatureManager design tree.

1. Show or Hide FeatureManager items.

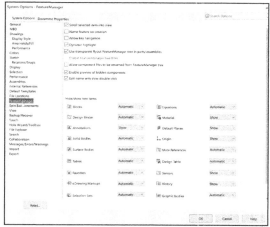

Click **Options** ⚙ from the Menu bar. Click **FeatureManager** from the System Options tab. Customize your FeatureManager from the Hide/Show tree Items dialog box.

2. Filter the FeatureManager design tree. Enter information in the filter field. You can filter by *Type of features, Feature names, Sketches, Folders, Mates, User-defined tags* and *Custom properties*.

Tags are keywords you can add to a SOLIDWORKS document to make them easier to filter and to search. The Tags ⊘ icon is located in the bottom right corner of the Graphics window.

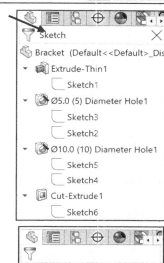

Collapse all items in the FeatureManager, **right-click** and select **Collapse items**, or press the **Shift + C** keys.

The FeatureManager design tree and the Graphics window are dynamically linked. Select sketches, features, drawing views, and construction geometry in either pane.

Split the FeatureManager design tree and either display two FeatureManager instances, or combine the FeatureManager design tree with the ConfigurationManager or PropertyManager.

Move between the FeatureManager design tree, PropertyManager, ConfigurationManager, DimXpertManager, DisplayManager and others by selecting the tab at the top of the menu.

Split

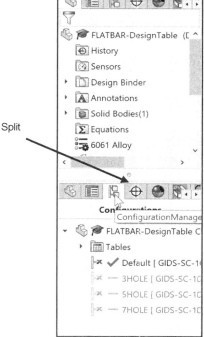

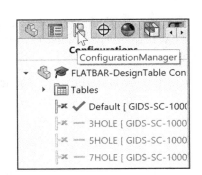

The ConfigurationManager tab is located to the right of the PropertyManager tab. Use the ConfigurationManager to create, select and view multiple configurations of parts and assemblies.

The icons in the ConfigurationManager denote whether the configuration was created manually or with a design table.

The DimXpertManager ⊕ tab provides the ability to insert dimensions and tolerances manually or automatically. The options are: **Auto Dimension Scheme** ⊕, **Auto Pair Tolerance** ⊞, **Basic, Location Dimension** ⊢ㅁ⊣, **Basic Size Dimension** ↘, **General Profile Tolerance** ⊗, **Show Tolerance Status** ±⊙, **Copy Scheme** ⊕, **Import Scheme** ⊕, **TolAnalyst Study** ⨎ and **Datum Target** ⊗.

💡 TolAnalyst is available in SOLIDWORKS Premium.

Fly-out FeatureManager

The fly-out FeatureManager design tree provides the ability to view and select items in the PropertyManager and the FeatureManager design tree at the same time.

Throughout the book, you will select commands and command options from the drop-down menu, fly-out FeatureManager, Context toolbar, or from a SOLIDWORKS toolbar.

💡 Another method for accessing a command is to use the accelerator key. Accelerator keys are special key strokes, which activate the drop-down menu options. Some commands in the menu bar and items in the drop-down menus have an underlined character.

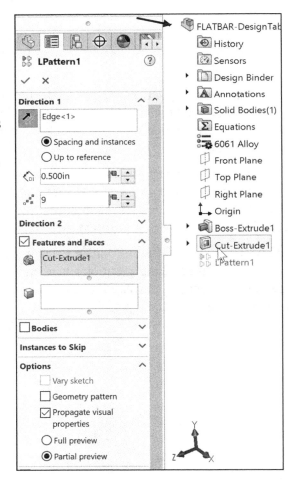

Task Pane

The Task Pane is displayed when a SOLIDWORKS session starts. You can show, hide, and reorder tabs in the Task Pane. You can also set a tab as the default so it appears when you open the Task Pane, pin or unpin to the default location.

The Task Pane contains the following default tabs:

- *SOLIDWORKS Resources* 🏠.

- *Design Library* 📖.

- *File Explorer* 📁.

- *View Palette* 🖼.

- *Appearances, Scenes and Decals* 🔵.

- *Custom Properties* 📋.

- *SOLIDWORKS Forum* 💬.

💡 Additional tabs are displayed with Add-Ins.

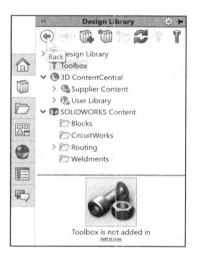

Use the **Back** and **Forward** buttons in the Design Library tab and the Appearances, Scenes, and Decals tab of the Task Pane to navigate in folders.

SOLIDWORKS Resources

The basic SOLIDWORKS Resources 🏠 menu displays the following default selections:

- *Welcome to SOLIDWORKS.*

- *SOLIDWORKS Tools.*

- *Online Resources.*

- *Subscription Services.*

Other user interfaces are available during the initial software installation selection: *Machine Design, Mold Design, Consumer Products Design, etc.*

Design Library

The Design Library ⬚ contains reusable parts, assemblies, and other elements including library features.

The Design Library tab contains four default selections. Each default selection contains additional sub categories.

The default selections are:

- *Design Library.*

- *Toolbox.*

- *3D ContentCentral (Internet access required).*

- *SOLIDWORKS Content (Internet access required).*

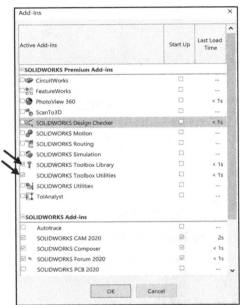

💡 Activate the SOLIDWORKS Toolbox. Click Tools, Add-Ins.., from the Main menu. Check the SOLIDWORKS Toolbox Library and SOLIDWORKS Toolbox Utilities box from the Add-ins dialog box or click SOLIDWORKS Toolbox from the SOLIDWORKS Add-Ins tab.

To access the Design Library folders in a non-network environment, click Add File Location ⬚ and browse to the needed path. Paths may vary depending on your SOLIDWORKS version and window setup. In a network environment, contact your IT department for system details.

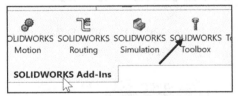

File Explorer

File Explorer 📁 duplicates Windows Explorer from your local computer and displays:

- *Recent Documents.*

- *Samples.*

- *Open in SOLIDWORKS*

- *Desktop.*

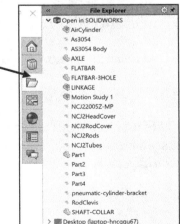

Search

The SOLIDWORKS Search box is displayed in the upper right corner of the SOLIDWORKS Graphics window (Menu Bar toolbar). Enter the text or key words to search.

New search modes have been added to SOLIDWORKS Search as illustrated.

View Palette

The View Palette tool located in the Task Pane provides the ability to insert drawing views of an active document, or click the Browse button to locate the desired document.

Click and drag the view from the View Palette into an active drawing sheet to create a drawing view.

The selected model is LINKAGE in the illustration.

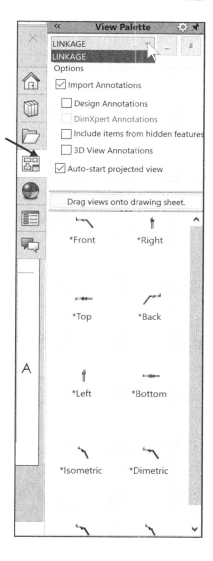

Appearances, Scenes, and Decals

Appearances, Scenes, and Decals ⬤ provide a simplified way to display models in a photo-realistic setting using a library of Appearances, Scenes, and Decals.

An appearance defines the visual properties of a model, including color and texture. Appearances do not affect physical properties, which are defined by materials.

Scenes provide a visual backdrop behind a model. In SOLIDWORKS they provide reflections on the model. PhotoView 360 is an Add-in. Drag and drop a selected appearance, scene or decal on a feature, surface, part or assembly.

Custom Properties

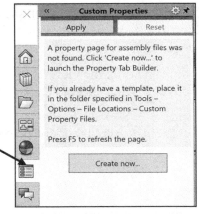

The Custom Properties ▦ tool provides the ability to enter custom and configuration specific properties directly into SOLIDWORKS files.

SOLIDWORKS Forum

Click the SOLIDWORKS Forum 💬 icon to search directly within the Task Pane. An internet connection is required. You are required to register and to log in for postings and discussions.

User Interface for Scaling High Resolution Screens

The SOLIDWORKS software supports high-resolution, high-pixel density displays. All aspects of the user interface respond to the Microsoft Windows® display scaling setting. In dialog boxes, PropertyManagers, and the FeatureManager design tree, the SOLIDWORKS software uses your display scaling setting to display buttons and icons at an appropriate size. Icons that are associated with text are scaled to a size appropriate for the text. In addition, for toolbars, you can display Small, Medium, or Large buttons. Click the **Options drop-down arrow** from the Standard Menu bar, and click Button size to size the icons.

Motion Study tab

Motion Studies are graphical simulations of motion for an assembly. Access the MotionManager from the Motion Study tab. The Motion Study tab is located in the bottom left corner of the Graphics window.

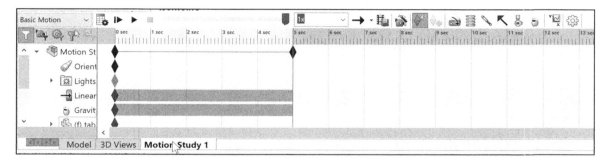

Incorporate visual properties such as lighting and camera perspective. Click the Motion Study tab to view the MotionManager. Click the Model tab to return to the FeatureManager design tree.

The MotionManager displays a timeline-based interface and provides the following selections from the drop-down menu as illustrated:

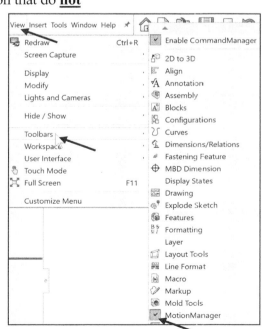

- *Animation:* Apply Animation to animate the motion of an assembly. Add a motor and insert positions of assembly components at various times using set key points. Use the Animation option to create animations for motion that do **not** require accounting for mass or gravity.

- *Basic Motion:* Apply Basic Motion for approximating the effects of motors, springs, collisions and gravity on assemblies. Basic Motion takes mass into account in calculating motion. Basic Motion computation is relatively fast, so you can use this for creating presentation animations using physics-based simulations. Use the Basic Motion option to create simulations of motion that account for mass, collisions or gravity.

⛯ If the Motion Study tab is not displayed in the Graphics window, click **View ➤ Toolbars ➤ MotionManager** from the Menu bar.

3D Views tab

SOLIDWORKS MBD (Model Based Definition) lets you create models without the need for drawings giving you an integrated manufacturing solution. MBD helps companies define, organize, and publish 3D product and manufacturing information (PMI), including 3D model data in industry standard file formats.

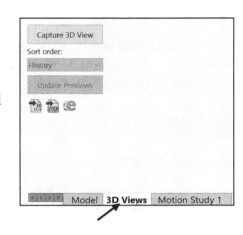

Create 3D drawing views of your parts and assemblies that contain the model settings needed for review and manufacturing. This lets users navigate back to those settings as they evaluate the design.

Use the tools in the MBD Dimensions CommandManager to set up your model with selected configurations, including explodes and abbreviated views, annotations, display states, zoom level, view orientation and section views. Capture those settings so that you and other users can return to them at any time using the 3D view palette.

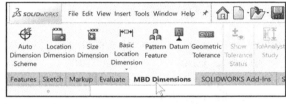

To access the 3D View palette, click the 3D Views tab at the bottom of the SOLIDWORKS window or the SOLIDWORKS MBD tab in the CommandManager. The Capture 3D View button opens the Capture 3D View PropertyManager, where you specify the 3D view name, and the configuration, display state and annotation view to capture. See SOLIDWORKS help for additional information.

Dynamic Reference Visualization (Parent/Child)

Dynamic Reference Visualization provides the ability to view the parent/child relationships between items in the FeatureManager design tree. When you hover over a feature with references in the FeatureManager design tree, arrows display showing the relationships. If a reference cannot be shown because a feature is not expanded, the arrow points to the feature that contains the reference and the actual reference appears in a text box to the right of the arrow.

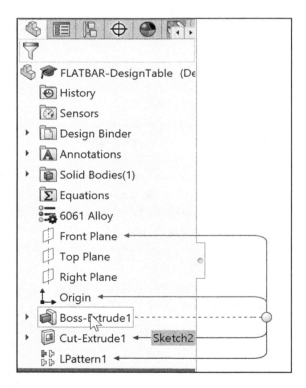

Use Dynamic reference visualization for a part, assembly and mates.

To display the Dynamic Reference Visualization, click **View ➢ User Interface ➢ Dynamic Reference Visualization Parent/Child)** from the Main menu bar.

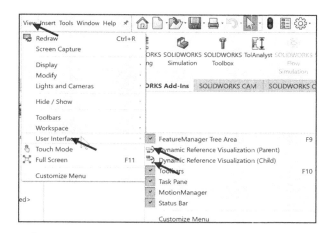

Mouse Movements

A mouse typically has two buttons: a primary button (usually the left button) and a secondary button (usually the right button). Most mice also include a scroll wheel between the buttons to help you scroll through documents and to Zoom in, Zoom out and rotate models in SOLIDWORKS. It is highly recommended that you use a mouse with at least a Primary, Scroll and Secondary button.

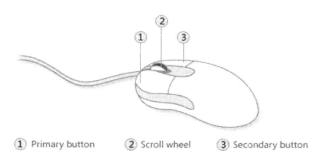

① Primary button   ② Scroll wheel   ③ Secondary button

Single-click

To click an item, point to the item on the screen, and then press and release the primary button (usually the left button). Clicking is most often used to select (mark) an item or open a menu. This is sometimes called single-clicking or left-clicking.

Double-click

To double-click an item, point to the item on the screen, and then click twice quickly. If the two clicks are spaced too far apart, they might be interpreted as two individual clicks rather than as one double-click. Double-clicking is most often used to open items on your desktop. For example, you can start a program or open a folder by double-clicking its icon on the desktop.

Right-click

To right-click an item, point to the item on the screen, and then press and release the secondary button (usually the right button). Right-clicking an item usually displays a list of things you can do with the item. Right-click in the open Graphics window or on a command in SOLIDWORKS, and additional pop-up context is displayed.

Scroll wheel

Use the scroll wheel to zoom-in or to zoom-out of the Graphics window in SOLIDWORKS. To zoom-in, roll the wheel backward (toward you). To zoom-out, roll the wheel forward (away from you).

Summary

The SOLIDWORKS (UI) is designed to make maximum use of the Graphics window for your model. Displayed toolbars and commands are kept to a minimum.

The SOLIDWORKS User Interface and CommandManager consist of the following main options: Menu bar toolbar, Menu bar menu, Drop-down menus, Context toolbars, Consolidated fly-out menus, System feedback icons, Confirmation Corner and Heads-up View toolbar.

The Part CommandManager controls the display of tabs: *Features*, *Sketch*, *Markup*, *Evaluate*, *MBD Dimensions* and various *SOLIDWORKS Add-Ins*.

The FeatureManager consists of various tabs:

- *FeatureManager design tree* tab.

- *PropertyManager* tab.

- *ConfigurationManager* tab.

- *DimXpertManager* tab.

- *DisplayManager* tab.

- *CAM FeatureManager tree* tab.

- *CAM Operation tree* tab.

- *CAM Tools tree* tab.

Click the direction arrows to expand or collapse the FeatureManager design tree.

CommandManager and FeatureManager tabs and file folders will vary depending on system setup and Add-ins.

You learned about creating a new SOLIDWORKS part and opening an existing SOLIDWORKS part along with using the Rollback bar to view the sketches and features.

If you modify a document property from an Overall drafting standard, a modify message is displayed as illustrated.

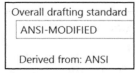

Overall drafting standard
ANSI-MODIFIED
Derived from: ANSI

🔅 Use the Search box, in the upper left corner of the Materials dialog box, to search through the entire materials library.

Templates are part, drawing and assembly documents which include user-defined parameters. Open a new part, drawing or assembly. Select a template for the new document.

In Chapter 2, obtain the working familiarity of the following SOLIDWORKS sketch and feature tools: *Line, Circle, Centerpoint Straight Slot, Smart Dimension, Extruded Boss/Base, Extruded Cut and Linear Pattern.*

Create three individual parts: AXLE, SHAFT-COLLAR and FLATBAR.

Create the assembly, LINKAGE using the three created parts and the downloaded sub-assembly - AirCylinder.

🔅 The book provides information on creating and storing special Part templates, Assembly templates and Drawing templates in the MY-TEMPLATES folder. The MY-TEMPLATES folder is added to the New SOLIDWORKS Document dialog box. Talk to your IT department *before you set* any new locations on a network system. The procedure in the book is designed for your personal computer.

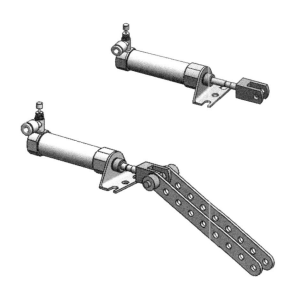

🔅 If you do not create the MY-TEMPLATE tab or the special part, drawing, or assembly templates, use the standard SOLIDWORKS default template and apply all of the needed document properties and custom properties.

Notes:

Chapter 2

Parts and Assembly Creation

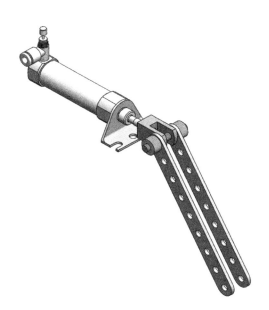

Below are the desired outcomes and usage competencies based on the completion of Chapter 2.

Desired Outcomes:	Usage Competencies:

• Create three parts:
 	
   ○ AXLE
 	
   ○ SHAFT-COLLAR
 	
   ○ FLATBAR
	• Create 2D sketch profiles on the correct Sketch plane.
	
	• Apply the following 3D features: Extruded Boss/Base, Extruded Cut and Linear Pattern.

• Create an assembly:
 	
   ○ LINKAGE assembly
 	
   ○ Apply Standard mates
 	
   ○ Motion Study
	• Understand the Assembly toolbar.
	
	• Insert components into an assembly.
	
	• Apply the following Standard mates: Concentric, Coincident and Parallel.
	
	• Understand the Quick mate procedure.

Notes:

Chapter 2 - Parts and Assembly Creation

Chapter Objective

SOLIDWORKS is a design software application used to model and create 2D and 3D sketches, 3D parts, 3D assemblies and 2D drawings. The chapter objective is to obtain the working familiarity of the following SOLIDWORKS sketch and feature tools: Line, Circle, Centerpoint, Straight Slot, Smart Dimension, Extruded Boss/Base, Extruded Cut, and Linear Pattern.

Create three individual parts: AXLE, SHAFT-COLLAR, and FLATBAR.

Create the assembly, LINKAGE, using the three created parts and the downloaded sub-assembly AirCylinder.

On the completion of this chapter, you will be able to:

- Set overall drafting standard, units, precision, and dimensioning standards for a SOLIDWORKS Part and Assembly document.

- Create a 2D sketch.

- Understand and apply the following Sketch tools:

  o Circle, Smart dimension and Centerpoint Straight Slot.

- Identify the correct Sketch plane for the 2D sketch:

  o Front, Top, Right.

- Insert geometric relations and dimensions:

  o Equal, Vertical, Horizontal, Parallel, Perpendicular, Coincident and MidPoint.

- Modify sketch dimensions and geometric relations.

- Create a 3D model.

- Understand and apply the following SOLIDWORKS features:

  o Extruded Boss/Base, Extruded Cut and Linear Pattern.

- Download an assembly document (with references) in SOLIDWORKS.

- Create an assembly document.

- Insert components:

    o Download and insert components.

- Understand and apply various Assembly feature tools.

- Apply the following Standard mates: Coincident, Concentric and Parallel.

- Utilize Standard and Quick mate procedure.

Chapter Overview

SOLIDWORKS is a 3D solid modeling CAD software package used to produce and model parts, assemblies, and drawings.

SOLIDWORKS provides design software to create 3D models and 2D drawings.

Create three parts in this chapter:

- AXLE

- SHAFT-COLLAR

- FLATBAR

Create the LINKAGE assembly. Download the AirCylinder assembly from the Pneumatic Components folder.

Combine the created parts with the AirCylinder sub-assembly to create the LINKAGE assembly.

Illustrations in the book display the default SOLIDWORKS user interface for 2020 SP0.

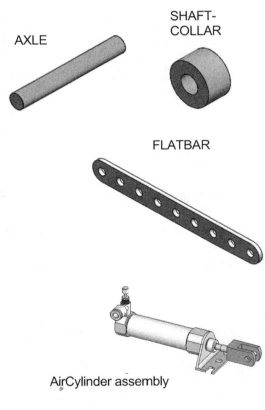

AXLE

SHAFT-COLLAR

FLATBAR

AirCylinder assembly

LINKAGE assembly

AXLE Part

The AXLE is a cylindrical rod. The AXLE supports the two FLATBAR parts.

Tangent edges and origins are displayed for educational purposes in this book.

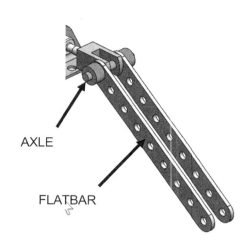

The AXLE rotates about its axis. The dimensions for the AXLE are determined from other components in the LINKAGE assembly.

AXLE

FLATBAR

Start a new SOLIDWORKS session.

Create the AXLE part.

Apply the feature to create the part. Features are the building blocks that add or remove material.

AXLE

Utilize the Extruded Boss/Base 🔲 tool from the Features toolbar to create a Boss-Extrude1 feature. The Extruded Boss/Base feature adds material. The Base feature (Boss-Extrude1) is the first feature of the part. The Base feature is the foundation of the part. Keep the Base feature <u>simple</u>!

The Base feature geometry for the AXLE is a simple extrusion. How do you create a solid Extruded Boss/Base feature for the AXLE?

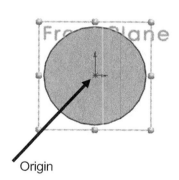

Origin

- Select the Front Plane as the Sketch plane.

- Sketch a circular 2D profile on the Front Plane, centered at the Origin as illustrated.

- Apply the Extruded Boss/Base Feature. Extend the profile perpendicular (⊥) to the Front Plane.

Utilize symmetry. Extrude the sketch with the Mid Plane End Condition in Direction 1. The Extruded Boss/Base feature is centered on both sides of the Front Plane.

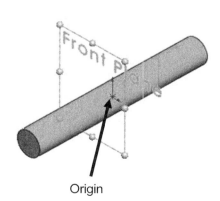

Origin

Create a new part. Select File, New from the Menu bar toolbar or click New ⬜ from the Menu bar menu. There are two options for new documents: **Novice** and **Advanced**. Select the Advanced option. Select the default Part document.

Activity: Start a SOLIDWORKS Session. Create a New Part.

Start a SOLIDWORKS 2020 session.

1) Click the **SOLIDWORKS 2020** application (or if available, **double-click** the SOLIDWORKS icon on the Desktop). View the Welcome dialog box. The Welcome dialog box provides a convenient means to open documents, view folders, access SOLIDWORKS resources, and stay updated on SOLIDWORKS news.

2) **Close** the SOLIDWORKS Welcome dialog box. The SOLIDWORKS program is displayed.

3) **Hover** the mouse pointer over the SOLIDWORKS icon.

4) **Pin** the Menu Bar toolbar. **View** your options.

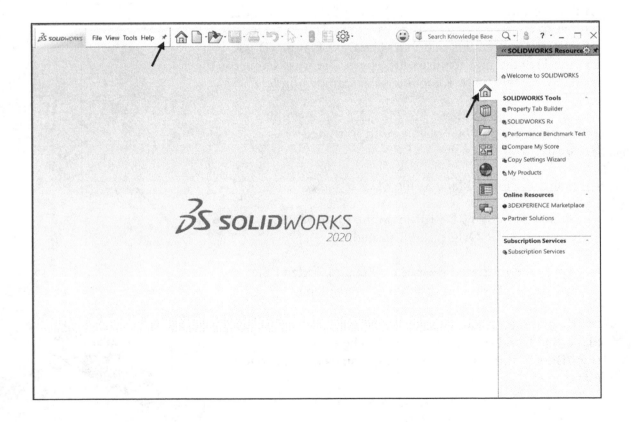

There are two default modes in the New SOLIDWORKS Document dialog box: Novice and Advanced. The *Novice* option is the default option with three templates. The *Advanced* mode contains access to additional templates and tabs that you create in system options. Use the *Advanced* mode in this book.

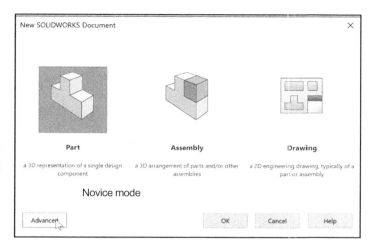

Create a New part.

5)   Click **New** ⬜ from the Menu bar. The New SOLIDWORKS Document dialog box is displayed.

Select Advanced Mode.

6)   Click the **Advanced** button to display the New SOLIDWORKS Document dialog box in Advanced mode.

7)   Click the **Templates** tab.

8)   Click **Part**. Part is the default template from the New SOLIDWORKS Document dialog box.

9)   Click **OK**.

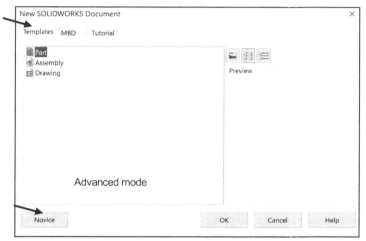

The *Advanced* mode remains selected for all new documents in the current SOLIDWORKS session. When you exit SOLIDWORKS, the *Advanced* mode setting is saved.

The default SOLIDWORKS installation contains three tabs in the New SOLIDWORKS Document dialog box: *Templates, Tutorial* and *MBD*. The *Templates* tab corresponds to the default SOLIDWORKS templates. The *Tutorial* tab corresponds to the templates utilized in the SOLIDWORKS Tutorials. The *MBD* tab corresponds to the templates utilized in the SOLIDWORKS (Model Based Definition).

🔆 During the initial SOLIDWORKS installation, you are requested to select either the ISO or ANSI drafting standard. ISO is typically a European drafting standard and uses First Angle Projection. The book is written using the ANSI (US) overall drafting standard and Third Angle Projection for all drawing documents.

Part1 is displayed in the FeatureManager and is the name of the document. Part1 is the default part window name.

The Part Origin ⊥ is displayed in blue in the center of the Graphics window. The Origin represents the intersection of the three default reference planes: *Front Plane*, *Top Plane* and *Right Plane*. The positive X-axis is horizontal and points to the right of the Origin in the Front view. The positive Y-axis is vertical and points upward in the Front view. The FeatureManager contains a list of features, reference geometry, and settings utilized in the part.

Edit the document units directly from the Graphics window as illustrated.

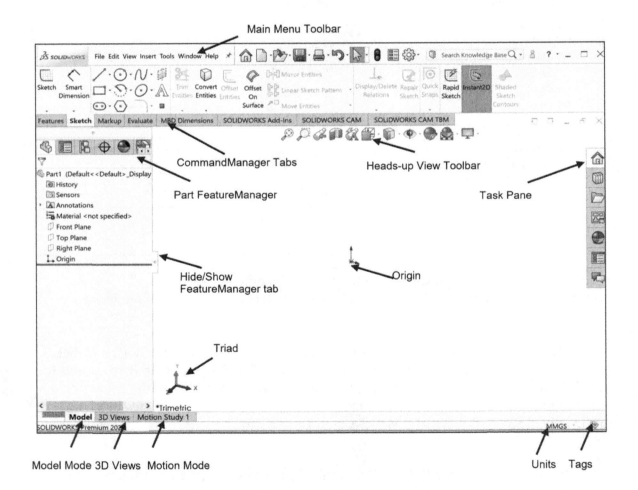

Activity: Set Document Properties - AXLE Part

Set Document Properties. Set drafting standard, units, and precision.

10) Click **Options** ⚙ from the Menu bar. The System Options General dialog box is displayed.

11) Click the **Document Properties** tab.

12) Select **ANSI** from the Overall drafting standard drop-down menu.

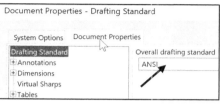

The Overall drafting standard determines the display of dimension text, arrows, symbols, and spacing.

Units are the measurement of physical quantities. Millimeter dimensioning and decimal inch dimensioning are the two most common unit types specified for engineering parts and drawings.

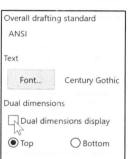

The primary units in this book are provided in IPS (inch, pound, second). The optional secondary units are provided in MMGS (millimeters, grams, second) and are indicated in brackets [ ].

Set document units and precision.

13) Click **Units**.

14) Click **IPS** (inch, pound, second) **[MMGS]** for Unit system.

15) Select **.123, [.12]** (three decimal places) for Length basic units.

16) Select **None** for Angle decimal places.

17) Click **OK** from the Document Properties - Units dialog box. The Part FeatureManager is displayed.

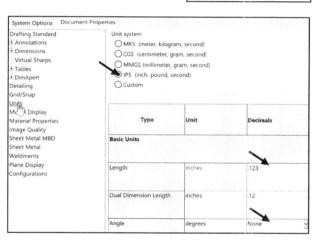

☀ The first sketch of a part is called the Base Sketch. The Base sketch is the foundation for the 3D model. The book focuses on 2D sketches and 3D features.

☀ If you modify a document property from an Overall drafting standard, a modify message is displayed as illustrated.

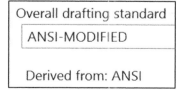

Activity: AXLE Part - Base Sketch - Extruded Base Feature

Create the Base sketch for the Extruded Base feature.

18) Right-click **Front Plane** from the FeatureManager. This is your Sketch plane. The Context toolbar is displayed.

19) Click **Sketch** 📑 from the Context toolbar as illustrated. The Sketch toolbar is displayed. Front Plane is your Sketch plane.

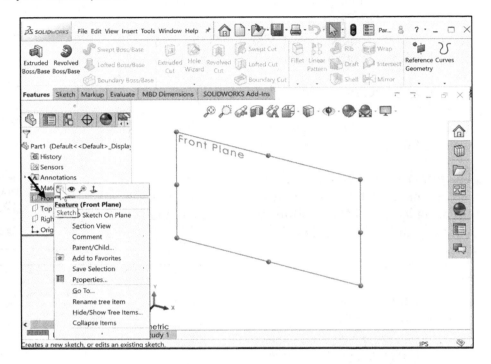

20) Click the **Circle** ⊙ tool from the Sketch toolbar. The Circle PropertyManager is displayed. The Circle-based tool uses a Consolidated Circle PropertyManager. The SOLIDWORKS application defaults to the last used tool type.

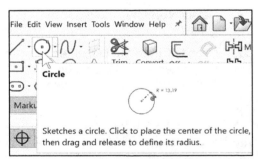

21) Drag the **mouse pointer** into the Graphics window. The cursor displays the Circle icon.

22) Click the **Origin** ⌞ of the circle. The cursor displays the Coincident to point feedback symbol.

23) Drag the **mouse pointer** to the right of the Origin to create the circle as illustrated. The center point of the circle is positioned at the Origin.

24) Click a **position** to create the circle. The activated circle is displayed in blue.

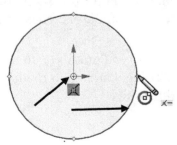

Add a dimension. Use the Smart dimension tool.

25) Click **Smart Dimension** from the Sketch toolbar.

The cursor displays the Smart Dimension icon.

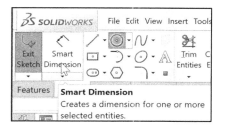

26) Click the **circumference** of the circle.

27) Click a **position** diagonally above the circle in the Graphics window.

28) Enter **.188**in, [**4.78**] in the Modify dialog box. The Dimension Modify dialog box provides the ability to select a unit drop-down menu to modify units in a sketch or feature.

29) Click the **Green Check mark** in the Modify dialog box. The diameter of the circle is .188 inches.

The circular sketch is centered at the Origin. The dimension indicates the diameter of the circle. Sketch Relations are displayed in the model.

Press the f key to fit the part document to the Graphics window.

Add relations, then dimensions. This keeps the user from having too many unnecessary dimensions. This also helps to show the design intent of the model. Dimension what geometry you intend to modify or adjust.

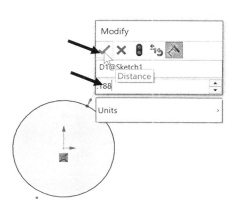

The primary units in this book are provided in IPS (inch, pound, second). The optional secondary units are provided in MMGS (millimeters, grams, second) and are indicated in brackets [ ].

When you create a new part or assembly, the three default Planes (Front, Right and Top) are aligned with specific views.

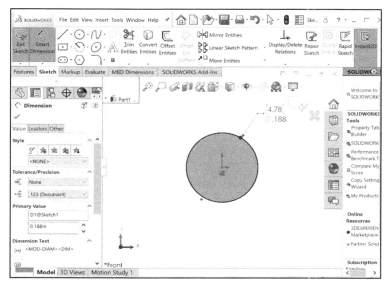

Extrude the sketch to create the Base Feature.

30) Click the **Features** tab from the CommandManager.

31) Click the **Extruded Boss/Base** Features tool. The Boss-Extrude PropertyManager is displayed. Blind is the default End Condition in Direction 1.

32) Select **Mid Plane** for End Condition in Direction 1. The Mid Plane End Condition in the Direction 1 box extrudes the sketch equally on both sides of the Sketch plane. The depth defines the extrude distance.

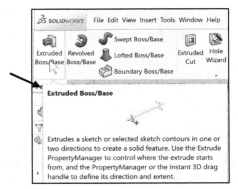

33) Enter **1.375**in, [**34.93**] for Depth in Direction 1. Accept the default conditions.

34) Click **OK** from the Boss-Extrude PropertyManager. Boss-Extrude1 is displayed in the FeatureManager.

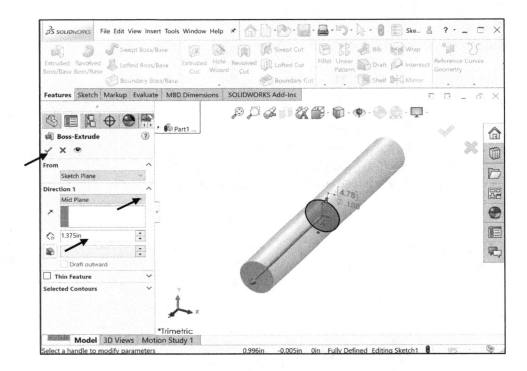

Fit the model to the Graphics window.
35) Press the **f** key. Note the location of the Origin in the model.

Use Symmetry. When possible and if it makes sense, model objects symmetrically about the origin.

Display an Isometric view of the model. Press the **space bar** to display the Orientation dialog box. Click the **Isometric view** icon.

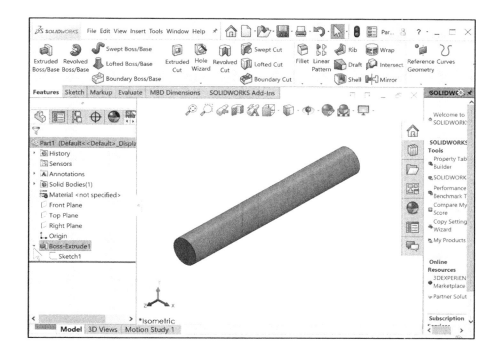

The Boss-Extrude1 feature name is displayed in the FeatureManager. The FeatureManager lists the features, planes, and other geometry that constructs the part. Extrude features add material. Extrude features require the following: *Sketch Plane*, *Sketch* and *depth*.

CommandManager and FeatureManager tabs and tree folders will vary depending on system setup and Add-ins.

Activity: AXLE Part - Save

Save the AXLE part. Enter name. Enter description. Save all parts in the SW-TUTORIAL-2020 folder.

36) Click **Save As** from the drop-down menu.

Create a new folder if needed.

37) Click **New** Folder. Enter **SW-TUTORIAL-2020** for the file folder name. In this book all models, assemblies and drawings are saved to the SW-TUTORIAL-2020 folder.

38) Double-click the **SW-TUTORIAL-2020** file folder. SW-TUTORIAL-2020 is the Save in file folder name.

39) Enter **AXLE** for the File name.

Enter **AXLE ROD** for the Description. Click **Save**. The AXLE FeatureManager is displayed.

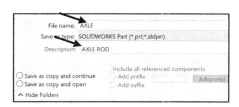

Activity: AXLE Part - Edit Appearance

Modify the color of the AXLE part.

40) Right-click the **AXLE** icon at the top of the FeatureManager.

41) Click the **Appearances** drop-down arrow.

42) Click **AXLE** as illustrated. The Color PropertyManager is displayed. AXLE is displayed in the Selection box.

43) Select a **light blue** color from the Color box. View your options.

44) Click **OK** ✔ from the Color PropertyManager. View the AXLE in the Graphics window.

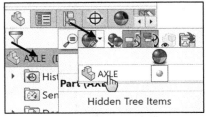

Use the Appearances PropertyManager to apply colors, material appearances, and transparency to parts and assembly components. For sketches or curves only, use the Sketch/Curve Color PropertyManager to apply colors.

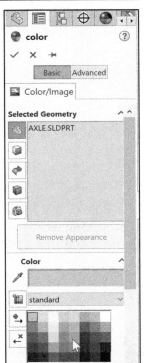

The Advanced tab includes the Illumination and Surface Finish tabs, and additional options in the Color/Image and Mapping tabs. To display the simplified Color/Image or Mapping interfaces, click the Basic tab.

Download all needed model files (SW-TUTORIAL-2020 folder) from the SDC Publication website (www.SDCpublications.com/downloads/978-1-63057-317-1).

Sketching in SOLIDWORKS is the basis for creating features. Features are the basis for creating parts, which can be put together into assemblies.

The sketch status appears in the window status bar and in the FeatureManager. Colors indicate the state of individual sketch entities. Sketches are generally in one of the following states:

1.) *(+) Over defined.* The sketch is displayed in orange - red.

2.) *(-) Under defined.* The sketch is displayed in blue.

3.) *No prefix.* The sketch is fully defined. This is the ideal sketch state. A fully defined sketch has complete information (manufacturing and inspection) and is displayed in black.

💡 The SketchXpert PropertyManager provides the ability to diagnose an over defined sketch to create a fully defined sketch. If you have an over defined sketch, click Over Defined at the bottom of the Graphics window toolbar. The SketchXpert PropertyManager is displayed. Click the Diagnose button.

Select the desired solution and click the Accept button from the Results box.

Activity: AXLE Part - View Modes

Orthographic projection is the process of projecting views onto Parallel planes with ⊥ projectors.

The default reference planes are the Front, Top and Right Planes.

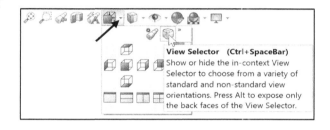

The Isometric view displays the part in 3D with two equal projection angles.

💡 Click **View**, **Hide/Show**, **Origins** from the Menu bar menu to display the Origin in the Graphics window.

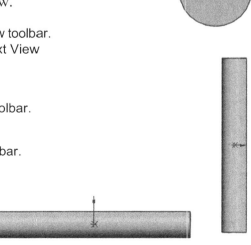

Origin

Display the various view modes using the Heads-up View toolbar.
45) **Deactivate** the View Selector. Hide the in-context View Selector box.

Display a Front view.
46) Click **Front view** 🔲 from the Heads-up View toolbar.

Display a Top view.
47) Click **Top view** 🔲 from the Heads-up View toolbar.

Display a Right view.
48) Click **Right view** 🔲 from the Heads-up View toolbar.

Display an Isometric view.

49)   Click **Isometric view** 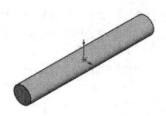 from the Heads-up View toolbar.

View modes manipulate the model in the Graphics window.

Zoom out/Zoom in.

50)   Press the lowercase **z** key to zoom out. Note: you can also use the middle mouse button.

51)   Press the upper-case **Z** key to zoom in. Note: you can also use the middle mouse button.

View the available view tools in the Graphics window.

52)   **Right-click** in the Graphics window. View the available view tools.

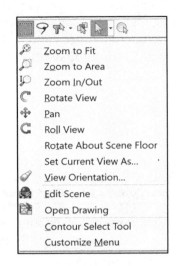

Deselect the view menu.

53)   Click **inside** the Graphics window.

Rotate the model.

54)   Click the **middle mouse** button and move your mouse. The model rotates. The Rotate icon is displayed.

55)   Press the **up arrow** on your keyboard. The arrow keys rotate the model in 15 degree increments.

 View modes remain active until deactivated from the View toolbar or unchecked from the pop-up menu.

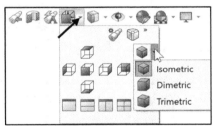

Utilize the center wheel of the mouse to Zoom In/Zoom Out and Rotate the model in the Graphics window.

View the various Display Styles.

56)   Click **Isometric view** 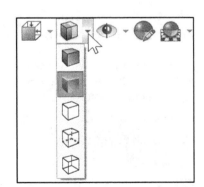 from the Heads-up View toolbar.

57)   Click the **drop-down arrow** from the Display Styles box from the Heads-up Views toolbar as illustrated. SOLIDWORKS provides five key Display Styles:

- *Shaded* . Displays a shaded view of the model with no edges.

- *Shaded With Edges* . Displays a shaded view of the model, with edges.

- *Hidden Lines Removed* . Displays only those model edges that can be seen from the current view orientation.

- *Hidden Lines Visible* . Displays all edges of the model. Edges that are hidden from the current view are displayed in a different color or font.

- *Wireframe* . Displays all edges of the model.

Return to Shaded With Edges. Save the AXLE part.

58)   Click the **Shaded With Edges** icon.

59)   Click **Save** . The AXLE part is complete. Later, apply material to the part.

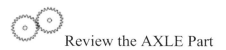   Review the AXLE Part

The AXLE part utilized the Extruded Boss/Base feature. The Extruded Boss/Base feature adds material. The Extruded feature required a Sketch Plane, sketch and depth. The AXLE Sketch plane was the Front Plane. The 2D circle was sketched centered at the Origin. A dimension defined the overall size of the sketch based on the dimensions of mating parts in the LINKAGE assembly.

The default name of the Base feature is Boss-Extrude1. Boss-Extrude1 utilized the Mid Plane End Condition. The Boss-Extrude1 feature is symmetrical about the Front Plane.

The Edit Color option modified the part color. Select the Part icon in the FeatureManager to modify the color of the part. Color and a prefix define the sketch status. A blue sketch is under defined. A black sketch is fully defined. A red sketch is over defined.

The default Reference planes are the Front, Top, and Right Planes. Utilize the Heads-up View toolbar to display the principle views of a part. The View Orientation and Display Style tools manipulate the model in the Graphics windows.

When you create a new part or assembly, the three default Planes (Front, Right and Top) are aligned with specific views. The Plane you select for the Base sketch determines the orientation of the part.

SHAFT-COLLAR Part

The SHAFT-COLLAR part is a hardened steel ring fastened to the AXLE part.

Two SHAFT-COLLAR parts are used to position the two FLATBAR parts on the AXLE.

Create the SHAFT-COLLAR part.

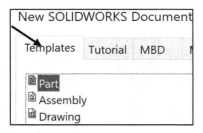

Utilize the Extruded Boss/Base 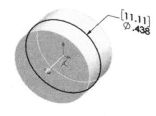 feature. The Extruded Boss/Base feature requires a 2D circular profile.

Utilize symmetry. Sketch a circle on the Front Plane centered at the Origin.

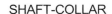

Extrude the sketch with the Mid Plane End Condition. The Extruded Boss/Base feature (Boss-Extrude1) is centered on both sides of the Front Plane.

The Extruded Cut  feature removes material. Utilize an Extruded Cut feature to create a hole. The Extruded Cut feature requires a 2D circular profile. Sketch a circle on the front face centered at the Origin.

SHAFT-COLLAR

The Through All End Condition extends the Extruded Cut feature from the front face through all existing geometry.

The book is designed to expose the new SOLIDWORKS user to many different tools, techniques and procedures.

Activity: SHAFT-COLLAR Part - Extruded Boss/Base Feature

Create a New part.

60) Click **New** from the Menu bar. The New SOLIDWORKS Document dialog box is displayed.

61) Click the **Templates** tab. Part is the default template from the New SOLIDWORKS Document dialog box.

62) Double-click **Part**. The Part FeatureManager is displayed.

Save the part. Enter name. Enter description.

63) Click **Save As**  from the drop-down menu.

64) Enter **SHAFT-COLLAR** for File name in the SW-TUTORIAL-2020 folder.

65) Enter **SHAFT-COLLAR** for Description.

66) Click **Save**. The SHAFT-COLLAR FeatureManager is displayed.

Set Document Properties. Set drafting standard, units, and precision.

67) Click **Options**  from the Main menu.

68) Click the **Document Properties** tab.

69) Select **ANSI** from the Overall drafting standard drop-down menu.

70) Click **Units**.

71) Click **IPS** (inch, pound, second), [**MMGS**] for Unit system.

72) Select **.123**, [**.12**] (three decimal places) for Length units Decimal places.

73) Select **None** for Angular units Decimal places.

74) Click **OK** from the Document Properties - Units dialog box.

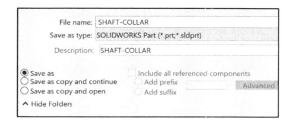

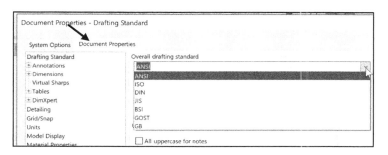

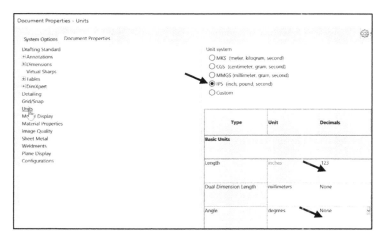

When you create a new part or assembly, the three default Planes (Front, Right and Top) are aligned with specific views. The Plane you select for the Base sketch determines the orientation of the part.

If you modify a document property from an Overall drafting standard, a modify message is displayed as illustrated.

Overall drafting standard
ANSI-MODIFIED

Derived from: ANSI

Insert a new sketch for the Extruded Base feature.

75) Right-click **Front Plane** from the FeatureManager. This is the Sketch plane. The Context toolbar is displayed.

76) Click **Sketch** 🖾 from the Context toolbar as illustrated. The Sketch toolbar is displayed.

77) Click the **Circle** ⊙ tool from the Sketch toolbar. The Circle PropertyManager is displayed.

78) Click the **Origin** ↳. The cursor displays the Coincident to point feedback symbol.

79) Drag the **mouse pointer** to the right of the Origin as illustrated.

80) Click a **position** to create the circle.

Add a dimension.

81) Click **Smart Dimension** ↖ from the Sketch toolbar.

82) Click the **circumference** of the circle. The cursor displays the diameter feedback symbol.

83) Click a **position** diagonally above the circle in the Graphics window.

84) Enter **.4375**in, [**11.11**] in the Modify dialog box.

85) Click the **Green Check mark** ✔ in the Modify dialog box. The black sketch is fully defined.

Note: Three decimal places are displayed. The diameter value .4375 rounds to .438.

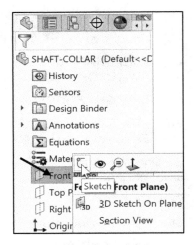

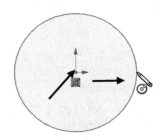

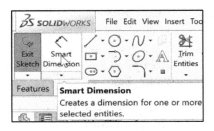

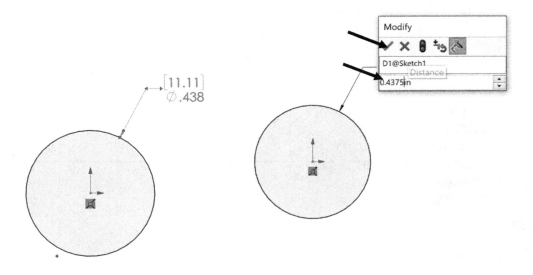

Extrude the sketch to create
the Base feature.

86) Click the **Features** tab
from the
CommandManager.

87) Click the **Extruded
Boss/Base** 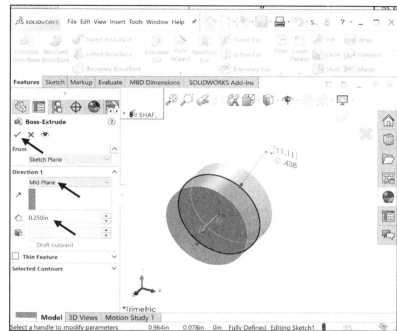 features tool. The
Boss-Extrude
PropertyManager is
displayed.

88) Select **Mid Plane** for
End Condition in
Direction 1.

89) Enter **.250**in, **[6.35]** for
Depth. Accept the
default conditions. View
the location of the
Origin.

90) Click **OK** ✔ from the
Boss-Extrude
PropertyManager.
Boss-Extrude1 is displayed in the FeatureManager.

Fit the model to the Graphics window. Display a trimetric view.
91) Press the **f** key.

92) Click **Trimetric** from the Heads-Up View toolbar.

Save the model.
93) Click **Save** .

Activity: SHAFT-COLLAR Part - Extruded Cut Feature

Insert a new sketch for the Extruded Cut feature.
94) Right-click the **front circular face** of the
Boss-Extrude1 feature for the Sketch plane. The
mouse pointer displays the face feedback icon.

95) Click **Sketch** from the Context toolbar as
illustrated. The Sketch toolbar is displayed. This
is your Sketch plane.

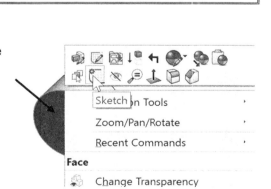

🔆 View the mouse pointer feedback icon for
the correct geometry: line, face, point, or vertex.

96) Click **Hidden Lines Removed** 🔲 from the Heads-up View toolbar.

97) Click the **Circle** ⊙ tool from the Sketch toolbar. The Circle PropertyManager is displayed.

98) Click the red **Origin** ↳. The cursor displays the Coincident to point feedback symbol.

99) Drag the **mouse pointer** to the right of the Origin.

100) Click a **position** to create the circle as illustrated.

Add a dimension.

101) Click the **Smart Dimension** ⟲ Sketch tool.

102) Click the **circumference** of the circle.

103) Click a **position** diagonally above the circle in the Graphics window.

104) Enter **.190**in, **[4.83]** in the Modify dialog box.

105) Click the **Green Check mark** ✔ in the Modify dialog box.

Insert an Extruded Cut feature.
106) Click the **Features** tab from the CommandManager.

107) Click **Extruded Cut** 🔲 from the Features toolbar. The Cut-Extrude PropertyManager is displayed.

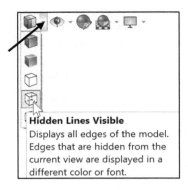

Hidden Lines Visible
Displays all edges of the model. Edges that are hidden from the current view are displayed in a different color or font.

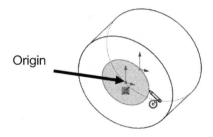

Origin

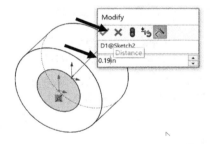

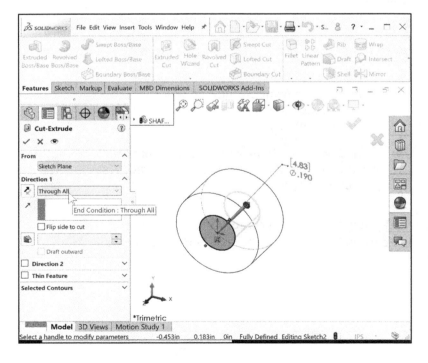

108) Select **Through All** for End Condition in Direction 1. The direction arrow points to the back. If needed, click the Reverse Direction button. Accept the default conditions.

109) Click **OK** ✔ from the Cut-Extrude PropertyManager. Cut-Extrude1 is displayed in the FeatureManager.

The Extruded Cut feature is named Cut-Extrude1. The Through All End Condition removes material from the Front Plane through the Boss-Extrude1 geometry.

 Model about the origin; this provides a point of reference.

Activity: SHAFT-COLLAR - Modify Dimensions and Edit Color

Modify the dimensions.

110) Click the **z** key a few times to Zoom in.

111) Double-click the **outside cylindrical face** of the SHAFT-COLLAR. The Boss-Extrude1 dimensions are displayed. Sketch dimensions are displayed in black. The Extrude depth dimensions are displayed in blue.

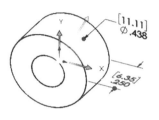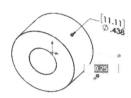

112) Double-click the **.250**in, [**6.35**] depth dimension.

113) Enter **.500**in, [**12.70**].

Rebuild the model.

114) Click **Rebuild** from the Menu bar.

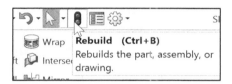

The Boss-Extrude1 feature and Cut-Extrude1 feature are modified.

Return to the original dimensions.

115) Click the **Undo** ↺ tool from the Menu bar.

Display Shaded With Edges.

116) Click **Shaded With Edges** from the Heads-up View toolbar. Modify the part color.

117) Right-click the **SHAFT-COLLAR Part** icon at the top of the FeatureManager.

118) Click the **Appearances** drop-down arrow.

119) Click **SHAFT-COLLAR** as illustrated. The Color PropertyManager is displayed. SHAFT-COLLAR is displayed in the Selection box.

120) Click the **Basic** tab.

121) Select a **light green** color from the Color box.

122) Click **OK** ✔ from the Color PropertyManager.

123) **View** the SHAFT-COLLAR in the Graphics window.

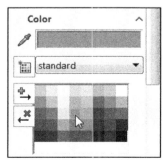

Save the SHAFT-COLLAR part.

124) Click **Save** 💾 . The SHAFT-COLLAR part is complete. All sketches should be fully defined. Later, apply material to the part.

 Review the SHAFT-COLLAR Part

The SHAFT-COLLAR utilized an Extruded Boss/Base 🔲 feature. The Extruded Boss/Base feature adds material. An Extruded feature required a Sketch Plane, sketch and depth.

The Sketch plane was the Front Plane. The 2D circle was sketched centered at the Origin. A dimension fully defined the overall size of the sketch. The default name of the feature was Boss-Extrude1. Boss-Extrude1 utilized the Mid Plane End Condition. The Boss-Extrude1 feature was symmetric about the Front Plane.

FLATBAR Part

The FLATBAR part fastens to the AXLE. The FLATBAR contains nine, ∅.190in holes spaced 0.5in apart.

The FLATBAR part is manufactured from .090inch 6061 alloy. Create the FLATBAR part. Utilize the Centerpoint Straight Slot Sketch 🔘 tool with an Extruded Boss/Base 🔲 feature.

The Extruded feature requires a 2D profile sketched on the Front Plane.

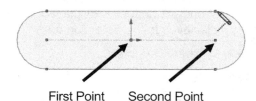

First Point   Second Point

The Centerpoint Straight Slot Sketch tool automatically applies design symmetry (Midpoint and Equal geometric relations). Create the 2D profile centered about the Origin. Relations control the size and position of entities with constraints.

Utilize an Extruded Cut 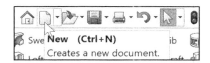 feature to create the first hole. This is the seed feature for the Linear Pattern.

Utilize a Linear Pattern 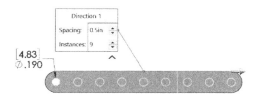 feature to create the remaining holes. A Linear Pattern creates an array of features in a specified direction.

Add relations, then dimensions. This keeps the user from having too many unnecessary dimensions. This also helps to show the design intent of the model.

Activity: FLATBAR Part - Extruded Base Feature

Create a New part.

125) Click **New** from the Menu bar. The New SOLIDWORKS Document dialog box is displayed.

126) Click the **Templates** tab.

127) Double-click **Part**. The Part FeatureManager is displayed.

Save the part. Enter name. Enter description.

128) Click **Save As** from the drop-down menu.

129) Enter **FLATBAR** for File name in the SW-TUTORIAL-2020 folder.

130) Enter **FLAT BAR 9 HOLES** for Description.

131) Click **Save**. The FLATBAR FeatureManager is displayed.

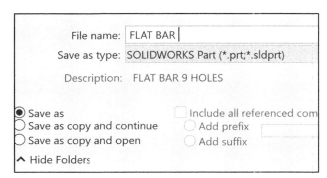

Set Document Properties. Set drafting standard, units, and precision.

132) Click **Options** ⚙ from the Main menu.

133) Click the **Document Properties** tab from the dialog box.

134) Select **ANSI** from the Overall drafting standard drop-down menu.

135) Click **Units**.

136) Click **IPS**, [**MMGS**] for Unit system.

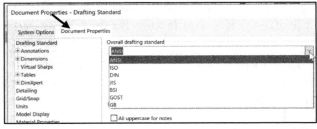

137) Select **.123**, [**.12**] for Length units Decimal places.

138) Select **None** for Angular units Decimal places.

139) Click **OK** to set the document units.

Insert a new sketch for the Extruded Base feature.

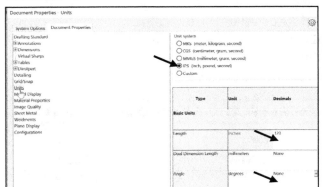

140) Right-click **Front Plane** from the FeatureManager. This is the Sketch plane.

141) Click **Sketch** ⬚ from the Context toolbar as illustrated. The Sketch toolbar is displayed.

Utilize the Consolidated Slot Sketch toolbar. Apply the Centerpoint Straight Slot Sketch tool.

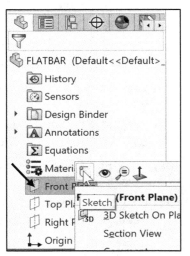

The Straight Slot Sketch tool provides the ability to sketch a straight slot from a center point. In this example, use the origin as your center point.

142) Click the **Centerpoint Straight Slot** ⬭ tool from the Sketch toolbar. The Slot PropertyManager is displayed.

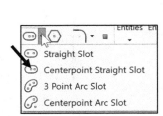

Create the Centerpoint Straight Slot with three points.
143) Click the **Origin**. This is your first point.

144) Click a **point** directly to the right of the Origin. This is your second point.

145) Click a **point** directly above the second point. This is your third point. The Centerpoint Straight Slot is displayed.

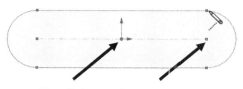

First Point   Second Point

146) Click **OK** ✔ from the Slot PropertyManager.

View Sketch relations.
147) Click **View**, **Hide/Show**, **Sketch Relations** from the Menu bar menu. View the sketch relations in the Graphics window.

Deactivate Sketch relations.
148) Click **View**, **Hide/Show**, un-check **Sketch Relations** from the Menu bar. The Straight Slot Sketch tool provides a midpoint relation with the Origin and Equal relations between the other sketch entities.

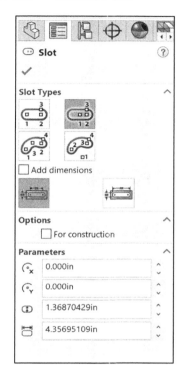

Add dimensions.
149) Click the **Smart Dimension** ✎ tool from the Sketch toolbar.

150) Click the **horizontal** centerline.

151) Click a **position** above the top horizontal line in the Graphics window.

152) Enter **4.000**in, **[101.6]** in the Modify dialog box.

153) Click the **Green Check mark** ✔ in the Modify dialog box.

154) Click the **right arc** of the FLATBAR.

155) Click a **position** diagonally to the right in the Graphics window.

156) Enter **.250**in, **[6.35]** in the Modify dialog box.

157) Click the **Green Check mark** ✔ in the Modify dialog box. The black sketch is fully defined.

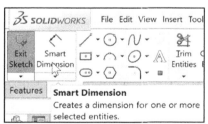

🔆 Model about the Origin; this provides a point of reference for your dimensions to fully define the sketch.

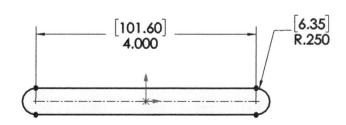

Extrude the sketch to create
the Base (Boss-Extrude1)
feature.

158) Click **Extruded**

   Boss/Base 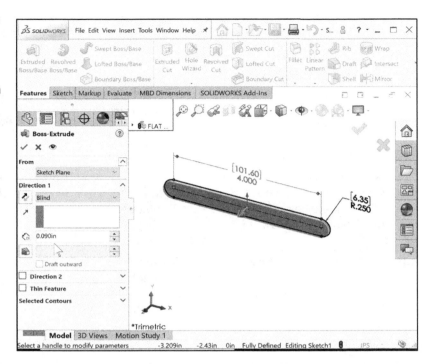 from
   the Features toolbar.
   The Boss-Extrude
   PropertyManager is
   displayed.

159) Enter **.090**in, **[2.29]**
   for Depth. Accept the
   default conditions.

160) Click **OK** ✔ from
   the Boss-Extrude
   PropertyManager.
   Boss-Extrude1 is
   displayed in the
   FeatureManager.

Fit the model to the
Graphics window.
161) Press the **f** key.

Save the FLATBAR part.

162) Click **Save** 💾.

Activity: FLATBAR Part - Extruded Cut Feature

Insert a new sketch for the Extruded Cut Feature.
163) Right-click the **front face** of the Boss-Extrude1
   feature in the Graphics window. This is the Sketch
   plane. Boss-Extrude1 is highlighted in the
   FeatureManager.

164) Click **Sketch** 📝 from the Context toolbar as
   illustrated. The Sketch toolbar is displayed.

Display a Front view - Hidden Lines Removed.
165) Click **Front view** 🔲 from the Heads-up View
   toolbar.

166) Click **Hidden Lines Removed** ◻ from the
   Heads-up View toolbar.

The process of placing the mouse pointer over
an existing arc to locate its center point is called
"wake up."

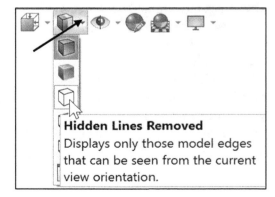

Wake up the Center point.

167) Click the **Circle** 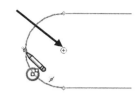 Sketch tool from the Sketch toolbar. The Circle PropertyManager is displayed.

168) Place the **mouse pointer** on the left arc. Do not click. The center point of the slot arc is displayed.

169) Click the **center point** of the arc. Click a **position** to the right of the center point to create the circle as illustrated.

Add a dimension.

170) Click the **Smart Dimension** Sketch tool.

171) Click the **circumference** of the circle. Click a **position** diagonally above and to the left of the circle in the Graphics window.

172) Enter **.190**in, [**4.83**] in the Modify box. Click the **Green Check mark** in the Modify dialog box.

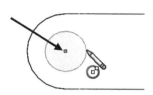

Display an Isometric view - Shaded With Edges.

173) Click **Isometric view** from the Heads-up View toolbar.

174) Click **Shaded With Edges** from the Heads-up View toolbar.

Insert an Extruded Cut feature.

175) Click the **Features** tab from the CommandManager.

176) Click **Extruded Cut** from the Features toolbar. The Cut-Extrude PropertyManager is displayed.

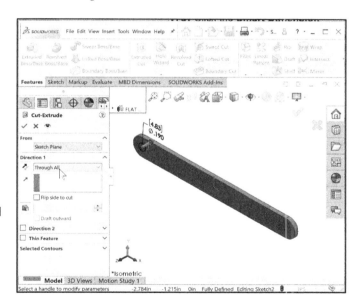

177) Select **Through All** for End Condition in Direction 1. The direction arrow points to the back. Accept the default conditions.

178) Click **OK** from the Cut-Extrude PropertyManager. The Cut-Extrude1 feature is displayed in the FeatureManager.

Save the FLATBAR part.

179) Click **Save**.

Think design intent. When do you use various End Conditions? What are you trying to do with the design? How does the component fit into an Assembly?

The blue Cut-Extrude1 icon indicates that the feature is selected.

Select features by clicking their icons in the FeatureManager or by selecting their geometry in the Graphics window.

When you create a new part or assembly, the three default Planes (Front, Right and Top) are aligned with specific views. The Plane you select for the Base sketch determines the orientation of the part.

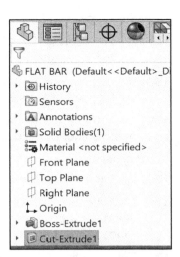

Activity: FLATBAR Part - Linear Pattern Feature

Create a Linear Pattern feature.

180) Click the **Linear Pattern** tool from the Features toolbar. The Linear Pattern PropertyManager is displayed. Cut-Extrude1 is displayed in the Features to Pattern box. Note: If Cut-Extrude1 is not displayed, click inside the Features to Pattern box. Click Cut-Extrude1 from the fly-out FeatureManager.

181) Click **inside** the Pattern Direction box.

182) Click the **top edge** of the Boss-Extrude1 feature for Direction1 in the Graphics window. Edge<1> is displayed in the Pattern Direction box.

183) Enter **0.5**in, **[12.70]** for Spacing.

184) Enter **9** for Number of Instances. Instances are the number of occurrences of a feature.

185) The Direction arrow should point to the right. Click the **Reverse Direction** button if required.

186) Check **Geometry pattern** from the Options box.

187) Click **OK** from the Linear Pattern PropertyManager. The LPattern1 feature is displayed in the FeatureManager.

Save the FLATBAR part.

188) Click **Save** 💾 . The FLATBAR part is complete. Later, apply material to the part.

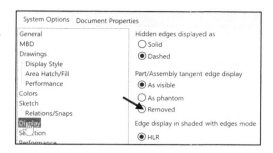

Close all documents.

189) Click **Windows, Close All** from the Menu bar.

 To remove Tangent edges, click **Display** from the System Options menu, and check the **Removed** box.

Review the FLATBAR Part

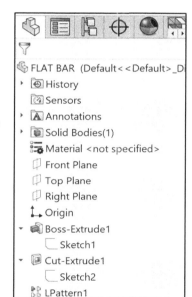

The FLATBAR part utilized an Extruded Boss/Base 🗔 feature as the first feature. The Sketch plane was the Front Plane. The 2D sketch utilized the Straight Slot Sketch tool to create the slot profile.

You added linear and radial dimensions to define your sketch. You applied the Extruded Boss/Base feature with a Blind End Condition in Direction 1. Boss-Extrude1 was created.

You created a circle sketch for the Extruded Cut feature on the front face of Boss-Extrude1. The front face was your Sketch plane for the Extruded Cut feature. The Extruded Cut 🗔 feature removed material to create the hole. The Extruded Cut feature default name was Cut-Extrude1. The Through All End Condition option in Direction 1 created the Cut-Extrude1 feature. The Cut-Extrude1 feature is the seed feature for the Linear Pattern of holes.

The Linear Pattern 🔢 feature created an array of 9 holes, equally spaced along the length of the FLATBAR part. Enter material later in an exercise.

LINKAGE Assembly

An assembly is a document that contains two or more parts. An assembly inserted into another assembly is called a sub-assembly. A part or sub-assembly inserted into an assembly is called a component. The LINKAGE assembly consists of the following components: AXLE, SHAFT-COLLAR, FLATBAR, and AirCylinder sub-assembly.

Establishing the correct component relationship in an assembly requires forethought on component interaction. Mates are geometric relationships that align and fit components in an assembly. Mates remove degrees of freedom from a component.

Mate Types

Mates reflect the physical behavior of a component in an assembly. The components in the LINKAGE assembly utilize Standard mate types. Review Standard, Advanced and Mechanical mate types.

Standard Mates:

Components are assembled with various mate types. The Standard mate types are:

Coincident mate: Locate the selected faces, edges, points or planes so they use the same infinite line. A Coincident mate positions two vertices for contact.

Parallel mate: Locate the selected items to lie in the same direction and to remain a constant distance apart. A parallel mate permits only translational motion of a single part with respect to another. No rotation is allowed.

Perpendicular mate: Locate the selected items at a 90 degree angle to each other. The perpendicular mate allows both translational and rotational motion of one part with respect to another.

Tangent Mate: Locates the selected items in a tangent mate. At least one selected item must be either a conical, cylindrical, or spherical face.

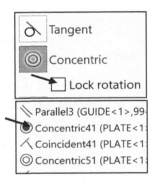

Concentric mate: Locates the selected items so they can share the same center point. You can prevent the rotation of components that are mated with concentric mates by selecting the Lock Rotation option. Locked concentric mates are indicated by an icon in the FeatureManager design tree.

Lock Mate: Maintains the position and orientation between two components.

Distance Mate: Locates the selected items with a specified distance between them. Use the drop-down arrow box or enter the distance value directly.

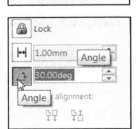

Angle Mate: Locates the selected items at the specified angle to each other. Use the drop-down arrow box or enter the angle value directly.

There are two Mate Alignment options. The Aligned option positions the components so that the normal vectors from the selected faces point in the same direction. The Anti-Aligned option positions the components so that the normal vectors from the selected faces point in opposite directions.

Advanced Mates:

The Advanced mate types are:

Profile Center: Mate to center automatically center-aligns common component types such as rectangular and circular profiles to each other.

Symmetric: Force two similar entities to be symmetric about a planar face or plane.

Width: Center a tab within the width of a groove. The Width mate is used to replace the Symmetric mate where components have tolerance and a gap rather than a tight fit.

Path Mate: Enable any point on a component to be set to follow a defined path.

Linear/Linear Coupler: Establish a relationship between the translation of one component and the translation of another component.

Limit: Allow components to move within a range of values for distance and angle. Select the angle and distance from the provided boxes. Specify a starting distance or angle as well as a maximum and minimum value.

Distance: Locate the selected items with a specified distance between them. Use the drop-down arrow box or enter the distance value directly (Limit mate).

Angle: Locate the selected items at the specified angle to each other. Use the drop-down arrow box or enter the angle value directly.

Mate alignment: Toggle the mate alignment as necessary. The options are:

Aligned: Locate the components so the normal or axis vectors for the selected faces point in the same direction.

Anti-Aligned: Locate the components so the normal or axis vectors for the selected faces point in the opposite direction.

Mechanical Mates:

The Mechanical mate types are:

Cam: Force a plane, cylinder, or point to be tangent or coincident to a series of tangent extruded faces. Four conditions can exist with the Cam mate: *Coincident, Tangent, CamMateCoincident* and *CamMateTangent*. Create the profile of the cam from lines, arcs and splines as long as they are tangent and form a closed loop. View SOLIDWORKS help for additional information.

Slot: Constrain the movement of a bolt or a slot to within a slot hole. You can mate bolts to straight or arced slots and you can mate slots to slots. You can select an axis, cylindrical face, or a slot to create slot mates.

Hinge: Limit the movement between two components to one rotational degree of freedom. It has the same effect as adding a Concentric mate plus a Coincident mate. You can also limit the angular movement between the two components.

Gear: Force two components to rotate relative to one another around selected axes. The Gear mate provides the ability to establish gear type relations between components without making the components physically mesh. Align the components before adding the Mechanical gear mate.

Rack and Pinion: Linear translation of one component (the rack) causes circular rotation in another component (the pinion), and vice versa. You can mate any two components to have this type of movement relative to each other. The components do not need to have gear teeth.

Screw: Constrain two components to be concentric, and adds a pitch relationship between the rotation of one component and the translation of the other. Translation of one component along the axis causes rotation of the other component according to the pitch relationship. Likewise, rotation of one component causes translation of the other component. Align the components before adding the Mechanical screw mate.

Universal Joint: Permit the transfer of rotations from one rigid body to another. This mate is particularly useful to transfer rotational motion around corners, or to transfer rotational motion between two connected shafts that are permitted to bend at the connection (drive shaft on an automobile).

Example: Utilize a Concentric mate between the AXLE cylindrical face and the FLATBAR Extruded Cut feature (hole). Utilize a Coincident mate between the SHAFT-COLLAR back face and the FLATBAR front flat face. The LINKAGE assembly requires the AirCylinder assembly. The AirCylinder assembly is located in the Pneumatic Components folder.

Activity: AirCylinder Assembly - Open and Save As option

Download all needed folders and files to your local hard drive.

190) Download the folders and files to your SW-TUTORIAL-2020 folder. Work from a local hard drive.

Create a New assembly.

191) Click **New** ⬜ from the Menu bar. The New SOLIDWORKS Document dialog box is displayed.

192) Click the **Templates** tab. Double-click **Assembly** from the New SOLIDWORKS Document dialog box. The Begin Assembly PropertyManager is displayed. Open models are displayed in the Open documents box. If there are no Open SOLIDWORKS models, the Window Open dialog is also displayed.

193) Browse to the **SW-TUTORIAL-2020\Pneumatic Components** folder on your hard drive. Double-click the **AirCylinder** assembly from the SW-TUTORIAL-2020\Pneumatic Components folder.

Graph paper
LOGO
MY-TEMPLATES
Pneumatic Components
SOLIDWORKS CSWA Model Folder

194) Click **OK** ✔ from the Begin Assembly PropertyManager. The AirCylinder assembly is displayed in the Graphics window. The (f) symbol is placed in front of the AirCylinder name in the FeatureManager. It is fixed to the origin.

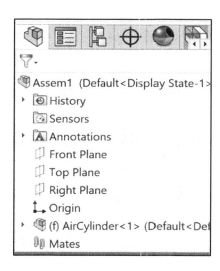

Mate to the first component added to the assembly. If you mate to the first component or base component of the assembly and decide to change its orientation later, all the components will move with it.

💡 Determine the static and dynamic behavior of mates in each sub-assembly before creating the top-level assembly.

195) If required, click **Yes** to Rebuild.

196) Click **Save As** 📋.

197) Select **SW-TUTORIAL-2020** on your hard drive for Save in folder.

198) Enter **LINKAGE** for file name.

199) Click the Include all referenced components box.

200) Click the **Advanced** button. An assembly and drawing document has part\component references. You must save the assembly document and all of the reference components to the same folder. Note your options.

201) Click the **Browse** button from the Save As with References folder.

202) Select the **SW-TUTORIAL-2020** folder on your hard drive.

203) Click **Save All**. The LINKAGE assembly FeatureManager is displayed.

You can also use the Pack and Go option to save an assembly or drawing document with references, toolbox components, studies, etc.

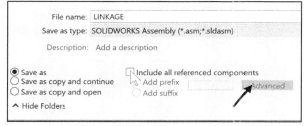

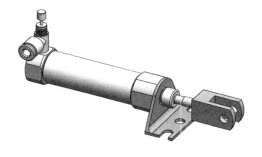

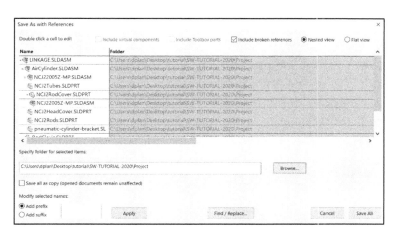

Display the RodClevis component in the FeatureManager.

204) Expand the AirCylinder assembly in the FeatureManager.

205) Click **RodClevis<1>** from the FeatureManager. The RodClevis is highlighted in the Graphics window.

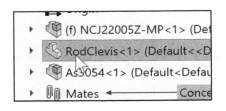

If required hide the Origins.

206) Click **View, Hide/Show,** un-check **Origins** from the Menu bar.

The AirCylinder is the first component in the LINKAGE assembly and is fixed (f) to the LINKAGE assembly Origin.

Display an Isometric view.

207) Click **Isometric view** .

Insert the AXLE part.

208) Click the **Assembly** tab in the CommandManager.

209) Click the **Insert Components** Assembly tool. The Insert Component PropertyManager is displayed. Note: Open SOLIDWORKS models are displayed in the Open documents box. If there are no Open models, the Window Open dialog is also displayed.

210) Browse to the **SW-TUTORIAL-2020** folder. Note if AXLE is active, double-click AXLE from the Open documents box.

211) Double-click **AXLE** from the SW-TUTORIAL-2020 folder.

212) Click a **position** to the front of the AirCylinder assembly as illustrated.

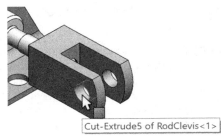

Cut-Extrude5 of RodClevis<1>

Enlarge the view.

213) Zoom in on the RodClevis and the AXLE.

Insert a Concentric mate. Use the Quick mate procedure.

214) Click the **inside front hole face** of the RodClevis. The cursor displays the face feedback symbol.

215) Hold the **Ctrl** key down.

216) Click the **long cylindrical face** of the AXLE. The cursor displays the face feedback symbol.

217) Release the **Ctrl** key. The Mate pop-up menu is displayed.

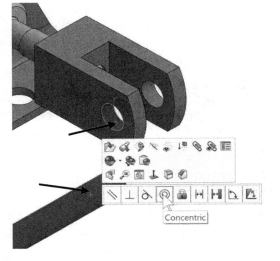

218) Click **Concentric** from the Mate pop-up menu.
The AXLE is positioned concentric to the
RodClevis hole.

Move the AXLE.

219) Click and drag the **AXLE** left to right. The AXLE
translates in and out of the RodClevis holes.

🔅 Position the mouse pointer in the middle of
the face to select the entire face. Do not position
the mouse pointer near the edge of the face. If the
wrong face or edge is selected, click inside the
Graphics window to deselect the entity.

Display a Top view.

220) Click **Top view** 📦.

Insert a Coincident mate. Expand the LINKAGE
assembly and components in the FeatureManager.

221) **Expand** the LINKAGE assembly
FeatureManager.

222) **Expand** the AirCylinder assembly.

223) Click the **Front Plane** of the AirCylinder
assembly.

224) **Expand** the AXLE part from
the FeatureManager.

225) Hold the **Ctrl** key down.

226) Click the **Front Plane** of the
AXLE part.

227) Release the **Ctrl** key. The
Mate pop-up menu is
displayed.

228) Click **Coincident mate**
from the pop-up menu.

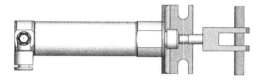

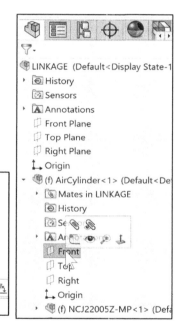

The AirCylinder Front Plane and the AXLE Front
Plane are Coincident. The AXLE is centered in the
RodClevis.

If you delete a mate and then recreate it, the mate
numbers will be in a different order.

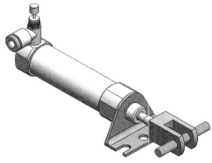

🔅 To fix the first component to the Origin in an
assembly, click OK ✔ from the Begin Assembly
PropertyManager or click the Origin in the Graphics window.

Display an Isometric view.

229) Click **Isometric view** .

Display the Mates in the folder.
230) Expand the Mates folder in the FeatureManager. View the created mates.

Activity: LINKAGE Assembly - Insert FLATBAR Part

Insert the FLATBAR part.

231) Click the **Insert Components** 📦 Assembly tool. The Insert Component PropertyManager is displayed. Note: Open SOLIDWORKS models are displayed in the Open documents box. If there are no Open models, the Window Open dialog is also displayed.

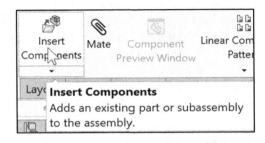

232) Double-click **FLATBAR** from the SW-TUTORIAL-2020 folder.

Place the component in the assembly.
233) Click a **position** in the Graphics window as illustrated.

Enlarge the view.
234) Zoom in on the AXLE and the left side of the FLATBAR to enlarge the view.

Insert a Concentric mate.
235) Click the inside **left front hole face** of the FLATBAR.

236) Hold the **Ctrl** key down.

237) Click the **long cylindrical face** of the AXLE.

238) Release the **Ctrl** key. The Mate pop-up menu is displayed.

239) Click **Concentric** from the Mate pop-up menu.

Fit the model to the Graphics window.
240) Press the **f** key.

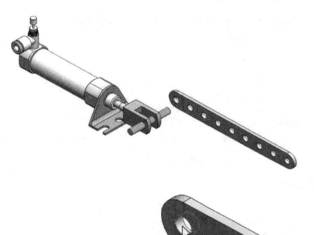

Move the FLATBAR.

241) Click and drag the **FLATBAR**. The FLATBAR translates and rotates along the AXLE.

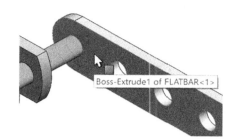

Insert a Coincident mate.

242) Click the **front face** of the FLATBAR.

243) **Rotate** the model to view the back face of the RodClevis.

244) Hold the **Ctrl** key down.

245) Click the **back face** of the RodClevis as illustrated.

246) Release the **Ctrl** key. The Mate pop-up menu is displayed.

247) Click **Coincident** from the Mate pop-up menu.

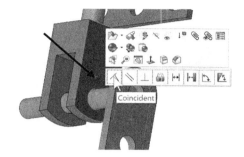

Display an Isometric view.

248) Click **Isometric view** .

Insert the second FLATBAR component.

249) Click the **Insert Components** Assembly tool.

250) Browse to the **SW-TUTORIAL-2020** folder.

251) Double-click **FLATBAR**.

252) Click a **position** to the front of the AirCylinder in the Graphics window as illustrated.

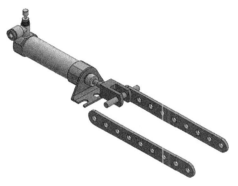

Enlarge the view.

253) **Zoom in** on the second FLATBAR and the AXLE.

Insert a Concentric mate.

254) Click the **left front inside hole face** of the second FLATBAR.

255) Hold the **Ctrl** key down.

256) Click the **long cylindrical face** of the AXLE.

257) Release the **Ctrl** key. The Mate pop-up menu is displayed.

258) Click **Concentric** from the Mate pop-up menu.

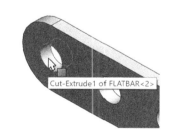

259) Click and drag the **second FLATBAR** to the front.

Fit the model to the Graphics window.

260) Press the **f** key.

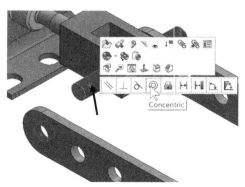

Insert a Coincident mate.

261) Rotate the model to view the back face of the second FLATBAR.

262) Click the **back face** of the second FLATBAR.

263) Rotate the model to view the front face of the RodClevis.

264) Hold the **Ctrl** key down.

265) Click the **front face** of the RodClevis.

266) Release the **Ctrl** key. The Mate pop-up menu is displayed.

267) Click **Coincident** from the Mate pop-up menu.

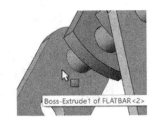

Insert a Parallel mate.

268) Press the **Shift-z** keys to Zoom in on the model.

269) Click the **top narrow face** of the first FLATBAR.

270) Hold the **Ctrl** key down.

271) Click the **top narrow face** of the second FLATBAR.

272) Release the **Ctrl** key.

273) Click **Parallel** ⟍ from the Mate pop-up menu.

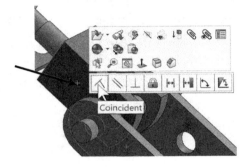

Display an Isometric view.

274) Click **Isometric view** ⬔ .

Move the two FLATBAR parts.

275) Click and drag the **second FLATBAR**. Both FLATBAR parts move together.

View the Mates folder.

276) Expand the Mates folder from the FeatureManager. View the created mates.

Determine the static and dynamic behavior of mates in each sub-assembly before creating the top level assembly.

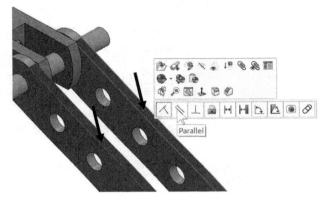

Activity: LINKAGE Assembly - Insert SHAFT-COLLAR Part

Insert the first SHAFT-COLLAR.

277) Click the **Insert Components** 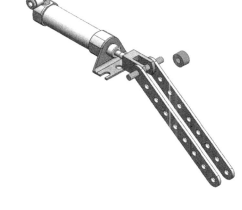 Assembly tool.

278) Double-click **SHAFT-COLLAR** from the SW-TUTORIAL-2020 folder.

279) Click a **position** to the back of the AXLE as illustrated.

Enlarge the view.

280) Click the **Zoom to Area** tool.

281) Zoom-in on the SHAFT-COLLAR and the AXLE component.

Deactivate the tool.

282) Click the **Zoom to Area** tool.

Insert a Concentric mate.
283) Click the **inside hole face** of the SHAFT-COLLAR.

284) Hold the **Ctrl** key down.

285) Click the **long cylindrical face** of the AXLE.

286) Release the **Ctrl** key. The Mate pop-up menu is displayed.

287) Click **Concentric** from the Mate pop-up menu.

Insert a Coincident mate.
288) Press the **Shift-z** keys to Zoom in on the model.

289) Click the **front face** of the SHAFT-COLLAR as illustrated.

290) Rotate the model to view the back face of the first FLATBAR.

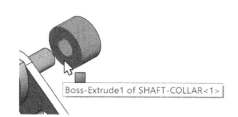

291) Hold the **Ctrl** key down.

292) Click the **back face** of the first FLATBAR.

293) Release the **Ctrl** key. The Mate pop-up menu is displayed.

294) Click **Coincident** from the Mate pop-up menu.

Display an Isometric view.

295) Click **Isometric view** .

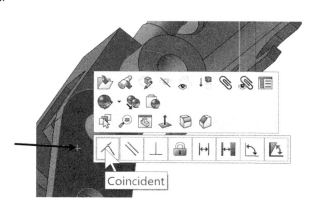

Insert the second SHAFT-COLLAR.

296) Click the **Insert Components** Assembly tool.

297) Double-click the **SHAFT-COLLAR** part from the SW-TUTORIAL-2020 folder.

298) Click a **position** near the AXLE as illustrated.

Enlarge the view.

299) Click the **Zoom to Area** tool.

300) Zoom-in on the second SHAFT-COLLAR and the AXLE to enlarge the view.

301) Click the **Zoom to Area** tool to deactivate the tool.

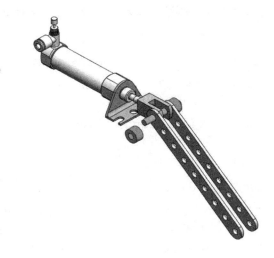

Insert a Concentric mate.
302) Click the **inside hole face** of the second SHAFT-COLLAR.

303) Hold the **Ctrl** key down.

304) Click the **long cylindrical face** of the AXLE.

305) Release the **Ctrl** key. The Mate pop-up menu is displayed.

306) Click **Concentric** from the Mate pop-up menu.

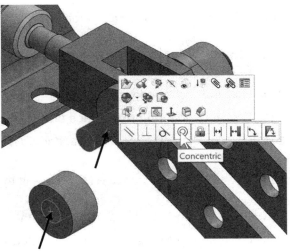

Insert a Coincident mate.
307) Click the **back face** of the second SHAFT-COLLAR.

308) Rotate to view the front face of the second FLATBAR.

309) Hold the **Ctrl** key down.

310) Click the **front face** of the second FLATBAR.

311) Release the **Ctrl** key. The Mate pop-up menu is displayed.

312) Click **Coincident** from the Mate pop-up menu.

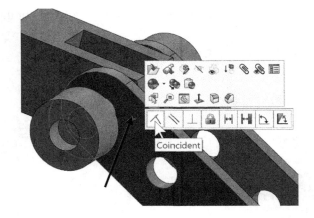

View the created Mates.
313) Expand the Mates folder. View the created mates.

Display an Isometric view.
314) Click **Isometric view** .

Fit the model to the Graphics window.
315) Press the **f** key.

Save the LINKAGE assembly.
316) Click **Save** .

317) Click **Rebuild and Save** the document. The LINKAGE assembly is complete.

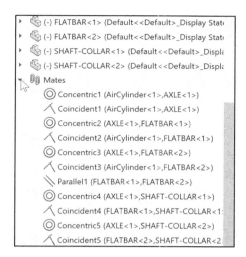

Use the Pack and Go option to save an assembly or drawing with references. The Pack and Go tool saves either to a folder or creates a zip file to e-mail. View SOLIDWORKS help for additional information.

 Review the LINKAGE Assembly

An assembly is a document that contains two or more parts. A part or sub-assembly inserted into an assembly is called a component. You created the LINKAGE assembly. The AirCylinder sub-assembly was the first component inserted into the LINKAGE assembly. The AirCylinder assembly was obtained from the Pneumatic Components folder.

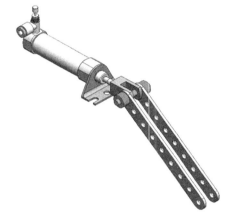

The AirCylinder assembly was fixed to the Origin. The Concentric and Coincident mates added Geometric relationships between the inserted components in the LINKAGE assembly.

To remove the fixed state, Right-click a component name in the FeatureManager. Click Float. The component is free to move.

To fix the first component to the Origin in an assembly, click OK from the Begin Assembly PropertyManager or click the Origin in the Graphics window.

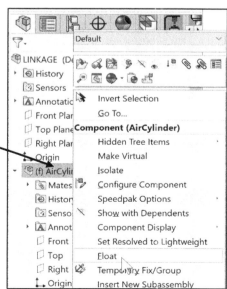

The AXLE part was the second component inserted into the LINKAGE assembly. The AXLE required a Concentric mate between the two cylindrical faces and a Coincident mate between the two Front Planes.

The FLATBAR part was the third component inserted into the LINKAGE assembly. The FLATBAR required a Concentric mate between the two cylindrical faces and a Coincident mate between the two flat faces.

A second FLATBAR was inserted into the LINKAGE assembly. A Parallel mate was added between the two FLATBARs.

Two SHAFT-COLLAR parts were inserted into the LINKAGE assembly. Each SHAFT-COLLAR required a Concentric mate between the two cylindrical faces and a Coincident mate between the two flat faces.

Motion Study - Basic Motion Tool

Motion Studies are graphical simulations of motion for assembly models. You can incorporate visual properties such as lighting and camera perspective into a motion study. Motion studies do not change an assembly model or its properties. They simulate and animate the motion you prescribe for your model. Use SOLIDWORKS mates to restrict the motion of components in an assembly when you model motion.

Create a Motion Study. Select the Basic Motion option from the MotionManager. The Basic Motion option provides the ability to approximate the effects of motors, springs, collisions and gravity on your assembly. Basic Motion takes mass into account in calculating motion. Note: The Animation option does not.

Activity: LINKAGE Assembly - Basic Motion

Insert a Rotary Motor using the Motion Study tab.

318) Click the **Motion Study 1** tab located in the bottom left corner of the Graphics window. The MotionManager is displayed.

319) Select **Basic Motion** for Type of study from the MotionManager drop-down menu as illustrated.

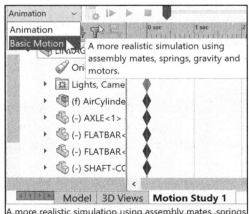

320) Click **Motor**  from the MotionManager. The Motor PropertyManager is displayed.

321) Click the **Rotary Motor** box.

322) Click the **FLATBAR front face** as illustrated. A red Rotary Motor icon is displayed. The red direction arrow points counterclockwise.

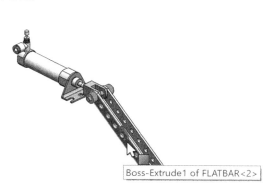

Boss-Extrude1 of FLATBAR<2>

323) Enter **150 RPM** for speed in the Motion box.

324) Click **OK** from the Motor PropertyManager.

Record the Simulation.

325) Click **Calculate** . The FLATBAR rotates in a counterclockwise direction for a set period of time.

326) Click **Play** . View the simulation.

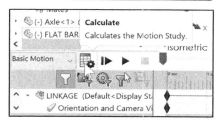

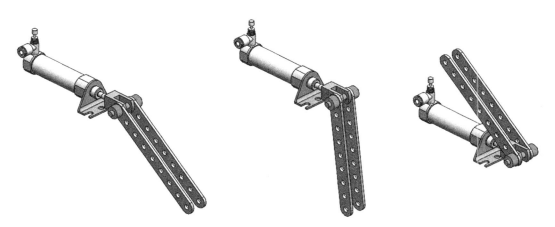

Save the simulation in an AVI file to the SW-TUTORIAL-2020 folder.

327) Click **Save Animation**.

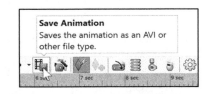

328) Click **Save** from the Save Animation to File dialog box. View your options.

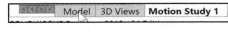

329) Click **OK** from the Video Compression box.

Close the Motion Study and return to SOLIDWORKS.

330) Click the **Model** tab location in the bottom left corner of the Graphics window.

Fit the assembly to the Graphics window.

331) Press the **f** key.

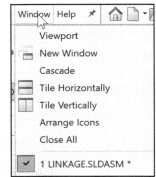

Save the LINKAGE assembly.

332) Click **Save** 💾.

Exit SOLIDWORKS.

333) Click **Windows**, **Close All** from the Menu bar.

The LINKAGE assembly chapter is complete.

Rename a feature or sketch for clarity. Slowly click the feature or sketch name twice and enter the new name when the old one is highlighted.

🔅 Use the Pack and Go option to save an assembly or drawing with references. The Pack and Go tool saves either to a folder or creates a zip file to e-mail. View SOLIDWORKS help for additional information.

 Review the Motion Study

The Rotary Motor Basic Motion tool combined Mates and Physical Dynamics to rotate the FLATBAR components in the LINKAGE assembly. The Rotary Motor was applied to the front face of the FLATBAR. You utilized the Calculate option to play the simulation. You saved the simulation in an .avi file.

🔍 Additional details on Motion Study, Assembly, mates, and Simulation are available in SOLIDWORKS help. Keywords: Motion Study and Basic Motion.

Chapter Summary

In this chapter, you created three parts (AXLE, SHAFT-COLLAR and FLATBAR), downloaded the AirCylinder assembly from the Pneumatic Components folder and created the LINKAGE assembly using the Standard and Quick mate procedure.

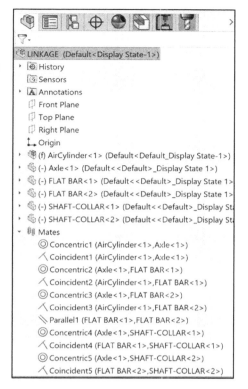

You developed an understanding of the SOLIDWORKS User Interface: Menus, Toolbars, Task Pane, CommandManager, FeatureManager, System feedback icons, Document Properties, Parts, and Assemblies.

You created 2D sketches and addressed the three key states of a sketch: *Fully Defined, Over Defined* and *Under Defined*. Note: Always review your FeatureManager for the proper sketch state.

You obtained the knowledge of the following SOLIDWORKS features: Extruded Boss/Base, Extruded Cut, and Linear Pattern. Features are the building blocks of parts. The Extruded Boss/Base feature required a Sketch plane, sketch, and depth.

The Extruded Boss/Base feature added material to a part. The Boss-Extruded1 feature was utilized in the AXLE, SHAFT-COLLAR and FLATBAR parts.

The Extruded Cut feature removed material from the part. The Extruded Cut feature was utilized to create a hole in the SHAFT-COLLAR and FLATBAR parts. Note: Both were Through All holes. We will address the Hole Wizard later in the book. The Linear Pattern feature was utilized to create an array of holes in the FLATBAR part.

When parts are inserted into an assembly, they are called components. You created the LINKAGE assembly by inserting the AirCylinder assembly, AXLE, SHAFT-COLLAR and FLATBAR parts.

Mates are geometric relationships that align and fit components in an assembly. Concentric, Coincident and Parallel mates were utilized to assemble the components.

You created a Motion Study. The Rotary Motor Basic Motion tool combined Mates and Physical Dynamics to rotate the FLATBAR components in the LINKAGE assembly.

Questions

1. Explain the steps in starting a SOLIDWORKS session.

2. Describe the procedure to begin a new 2D sketch.

3. Explain the steps required to modify units in a part document from inches to millimeters.

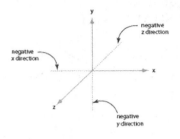

4. Describe the procedure to create a simple 3D cylinder on the Front Plane with an Extruded Boss/Base (Boss-Extrude1) feature.

5. Identify the three default Reference planes in SOLIDWORKS.

6. Describe a Base feature. Provide two examples from this chapter.

7. Describe the differences between an Extruded Boss/Base feature and an Extruded Cut feature.

8. The sketch color black indicates a sketch is _____ defined.

9. The sketch color blue indicates a sketch is _____ defined.

10. The sketch color red indicates a sketch is _____ defined.

11. Describe the procedure to "wake up" a center point in a sketch.

12. Define a Geometric relation. Provide three examples.

13. Describe the procedure to create a Linear Pattern feature.

14. Describe an assembly or sub-assembly.

15. What are mates and why are they important in assembling components?

16. In an assembly, each component has_____# degrees of freedom. Name them.

17. True or False. A fixed component cannot move in an assembly.

18. Review the Design Intent section in the book. Identify how you incorporated design intent into a part or assembly document.

Exercises

Exercise 2.1: Identify the Sketch plane for the Boss-Extrude1 (Base) feature as illustrated. Simplify the number of features.

A: Top Plane

B: Front Plane

C: Right Plane

D: Left Plane

Correct answer _____.

Create the part. Dimensions are arbitrary.

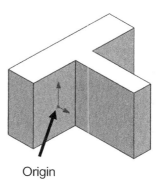

Origin

Exercise 2.2: Identify the Sketch plane for the Boss-Extrude1 (Base) feature as illustrated. Simplify the number of features.

A: Top Plane

B: Front Plane

C: Right Plane

D: Left Plane

Correct answer _____.

Create the part. Dimensions are arbitrary.

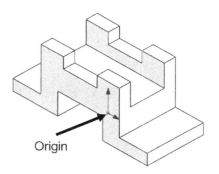

Origin

Exercise 2.3: Identify the Sketch plane for the Boss-Extrude1 (Base) feature as illustrated. Simplify the number of features.

A: Top Plane

B: Front Plane

C: Right Plane

D: Left Plane

Correct answer _____.

Create the part. Dimensions are arbitrary.

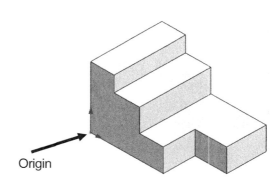

Origin

Exercise 2.4: FLATBAR - 3HOLE Part

Create an ANSI, IPS FLATBAR - 3HOLE part document.

- Utilize the Front Plane for the Sketch plane. Insert an Extruded Base (Boss-Extrude1) feature. Do not display Tangent Edges.

- Create an Extruded Cut feature. This is your seed feature. Apply the Linear Pattern feature. The FLATBAR - 3HOLE part is manufactured from 0.06in., [1.5mm] 6061 Alloy.

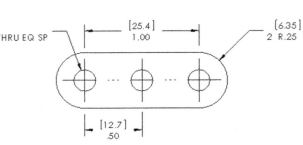

Exercise 2.5: FLATBAR - 5HOLE Part

Create an ANSI, IPS, FLATBAR - 5HOLE part as illustrated.

- Utilize the Front Plane for the Sketch plane. Insert an Extruded Base (Boss-Extrude1) feature.

- Create an Extruded Cut feature. This is your seed feature. Apply the Linear Pattern feature. The FLATBAR - 5HOLE part is manufactured from 0.06in, [1.5mm] 6061 Alloy.

- Calculate the required dimensions for the FLATBAR - 5HOLE part. Use the following information: Holes are .500in. on center, Radius is .250in., and hole diameter is .190in.

- Do not display Tangent edges in the final model.

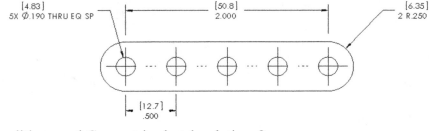

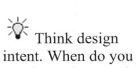 Think design intent. When do you use the various End Conditions and Geometric sketch relations? What are you trying to do with the design? How does the component fit into an assembly?

Exercise 2.6: Simple Block Part

Create the illustrated ANSI part. Note the location of the Origin in the illustration.

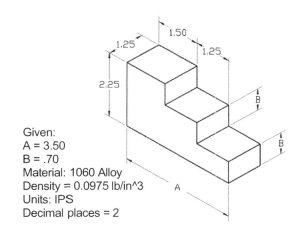

- Calculate the overall mass of the illustrated model.

- Apply the Mass Properties tool.

- Think about the steps that you would use to build the model.

- Review the provided information carefully.

- Units are represented in the IPS (inch, pound, second) system.

- A = 3.50in, B = .70in

Given:
A = 3.50
B = .70
Material: 1060 Alloy
Density = 0.0975 lb/in^3
Units: IPS
Decimal places = 2

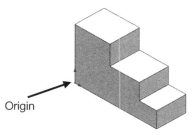

Origin

Exercise 2.7: Simple Block Part

Create the illustrated ANSI part. Note the location of the Origin in the illustration.

Create the sketch symmetric about the Front Plane. The Front Plane in this problem is **not** your Sketch Plane. Utilize the Blind End Condition in Direction 1.

Given:
A = 3.00
B = .75
Material: Copper
Density = 0.321 lb/in^3
Units: IPS
Decimal places = 2

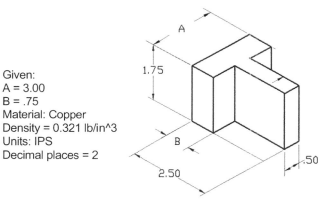

- Calculate the overall mass of the illustrated model.

- Apply the Mass Properties tool.

- Think about the steps that you would use to build the model.

- Review the provided information carefully. Units are represented in the IPS (inch, pound, second) system.

- A = 3.00in, B = .75in

Note: Sketch1 is symmetrical.

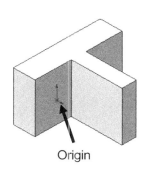

Origin

Exercise 2.8: Simple Block Part

Create an ANSI part from the illustrated model. Note the location of the Origin in the illustration.

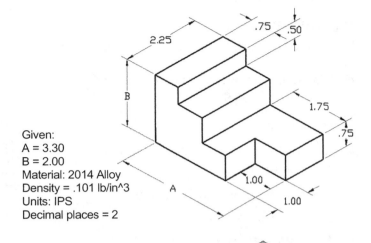

- Calculate the volume of the part and locate the Center of mass with the provided information.

Given:
A = 3.30
B = 2.00
Material: 2014 Alloy
Density = .101 lb/in^3
Units: IPS
Decimal places = 2

- Apply the Mass Properties tool.

- Think about the steps that you would use to build the model.

- Review the provided information carefully.

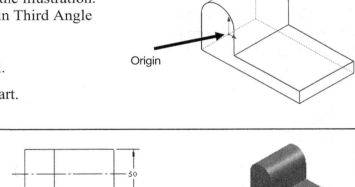

Origin

Exercise 2.9: Simple Block Part

Create an ANSI, MMGS part from the illustrated drawing: Front, Top, Right and Isometric views.

Note the location of the Origin in the illustration. The drawing views are displayed in Third Angle Projection.

- Apply 1060 Alloy for material.

- Calculate the Volume of the part.

- Locate the Center of mass.

Think about the steps that you would use to build the model. The part is symmetric about the Front Plane.

Origin

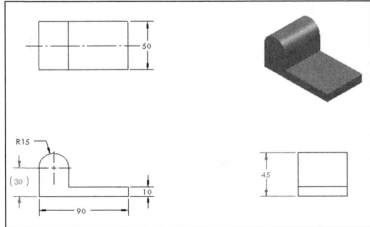

Exercise 2.10: Simple Block Part

Create the ANSI, MMGS part from the illustrated drawing: Front, Top, Right and Isometric views.

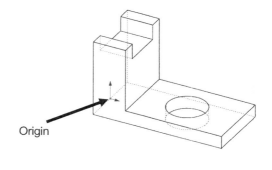

Origin

- Apply 1060 Alloy for material.

- The part is symmetric about the Front Plane.

- Calculate the Volume of the part and locate the Center of mass.

Think about the steps that you would use to build the model.

The drawing views are displayed in Third Angle Projection.

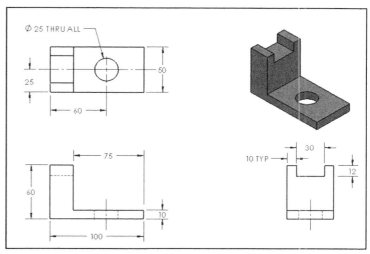

Exercise 2.11: LINKAGE-2 Assembly

Create the LINKAGE-2 assembly.

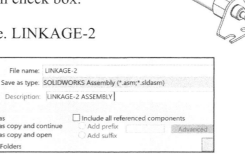

- Copy the LINKAGE assembly from the Chapter 2 Homework folder to your local hard drive.

- Open the assembly.

- Select Save As from the drop-down Menu bar.

- Check the Save as copy and open check box.

- Enter LINKAGE-2 for file name. LINKAGE-2 ASSEMBLY for description.

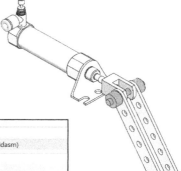

File name:	LINKAGE-2
Save as type:	SOLIDWORKS Assembly (*.asm;*.sldasm)
Description:	LINKAGE-2 ASSEMBLY

- Save as
- Save as copy and continue
- Save as copy and open
- Hide Folders
- ☐ Include all referenced components
- Add prefix
- Add suffix
- Advanced

The FLATBAR-3HOLE part was created in Exercise 2.4. Utilize two AXLE parts, four SHAFT COLLAR parts, and two FLATBAR-3HOLE parts to create the LINKAGE-2 assembly as illustrated.

- Insert the first AXLE part.

- Insert a Concentric mate.

- Insert a Coincident mate.

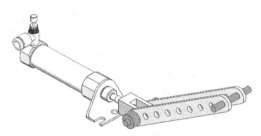

- Insert the first FLATBAR-3HOLE part.

- Insert a Concentric mate.

- Insert a Coincident mate.

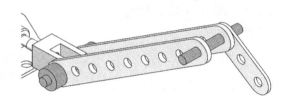

- Perform the same procedure for the second FLATBAR-3HOLE part.

- Insert a Parallel mate between the 2 FLATBAR-3HOLE parts. Note: The 2 FLATBAR-3HOLE parts move together.

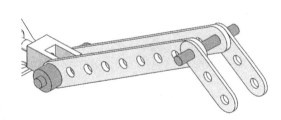

☀ When a component is in the Lightweight

🔷 state, only a subset of its model data is loaded into memory. The remaining model data is loaded on an as-needed basis.

☀ When a component is fully resolved, all its model data is loaded into memory.

- Insert the second AXLE part.

- Insert a Concentric mate.

- Insert a Coincident mate.

- Insert the first SHAFT-COLLAR part.

- Insert a Concentric mate.

- Insert a Coincident mate.

- Perform the same tasks to insert the other three required SHAFT-COLLAR parts as illustrated.

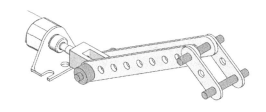

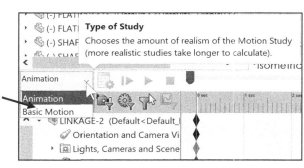

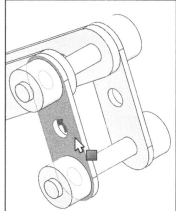

Exercise 2.12: LINKAGE-2 Assembly Motion Study

Create a Motion Study using the LINKAGE-2 Assembly that was created in the previous exercise.

- Create a Basic Motion Study.

- Apply a Rotary Motor to the front FLATBAR-3HOLE as illustrated.

- Play and Save the Simulation.

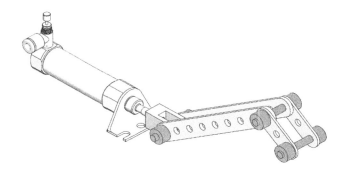

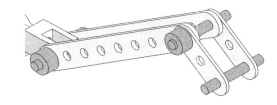

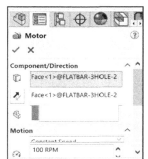

Exercise 2.13: ROCKER Assembly

Create a ROCKER assembly. The ROCKER assembly consists of two AXLE parts, two FLATBAR-5HOLE parts, and two FLATBAR-3HOLE parts.

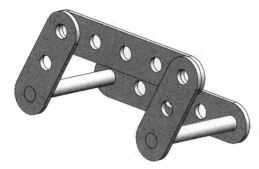

The FLATBAR-3HOLE parts are linked together with the FLATBAR-5HOLE.

The three parts rotate clockwise and counterclockwise, above the Top Plane. Create the ROCKER assembly.

- Insert the first FLATBAR-5HOLE part. The FLATBAR-5HOLE is fixed to the Origin of the ROCKER assembly.

- Insert the first AXLE part.

- Insert a Concentric mate.

- Insert a Coincident mate.

- Insert the second AXLE part.

- Insert a Concentric mate.

- Insert a Coincident mate.

- Insert the first FLATBAR-3HOLE part.

- Insert a Concentric mate.

- Insert a Coincident mate.

- Insert the second FLATBAR-3HOLE part.

- Insert a Concentric mate.

- Insert a Coincident mate.

- Insert the second FLATBAR-5HOLE part.

- Insert the required mates.

Note: The end holes of the second FLATBAR-5HOLE are concentric with the end holes of the FLATBAR-3HOLE parts.

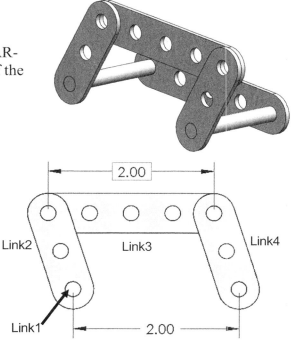

Note: In mechanical design, the ROCKER assembly is classified as a mechanism. A Four-Bar Linkage is a common mechanism comprised of four links.

Link1 is called the Frame.

The AXLE part is Link1.

Link2 and Link4 are called the Cranks.

The FLATBAR-3HOLE parts are Link2 and Link4. Link3 is called the Coupler. The FLATBAR-5HOLE part is Link3.

If an assembly or component is loaded in a Lightweight state, right-click the assembly name or component name from the FeatureManager. Click Set Lightweight to Resolved.

Determine the static and dynamic behavior of mates in each sub-assembly before creating the top-level assembly.

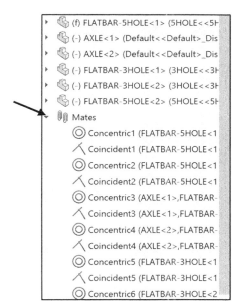

Exercise 2.14: 4 Bar linkage

Create the 4-bar linkage assembly as illustrated. Create the five components. Assume dimensions.

In an assembly, fix (f) the first component to the origin or fully define it to the three default reference planes.

Insert all needed mates to simulate the movement of a 4-bar linkage assembly.

Read the section on Coincident, Concentric and Distance mates in SOLIDWORKS help and in the SOLIDWORKS 2020 Tutorial book.

Create a base with text for extra credit.

Below are a few sample models from my Freshman Engineering class.

Note the different designs to maintain the proper movement of a 4-bar linkage using a base.

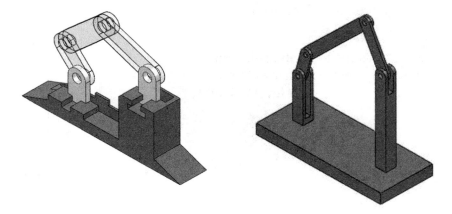

Exercise 2.15: Fill Pattern 2-1

Create a Polygon Layout Fill Pattern feature. Apply the seed cut option.

1. Open **Fill Pattern 2-1** from the Chapter 2 Homework folder.

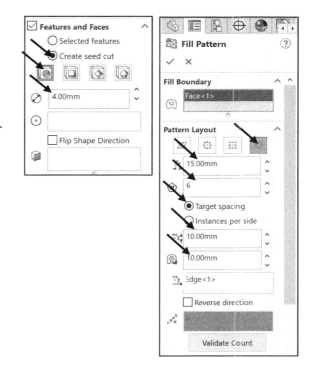

2. Click the **Fill Pattern** 🔠 Features tool. The Fill Pattern PropertyManager is displayed.

3. Click the **Front face** of Extrude1. Face<1> is displayed in the Fill Boundary box. The direction arrow points to the right.

4. Click **Polygon** for Pattern Layout.

5. Click **Target spacing**.

6. Enter **15**mm for Loop Spacing.

7. Enter **6** for Polygon sides.

8. Enter **10**mm for Instances Spacing.

9. Enter **10**mm for Margins. View the direction arrow.

10. Click **Create seed cut**.

11. Click **Circle** for Features to Pattern.

12. Enter **4**mm for Diameter.

13. Click **OK** ✅ from the Fill Pattern PropertyManager. Fill Pattern1 is created and is displayed in the FeatureManager.

14. **View** the results.

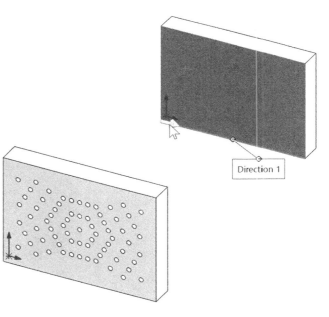

Exercise 2.16: Fill Pattern 2-2

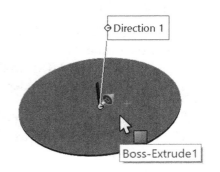

Create a Circular Layout Fill Pattern feature.

1. **Copy** the Chapter 2 Homework folder to your hard drive.

2. Open **Fill Pattern 2-2** from the Chapter 2 Homework folder.

3. Click the **Fill Pattern** 🖼 Features tool. The Fill Pattern PropertyManager is displayed.

4. Click the **top face** of Boss-Extrude1. Face<1> is displayed in the Fill Boundary box. The direction arrow points to the back.

5. Click **Circular** for Pattern Layout.

6. Enter **.10**in for Loop Spacing.

7. Click **Target spacing**.

8. Enter .**10**in Instance Spacing.

9. Enter **.05**in for Margins. Edge<1> is selected for Pattern Direction.

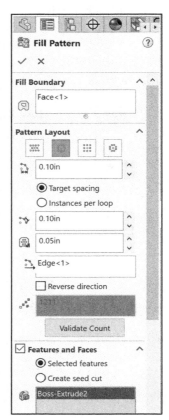

10. Click **inside** the Features to Pattern box.

11. Click **Boss-Extrude2** from the fly-out FeatureManager. Boss-Extrude2 is displayed in the Features to Pattern box.

12. Click **OK** ✔ from the Fill Pattern PropertyManager. Fill Pattern1 is created and is displayed in the FeatureManager.

13. **View** the results.

Exercise 2.17: Tangent Mate

Create a Tangent mate between the roller-wheel and the cam on both sides to simulate movement.

1. **Copy** the Chapter 2 Homework folder to your hard drive.

2. Open **Tangent** from the Chapter 2 Homework\Tangent folder. The assembly is displayed.

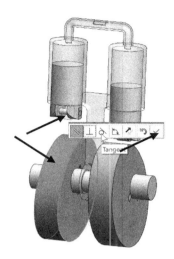

Create a Tangent mate between the roller-wheel and the cam on both sides. The cam was created with imported geometry (curve).

A Tangent mate places the selected items tangent to each other. At least one selected item must be either a conical, cylindrical, spherical face. An Angle mate places the selected items at the specified angle to each other.

3. Click the **Mate** ✎ tool from the Assembly tab.

4. **Pin** the Mate PropertyManager.

Create the first Tangent mate.

5. Click the **contact face** of the left cam as illustrated.

6. Click the **contact face** of the left roller-wheel as illustrated.

7. Click **Tangent** ○ mate from the PropertyManager.

8. Click **OK** ✔ from the Pop-up menu.

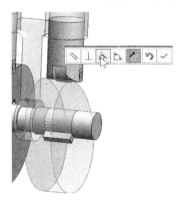

You can set an option in the Mate PropertyManager so that the first component you select from becomes transparent. Then selecting from the second component is easier, especially if the second component is behind the first. The option is supported for all mate types except those that might have more than one selection from the first component (width, symmetry, linear coupler, cam, and hinge).

Create the second Tangent mate.

9. Click the **contact face** of the right cam.

10. Click the **contact face** of the right roller-wheel.

11. Click **Tangent** ○ mate from the PropertyManager.

12. Click **OK** ✔ from the Pop-up menu. **Un-Pin** the Mate PropertyManager. Click **OK** ✔ from the Mate PropertyManager.

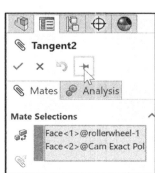

13. Display an **Isometric** view. **Move** the cam. View the results.

Exercise 2.18: Motion Study 1

Apply a motor to move components in an
assembly. Set up an animation motion study. Use
the motion study time line and the
MotionManager design tree to suppress mates in
the motion study.

1. **Copy** the Chapter 2 Homework folder to
   your hard drive.

2. Open **Motion Study 1** from the Chapter 2
   Homework\Motion Study folder.

3. Click the **Motion Study 1** tab at the bottom
   of the Graphics window.

4. Select **Basic Motion** for Study type.

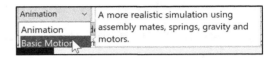

5. Click the **Motor** 🔧 tool from the
   MotionManager toolbar. The Motor
   PropertyManager is displayed.

6. Click **Rotary Motor** for Motor Type.

7. Click the **illustrated face** for Motor
   Direction in the Graphics window. The
   direction arrow points counterclockwise.

8. Select **Constant speed** for Motor Type.

9. Enter **30 RPM** for Speed.

10. Click **OK** ✔ from the Motor PropertyManager.

Set the duration and run your animation on the 4bar-Linkage.

11. Drag the key to **6** seconds as
    illustrated.

12. Click **Play from Start** ▮▶ from
    the MotionManager. View the
    results.

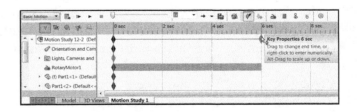

In the next section, suppress a
Concentric mate on a component.
View the results in the Motion Study.

13. **Expand** the Mate folder in the MotionManager Design tree.

14. Right-click at 2 seconds and select **Place Key** 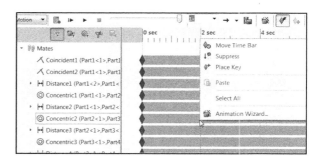 for the Concentric2 mate.

15. Right-click at 4 seconds and select **Place Key** for the Concentric2 mate.

16. Set the time bar to **2 seconds**.

17. Right-click **Concentric2** in the MotionManager design tree.

18. Click **Suppress**. The mate is suppressed between the 2 and 4 second marks.

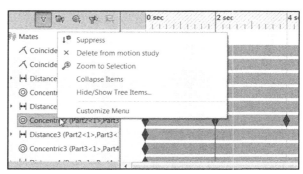

19. Click the **Calculate** tool. After calculating a motion study, you can view it again.

20. Click **Play from Start** ▶ from the MotionManager toolbar. View the results. Explore your options and the available tools. View SOLIDWORKS Help for additional information on a Motion Study.

Return to the FeatureManager.

21. Click the **Model** tab to return to SOLIDWORKS.

22. **Close** the model.

Notes:

Chapter 3

FRONT-SUPPORT Assembly

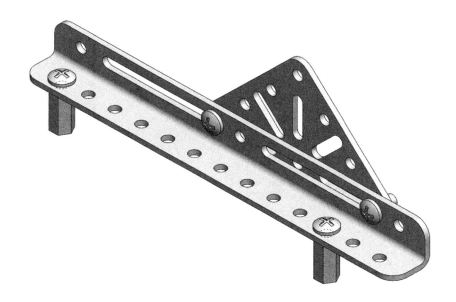

Below are the desired outcomes and usage competencies based on the completion of Chapter 3.

Desired Outcomes:	**Usage Competencies**:

• Create four parts:
 	
   o HEX-STANDOFF
 	
   o ANGLE-13HOLE
 	
   o TRIANGLE
 	
   o SCREW
	• Apply the following model features: Extruded Boss/Base, Extruded Thin, Extruded Cut, Revolved Boss/Base, Hole Wizard, Linear Pattern, Circular Pattern, Mirror, Fillet and Chamfer.
	
	• Apply sketch techniques with various sketch tools and Construction geometry.

• Create an assembly: o FRONT-SUPPORT assembly	• Comprehend the assembly process and insert the following mate types: Concentric, Coincident, Parallel, Distance, Gear, Screw, Hinge, and Cam.

Notes:

Chapter 3 - FRONT-SUPPORT Assembly

Chapter Objective

Create four new parts utilizing the Top, Front and Right Planes. Determine the Sketch plane for each feature. Obtain the knowledge of the following SOLIDWORKS features: Extruded Boss/Base, Extruded Thin, Extruded Cut, Revolved Boss/Base, Hole Wizard, Linear Pattern, Circular Pattern, Fillet, and Chamfer.

Apply sketch techniques with various Sketch tools: Line, Circle, Corner Rectangle, Centerline, Dynamic Mirror, Straight Slot, Trim Entities, Polygon, Tangent Arc, Sketch Fillet, Offset Entities, and Convert Entities.

Create four new parts:

1. HEX-STANDOFF
2. ANGLE-13HOLE
3. TRIANGLE
4. SCREW

Create the FRONT-SUPPORT assembly.

On the completion of this chapter, you will be able to:

- Select the correct Sketch plane.
- Generate a 2D sketch.
- Insert the required dimensions and Geometric relations.
- Apply the following SOLIDWORKS features:
  - Extruded Boss/Base
  - Extruded Cut
  - Extruded Thin
  - Revolved Base
  - Linear and Circular Pattern
  - Mirror
  - Fillet
  - Hole Wizard
  - Chamfer
- Apply the following mate types: Concentric, Coincident, Parallel, Distance, Gear, and Cam.

Chapter Overview

The FRONT-SUPPORT assembly supports various pneumatic components and is incorporated into the PNEUMATIC-TEST-MODULE.

Create four new parts in this chapter:

1. HEX-STANDOFF

2. ANGLE-13HOLE

3. TRIANGLE

4. SCREW

ANGLE-13HOLE

HEX-STANDOFF

Create the FRONT-SUPPORT assembly using the four new created parts.

The FRONT-SUPPORT assembly is used in the exercises at the end of this chapter and in later chapters of the book.

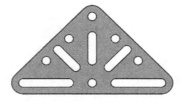

TRIANGLE

Simplified SCREW

Think design intent. When do you use various End Conditions and Geometric sketch relations? What are you trying to do with the design? How does the component fit into an assembly?

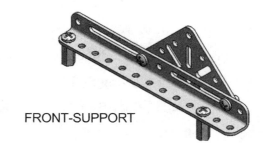

FRONT-SUPPORT

Download all needed model files from the SDC Publications website (www.sdcpublications.com) to a local hard drive.

Reference Planes and Orthographic Projection

The three default ⊥ Reference planes represent infinite 2D planes in 3D space:

- Front
- Top
- Right

Planes have no thickness or mass.

Orthographic projection is the process of projecting views onto parallel planes with ⊥ projectors.

The default ⊥ datum planes are:

- Primary
- Secondary
- Tertiary

These are the planes used in manufacturing:

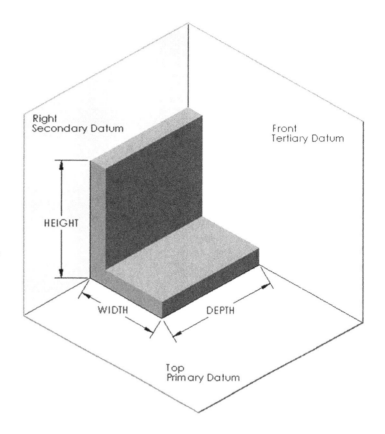

- Primary datum plane contacts the part at a minimum of three points.
- Secondary datum plane contacts the part at a minimum of two points.
- Tertiary datum plane contacts the part at a minimum of one point.

The part view orientation depends on the Base feature Sketch plane.

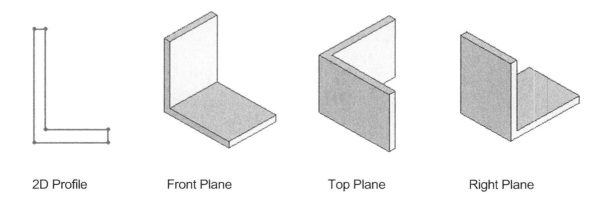

   2D Profile   Front Plane   Top Plane   Right Plane

The part view orientation is dependent on the Base feature Sketch plane. Compare the available default Sketch planes in the FeatureManager: Front, Top and Right Plane.

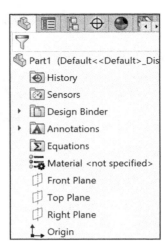

Each Boss-Extrude1 feature above was created with an L-shaped 2D Sketch profile. The six principle views of Orthographic projection listed in the ASME Y14.3standard are:

- Top

- Front

- Right side

- Bottom

- Rear (Back)

- Left side

SOLIDWORKS Standard view names correspond to these Orthographic projection view names.

ASME Y14.3M Principle View Name:	SOLIDWORKS Standard View:
Front	Front
Top	Top
Right side	Right
Bottom	Bottom
Rear	Back
Left side	Left

The standard drawing views in Third Angle Orthographic projection are:

- Front

- Top

- Right

- Isometric

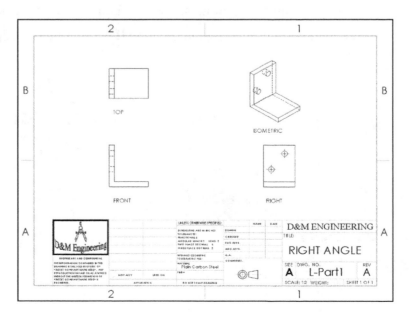

There are two Orthographic projection drawing systems. The first Orthographic projection system is called the Third Angle projection. The second Orthographic projection system is called the First Angle projection. The systems are derived from positioning a 3D object in the third or first quadrant.

Third Angle Projection

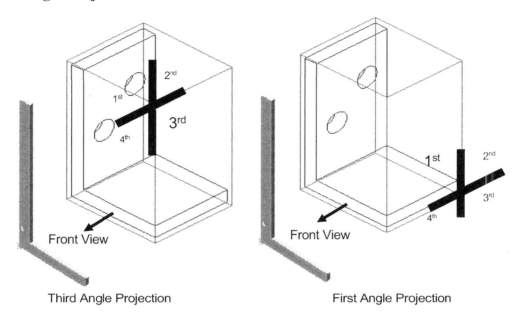

Third Angle Projection   First Angle Projection

The part is positioned in the third quadrant in third angle projection. The 2D projection planes are located between the viewer and the part. The projected views are placed on a drawing.

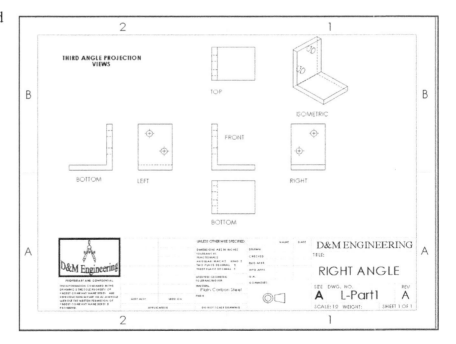

First Angle Projection

The part is positioned in the first quadrant in First Angle projection. Views are projected onto the planes located behind the part. The projected views are placed on a drawing. First Angle projection is primarily used in Europe and Asia.

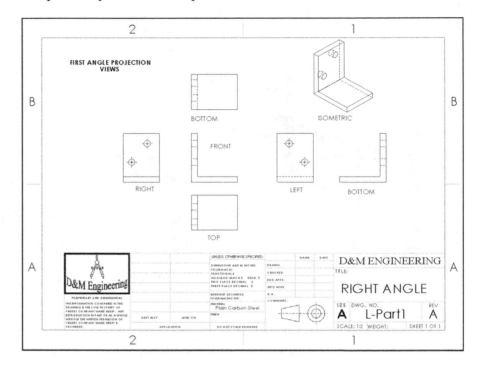

Third Angle projection is primarily used in the U.S. & Canada and is based on the ASME Y14.3M Multi and Sectional View Drawings standard. Designers should have knowledge and understanding of both systems.

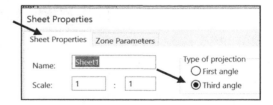

There are numerous multi-national companies. Example: A part is designed in the U.S., manufactured in Japan and destined for a European market.

Third Angle projection is used in this text. A truncated cone symbol appears on the drawing to indicate the Projection System:

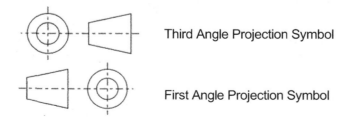

Third Angle Projection Symbol

First Angle Projection Symbol

Select the Sketch plane based on symmetry and orientation of the part in the FRONT-SUPPORT assembly. Utilize the standard views: *Front*, *Back*, *Right*, *Left*, *Top*, *Bottom* and *Isometric* to orient the part. Create the 2D drawings for the parts in Chapter 4.

HEX-STANDOFF Part

The HEX-STANDOFF part is a hexagonal shaped part utilized to elevate components in the FRONT-SUPPORT assembly. Machine screws are utilized to fasten components to the HEX-STANDOFF.

Create the HEX-STANDOFF part with the Extruded Boss/Base  feature. The Sketch plane for the HEX-STANDOFF Boss-Extrude1 feature is the Top Plane.

Create the HEX-STANDOFF in the orientation utilized by the FRONT-SUPPORT assembly.

HEX-STANDOFF

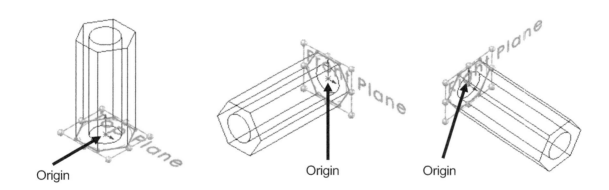

Origin   Origin   Origin

Note the location of the origin in the model. When you create a new part, the three default Planes (Front, Right and Top) are aligned with specific views. The Plane you select for the Base sketch determines the orientation of the part.

All sketches should be fully defined in the FeatureManager.

Insert Geometric relations first then dimensions in a sketch to maintain design intent.

The Boss-Extrude1 feature sketch consists of two profiles. The first sketch is a circle centered at the Origin on the Top Plane.

The second sketch is a polygon with 6 sides centered at the Origin. The polygon utilizes an inscribed circle to construct the geometry.

Geometric relations are constraints that control the size and position of the sketch entities. Apply a Horizontal relation in the polygon sketch.

Extrude the sketch perpendicular to the Top Plane. Utilize the Edit Sketch tool to modify the sketch.

The Hole Wizard feature creates complex and simple Hole features. Utilize the Hole Wizard feature to create a Tapped Hole.

The Tapped Hole depth and diameter are based on drill size and screw type parameters. Apply a Coincident relation to position the Tapped Hole aligned with the Origin.

Activity: HEX-STANDOFF Part - Extruded Boss/Base Feature

Create a New part.

1)   Click **New** 🗋 from the Menu bar.

2)   Click the **Templates** tab.

3)   Double-click **Part**. The Part FeatureManager is displayed.

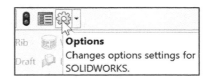

Set Document Properties. Set drafting standard.

4)   Click **Options** ⚙ from the Main menu.

5)   Click the **Document Properties** tab from the dialog box. Select **ANSI** from the Overall drafting standard box.

Set document units and precision.

6)   Click **Units**.

7)   Select **IPS**, [**MMGS**] for Unit system.

8)   Select **.123**, [**.12**] for Length units Decimal places.

9)   Select **None** for Angular units Decimal places.

10)   Click **OK**. The Part FeatureManager is displayed.

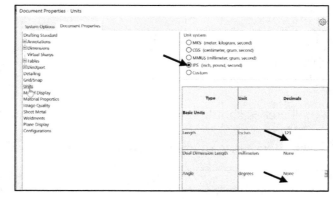

The primary units are provided in IPS (inch, pound, seconds). The optional secondary units are provided in MMGS (millimeter, gram, second) and are indicated in brackets [ ]. Illustrations are provided in inches and millimeters.

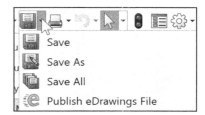

Save the part. Enter name. Enter description.

11) Click **Save As** .

12) Select the **SW-TUTORIAL-2020** folder.

13) Enter **HEX-STANDOFF** for File name.

14) Enter **HEX-STANDOFF 10-24** for Description. Click **Save**. The HEX-STANDOFF FeatureManager is displayed.

Select the Sketch plane.
15) Right-click **Top Plane** from the FeatureManager. This is the Sketch plane.

Insert a Boss-Extrude1 feature (Base feature) sketched on the Top Plane. Note: A plane is an infinite 2D area. The blue boundary is for visual reference.

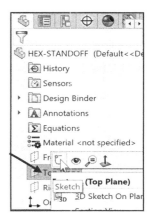

Insert a new sketch.
16) Click **Sketch** from the Context toolbar. The Sketch toolbar is displayed.

17) Click the **Circle** Sketch tool. The Circle PropertyManager is displayed.

18) Drag the **mouse pointer** into the Graphics window. The cursor displays the Circle icon.

19) Click the **Origin** .

20) Drag the **mouse pointer** to the right of the Origin.

21) Click a **position** to create the circle as illustrated.

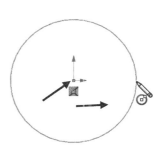

Insert a Polygon.

22) Click the **Polygon** Sketch tool. The Polygon PropertyManager is displayed. The cursor displays the Polygon icon.

23) Click the **Origin** as illustrated.

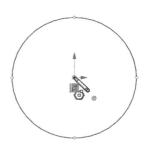

24)   Drag the **mouse pointer** horizontally to the right.

25)   Click a **position** to the right of the circle to create the hexagon as illustrated. **Display** Geometric relations.

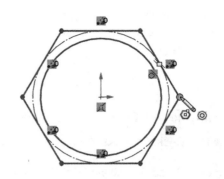

Add a dimension.

26)   Click the **Smart Dimension** Sketch tool.

27)   Click the **circumference** of the first circle.

28)   Click a **position** diagonally above the hexagon to locate the dimension.

29)   Enter .**150**in, [**3.81**] in the Modify box.

30)   Click the **Green Check mark** in the Modify dialog box.

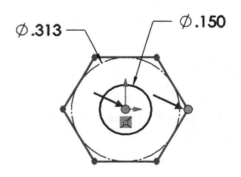

31)   Click the **circumference** of the inscribed circle. Click a **position** diagonally below the hexagon to locate the dimension.

32)   Enter .**313**in, [**7.95**] in the Modify box.

33)   Click the **Green Check mark** in the Modify dialog box. The black sketch is fully defined.

34)   **Deactivate** the Geometric relations.

Fit the model to the Graphics window.

35)   Press the **f** key to fit the model to the Graphics window.

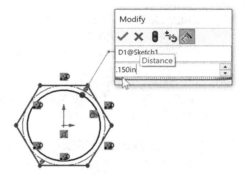

If required, click the arrowhead dot to toggle the direction of the dimension arrow.

36)   Click **OK** from the Dimension PropertyManager.

If needed, add a Horizontal relation.

37)   Click the **Origin** . Hold the **Ctrl** key down.

38)   Click the right most **point** of the hexagon. The Properties PropertyManager is displayed.

39)   Release the **Ctrl** key. Click **Horizontal** from the Add Relations box.

40)   Click **OK** from the Properties PropertyManager.

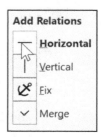

Extrude the sketch.

41)   Click the **Features** tab in the CommandManager.

42)   Click **Extruded Boss/Base** from the Features toolbar. The Boss-Extrude PropertyManager is displayed. Blind is the default End Condition in Direction1. The direction arrow points upward.

43)   Enter **.735**in, [**18.67**] for Depth.

44)   Click **OK** ✔ from the Boss-Extrude PropertyManager. Boss-Extrude1 is displayed in the FeatureManager.

Fit the model to the Graphics window.

45)   Press the **f** key.

The Boss-Extrude1 feature (Base feature) was sketched on the Top Plane. Changes occur in the design process. Edit the sketch of Boss-Extrude1. Delete the circle and close the sketch. Apply the Hole Wizard feature to create a Tapped Hole.

Edit Sketch1.

46)   **Expand** Boss-Extrude1 in the FeatureManager.

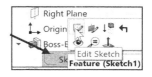

47)   Right-click **Sketch1** in the FeatureManager.

48)   Click **Edit Sketch** from the Context toolbar.

Delete the inside circle.

49)   Click the **circumference** of the inside circle as illustrated. The Circle PropertyManager is displayed.

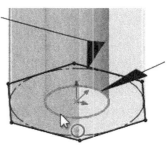

50)   Press the **Delete** key.

51)   Click **Yes** to the Sketcher Confirm Delete message. Both the circle geometry and its dimension are deleted.

Save and close the sketch.

52)   Click **Save** 💾.

53)   Click **Rebuild and Save**. The Boss-Extrude1 feature is updated. Note: Sketch1 is fully defined.

Fit the model to the Graphics window.

54)   Press the **f** key.

Activity: HEX-STANDOFF Part - Hole Wizard Feature

Insert a Tapped Hole with the Hole Wizard feature tool. Create a 2D Sketch.

55) Click **Hidden Lines Visible** ⬚ from the Heads-up View toolbar.

56) Click **Hole Wizard** from the Features toolbar. The Hole Specification PropertyManager is displayed.

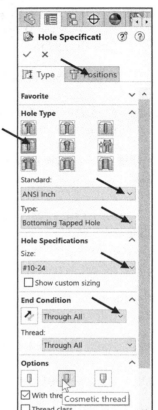

Note: For metric, utilize ANSI Metric and M5x0.8 for size.

57) Click **Straight Tap** for Hole Specification.

58) Select **ANSI Inch**, [**ANSI Metric**] for Standard.

59) Select **Bottoming Tapped Hole** for Type.

60) Select **#10-24**, [**Ø5**] for Size.

61) Select **Through All** for End Condition.

62) Click the **Cosmetic thread** box. Accept the default conditions.

63) Click the **Positions** tab.

64) Click the **top face** of Boss-Extrude1 to the right of the origin. Do not select the center of the part.

65) Click the **center of the part** to locate the center point of the hole as illustrated. The Straight Tapped hole is displayed in yellow. Yellow is a preview color. The Point Sketch tool is automatically selected. No other holes are required. You created a Coincident relation to the center point of the Tapped Hole in the Top view.

66) Right-click **Select** in the Graphics window to deselect the Point Sketch tool. Do not right-click on the model.

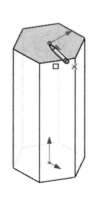

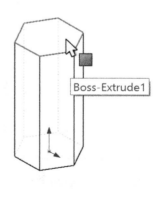

67)   Click **OK** ✔ from the Hole Position PropertyManager.

The #10-24 Tapped Hole1 feature is displayed in the
FeatureManager. Sketch2 determines the center point
location of the Tapped Hole. Sketch3 is the profile of
the Tapped Hole.

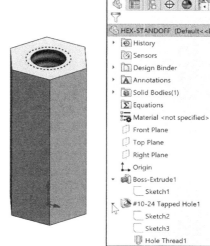

Save the HEX-STANDOFF part.

68)   Click **Shaded With Edges** 🔲 from the Heads-up
View toolbar.

69)   Click **inside** the Graphics window.

70)   Click **Save** 💾. The HEX-STANDOFF is complete.
Later, add material. View the threads.

To view the thread, right-click the Annotations folder, and click Details. Check the
Cosmetic thread box and the Shaded cosmetic threads box. Click OK.

 Review the HEX-STANDOFF Part.

The HEX-STANDOFF part utilized the Extruded Boss/Base feature. The
Boss-Extrude1 feature required a sketch on the Top Plane. The first profile
was a circle centered at the Origin on the Top Plane. The second profile
used the Polygon Sketch tool. You utilized the Edit Sketch tool to modify
the Sketch profile and to delete the circle.

The Hole Wizard feature created a Tapped Hole. The Hole Wizard feature
required the Boss-Extrude1 top face as the Sketch plane.

A Coincident relation located the center point of the Tapped Hole aligned
with respect to the Origin.

CommandManager and FeatureManager tabs and file folders will vary
depending on system setup and Add-ins.

ANGLE-13HOLE Part

The ANGLE-13HOLE part is an L-shaped
support bracket. The ANGLE-13HOLE part is
manufactured from 0.090in, [2.3] aluminum.

The ANGLE-13HOLE part contains fillets, holes,
and slot cuts.

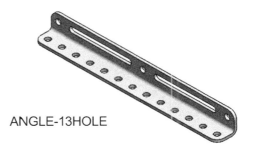

ANGLE-13HOLE

Simplify the overall design into seven features. Utilize symmetry and Linear Patterns.

The open L-Shaped profile is sketched on the Right Plane.

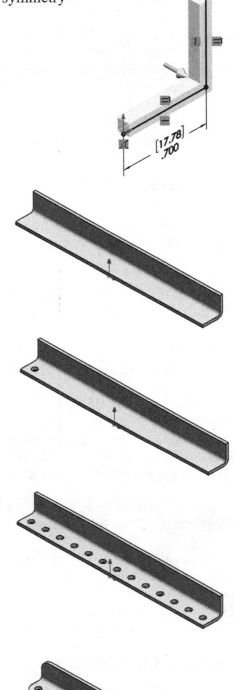

Utilize an Extruded Thin 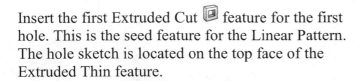 feature with the Mid Plane option to locate the part symmetrical to the Right Plane.

Insert the first Extruded Cut 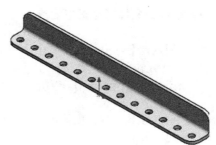 feature for the first hole. This is the seed feature for the Linear Pattern. The hole sketch is located on the top face of the Extruded Thin feature.

Insert a Linear Pattern feature to create an array of 13 holes along the bottom horizontal edge.

Insert a Fillet feature to round the four corners.

Model about the origin; this provides a point of reference.

Insert the second Extruded Cut 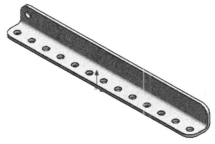 feature on the front face of the Extruded Thin feature. This is the seed feature for the second Linear Pattern.

Insert a Linear Pattern 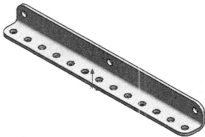 feature to create an array of 3 holes along the top horizontal edge.

Utilize the Sketch Mirror tool to create the slot profile. Use the Slot Sketch tool.

Insert the third Extruded Cut 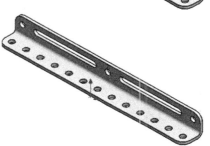 feature to create the two slots.

🔆 Select the Sketch plane for the Base feature that corresponds to the parts orientation in the assembly.

Activity: ANGLE-13HOLE Part - Documents Properties

Create a New part.

71) Click **New** ⬜ from the Menu bar.

72) Double-click **Part**. The Part FeatureManager is displayed.

Set Document Properties. Set drafting standard, units, and precision.

73) Click **Options** ⚙ from the Main menu.

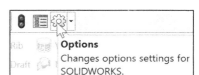

74) Click the **Document Properties** tab from the dialog box.

75) Select **ANSI** from the Overall drafting standard drop-down menu.

76) Click **Units**.

77) Select **IPS**, [**MMGS**] for Unit system.

78) Select **.123**, [**.12**] for Length units Decimal places.

79) Select **None** for Angular units Decimal places. Click **OK**.

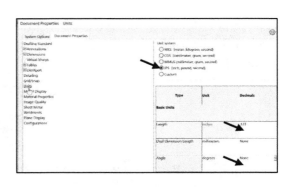

Save the part. Enter name and description.

80) Click **Save As** 🖫 . Select the **SW-TUTORIAL-2020** file folder.

81) Enter **ANGLE-13 HOLE** for File name.

82) Enter **ANGLE BRACKET-13 HOLE** for Description.

83) Click **Save**. The ANGLE-13 Hole FeatureManager is displayed.

Activity: ANGLE-13HOLE Part - Extruded Thin Feature

Insert an Extruded Thin feature sketched on the Right Plane.
Select the Sketch plane.

84) Right-click **Right Plane** from the FeatureManager.

Sketch a horizontal line.

85) Click **Sketch** ⊞ from the Context toolbar. The Sketch toolbar is displayed.

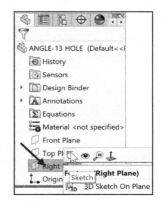

86) Click the **Line** ╱ Sketch tool from the Sketch toolbar.

87) Click the **Origin** ↳ as illustrated.

88) Click a **position** to the right of the Origin.

Sketch a vertical line.

89) Click a **position** directly above the right end point.

De-select the Line Sketch tool.

90) Right-click **Select** in the Graphics window.

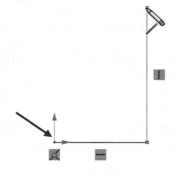

Add an Equal relation.

91) Click the **vertical** line. Hold the **Ctrl** key down.

92) Click the **horizontal line**. Release the **Ctrl** key.

93) Right-click **Make Equal** = from the Context toolbar.

94) Click **OK** ✓ from the Properties PropertyManager.

Add a dimension.

95) Click the **Smart Dimension** ✎ Sketch tool.

96) Click the **horizontal** line.

97) Click a **position** below the profile.

98) Enter **.700**in, **[17.78]** in the Modify box.

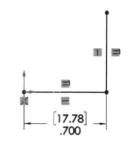

99) Click the **Green Check mark** ✔ . The black sketch is fully defined.

💡 Save rebuild time. Add relations and dimensions to fully define a sketch. Fully defined sketches are displayed in black.

Extrude the sketch.

100) Click **Extruded Boss/Base** 🗗 from the Features toolbar. The Boss-Extrude PropertyManager is displayed.

101) Select **Mid Plane** for End Condition in Direction 1.

102) Enter **7.000**in, **[177.8]** for Depth. Note: Thin Feature is checked.

103) If needed, click the **Reverse Direction Arrow** button for One-Direction. Material thickness is created above the Origin.

104) Enter **.090**in, **[2.3]** for Thickness.

105) Check the **Auto-fillet corners** box.

106) Enter **.090**in, **[2.3]** for Fillet Radius.

107) Click **OK** ✔ from the Boss-Extrude PropertyManager. Extrude-Thin1 is displayed in the FeatureManager. Sketch1 is fully defined.

Fit the model to the Graphics window.

108) Press the **f** key.

💡 Think design intent. When do you use the various End Conditions and Geometric sketch relations? What are you trying to do with the design? How does the component fit into an assembly?

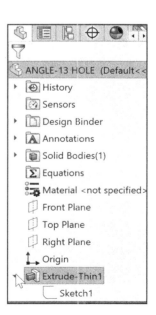

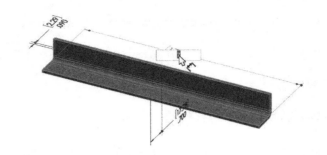

Clarify the Extrude-Thin1 feature direction and thickness options. Utilize multiple view orientations and Zoom In before selecting OK ✔ from the Boss-Extrude PropertyManager.

Modify feature dimensions.

109) Click **Extrude-Thin1** in the FeatureManager.

110) Click the **7.000**in, **[177.80]** dimension in the Graphics window.

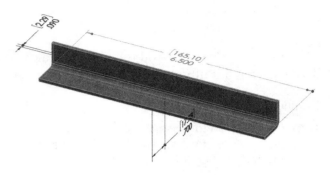

111) Enter **6.500**in, **[165.10]**.

112) Click **inside** the Graphics window.

Save the model.

113) Click **Save** 💾.

Activity: ANGLE-13HOLE Part - Extruded Cut Feature

Insert a new sketch for the Extruded Cut feature.

114) Right-click the **top face** of Extrude-Thin1 as illustrated. This is the Sketch plane.

115) Click **Sketch** ▦ from the Context toolbar. The Sketch toolbar is displayed.

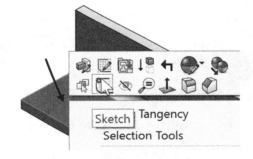

116) Click **Top view** ⬛ from the Heads-up View toolbar.

117) Click the **Circle** ⊙ Sketch tool.

118) Sketch a **circle** on the left side of the Origin as illustrated.

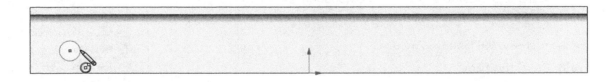

Add dimensions.

119) Click the **Smart Dimension** ✏ Sketch tool.

120) Click the **Origin** .

121) Click the **center point** of the circle.

122) Click a **position** below the horizontal profile line.

123) Enter **3.000**in, [**76.2**].

124) Click the **Green Check mark** ✓.

125) Click the **bottom horizontal line**.

126) Click the **center point** of the circle.

127) Click a **position** to the left of the profile.

128) Enter **.250**in, [**6.35**].

129) Click the **Green Check mark** ✓.

130) Click the **circumference** of the circle.

131) Click a **position** diagonally above the profile.

132) Enter **.190**in, [**4.83**].

133) Click the **Green Check mark** ✓.

Insert an Extruded Cut Feature.

134) Click **Extruded Cut**  from the Features toolbar. The Cut-Extrude PropertyManager is displayed.

135) Select **Through All** for End Condition in Direction 1. Accept the default conditions.

136) Click **OK** ✓ from the Cut-Extrude PropertyManager. Cut-Extrude1 is displayed in the FeatureManager. Cut-Extrude1 is the seed feature for the Linear Pattern feature of holes.

Display an Isometric view.

137) Click **Isometric view** from the Heads-up View toolbar.

Save the model.

138) Click **Save** .

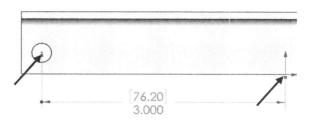

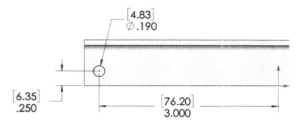

Activity: ANGLE-13HOLE Part - Linear Pattern Feature

Insert a Linear Pattern feature.

139) Click **Top View** from the
Heads-up View toolbar.

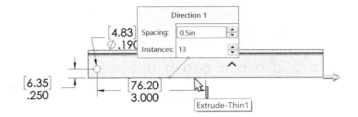

140) Click **Linear Pattern** from
the Features toolbar. The Linear
Pattern PropertyManager is
displayed. Cut-Extrude1 is
displayed in the Features to
Pattern box.

141) Click the **bottom horizontal edge** of the Extrude-Thin1 feature for
Direction1. Edge<1> is displayed in the Pattern Direction box. The
direction arrow points to the right. If required, click the Reverse
Direction button.

142) Enter **0.5**in, [**12.70**] for Spacing.

143) Enter **13** for Number of Instances. If needed, click inside the
Features and Faces box.

144) Click **Cut-Extrude1** from the fly-out FeatureManager.

145) Check the **Geometry pattern** box under Options.

146) Click **OK** from the Linear Pattern PropertyManager. LPattern1
is displayed in the FeatureManager.

Display an Isometric view.

147) Click **Isometric view** from the Heads-
up View toolbar.

Save the ANGLE-13HOLE part.

148) Click **Save** . Note: All sketches should
be fully defined in the FeatureManager.

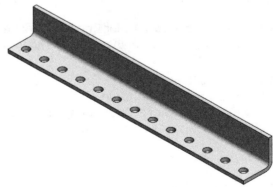

When possible and if it makes sense,
model objects symmetrically about the
origin. Even if the part is not symmetrical,
the way it attaches or is manufactured could
have symmetry.

Activity: ANGLE-13HOLE Part - Fillet Feature

Insert a Constant Radius Fillet Feature.

149) **Zoom in** on the right top edge as illustrated.

150) Click the **right top edge** of the Extrude-Thin1 feature.

151) Click **Fillet** 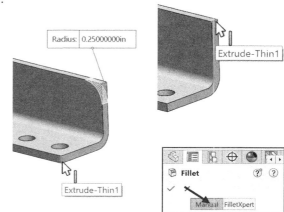 from the Features toolbar. The Fillet PropertyManager is displayed.

152) Click the **Manual** tab.

153) Click **Constant Radius Fillet Type**.

154) Enter **.250 [6.35]** for Radius.

155) Click the **right bottom edge**. Edge<1> and Edge<2> are displayed in the Items To Fillet box. Note the new fillet pop-up menu.

156) Click **OK** ✔ from the Fillet PropertyManager. Fillet1 is displayed in the FeatureManager.

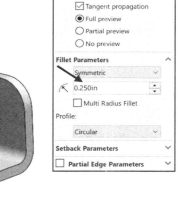

Two Fillet PropertyManager tabs are available. Use the Manual tab to control features for all Fillet types. Use the FilletXpert tab when you want SOLIDWORKS to manage the structure of the underlying features only for a Constant radius Fillet type. Click the ⑦ button for additional information.

Fit the model to the Graphics window.

157) Press the **f** key.

Edit the Fillet feature.

158) **Zoom in** on the left side of the Extrude-Thin1 feature.

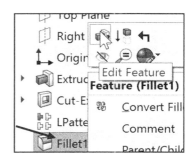

159) Right-click **Fillet1** from the FeatureManager.

160) Click **Edit Feature** from the Context toolbar. The Fillet1 PropertyManager is displayed.

161) Click the **left top edge** and **left bottom edge**. Edge<3> and Edge <4> are added to the Items To Fillet box.

162) Click **OK** ✔ from the Fillet1 PropertyManager. The four edges have a Fillet feature with a .250in radius.

Display an Isometric view.

163) Click **Isometric view** 🔲 from the Heads-up View toolbar.

Save the ANGLE-13HOLE part.

164) Click **Save** 💾.

The book is designed to expose the new SOLIDWORKS user to many different tools, techniques and procedures. It may not always use the most direct tool or process.

Activity: ANGLE-13HOLE Part - Second Extruded Cut/Linear Pattern

---

Insert a new sketch for the second Extruded Cut feature.

165) Right-click the **front face** of the Extrude-Thin1 feature in the Graphics window. The front face is the Sketch plane.

166) Click **Sketch** 🔲 from the Context toolbar. The Sketch toolbar is displayed.

167) Click **Front view** 🔲 from the Heads-up View toolbar.

168) Click **Wireframe** 🔲 from the Heads-up View toolbar to display LPattern1.

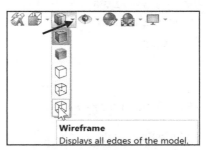

Note: Do not align the center point of the circle with the center point of the LPattern1 feature. Do not align the center point of the circle with the center point of the Fillet radius. Control the center point position with dimensions.

169) Click the **Circle** 🔾 Sketch tool. The cursor displays the Circle icon.

170) Sketch a **circle** on the left side of the Origin between the two LPattern1 holes as illustrated.

Add dimensions.

171) Click the **Smart Dimension** Sketch tool.

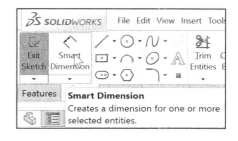

172) Click the **Origin**.

173) Click the **center point** of the circle.

174) Click a **position** below the horizontal profile line.

175) Enter **3.000**in, [**76.20**].

176) Click the **Green Check mark**.

177) Click the **top horizontal line**.

178) Click the **center point** of the circle.

179) Click a **position** to the left of the profile.

180) Enter **.250**in, [**6.35**].

181) Click the **Green Check mark**.

182) Click the **circumference** of the circle.

183) Click a **position** above the profile.

184) Enter **.190**in, [**4.83**].

185) Click the **Green Check mark**.

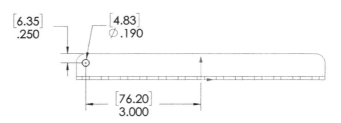

Insert an Extruded Cut Feature.

186) Click **Extruded Cut** from the Features toolbar. The Cut-Extrude PropertyManager is displayed.

187) Select **Through All** for End Condition in Direction1. Accept the default conditions.

188) Click **OK** from the Cut-Extrude PropertyManager. Cut-Extrude2 is displayed in the FeatureManager.

Create the second Linear Pattern Feature.

189) Click **Linear Pattern** from the Features toolbar. The Linear Pattern FeatureManager is displayed. Cut-Extrude2 is displayed in the Features to Pattern box.

190) Click inside the **Pattern Direction** box.

191) Click the **top horizontal edge** of the Extrude-Thin1 feature for Direction1. Edge<1> is displayed in the Pattern Direction box. The Direction arrow points to the right. If required, click the **Reverse Direction** button.

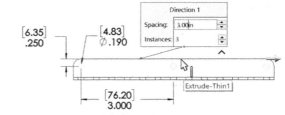

192) Enter **3.000**in, **[76.20]** for Spacing.

193) Enter **3** for Number of Instances. Note: Cut-Extrude2 (Seed feature) is displayed in the Features to Pattern box. Check the **Geometry pattern** box under Options.

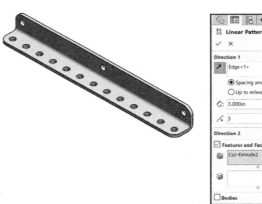

194) Click **OK** ✔ from the Linear Pattern PropertyManager. LPattern2 is displayed in the FeatureManager.

Display an Isometric view - Shaded With Edges.

195) Click **Isometric view** 🔲 from the Heads-up View toolbar.

196) Click **Shaded With Edges** 🔲 from the Heads-up View toolbar.

Save the ANGLE-13HOLE part.

197) Click **Save** 💾.

Activity: ANGLE-13HOLE Part - Third Extruded Cut

Insert a new sketch for the third Extruded Cut Feature.

198) Select the Sketch plane. Right-click the **front face** of Extrude-Thin1.

199) Click **Sketch** 🔲 from the Context toolbar. The Sketch toolbar is displayed. Click **Front view** 🔲 from the Heads-up View toolbar.

200) Click **Hidden Lines Removed** 🔲 from the Heads-up View toolbar.

Sketch a vertical centerline.

201) Click the **Centerline** ✏ Sketch tool. The Insert Line PropertyManager is displayed.

202) Click the **Origin** 📐.

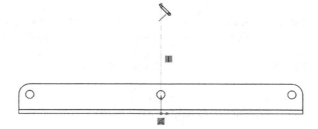

203) Click a **vertical position** above the top horizontal line.

204) Click **Tools**, **Sketch Tools**, **Dynamic Mirror** from the Menu bar.

205) Click the **centerline** in the Graphics window.

Sketch a rectangle.

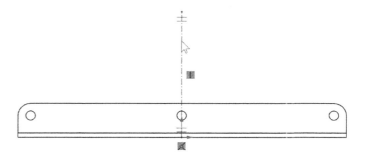

206) Click **Wireframe** ⬡ from the Heads-up View toolbar.

Apply the Straight Slot Sketch tool. Sketch a straight slot using two end points and a point for height.

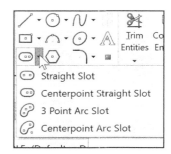

207) Click the **Straight Slot** ⬭ Sketch tool from the Consolidated Slot toolbar. Do not align the rectangle first point and second point to the center points of Lpattern1.

208) Click the **first point** of the rectangle to the left of the Origin as illustrated.

209) Click the **second point** directly to the right.

210) Click the **third point** to create the slot as illustrated. View the two slots on the model.

211) Click **OK** ✔ from the Slot PropertyManager.

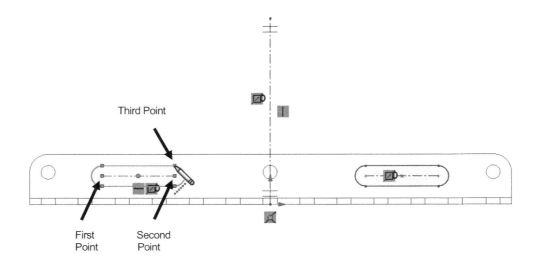

Deactivate the Dynamic Mirror tool.

212) Click **Tools, Sketch Tools, Dynamic Mirror** from the Menu bar.

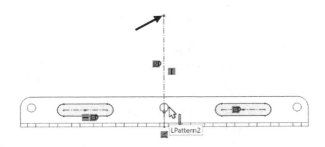

Add a Concentric relation.

213) Zoom in on the centerline.

214) Click the **top end point** of the centerline.

215) Hold the **Ctrl** key down.

216) Click the **circumference** of the circle.

217) Release the **Ctrl** key.

218) Click **Concentric** from the Add Relations box.

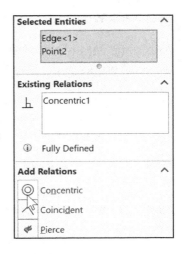

The endpoint of the centerline is positioned in the center of the circle. Note: Right-click Clear Selections to remove selected entities from the Add Relations box.

Add an Equal relation.

219) Click the **circumference** of the circle.

220) Hold the **Ctrl** key down.

221) Click the **left arc** of the first rectangle.

222) Release the **Ctrl** key.

223) Click **Equal** ﹦ from the Add Relations box. The arc radius is equal to the circle radius.

Add a Horizontal relation.

224) Click the **top end point** of the centerline as illustrated.

225) Hold the **Ctrl** key down.

226) Click the **center point** of the left arc of the first rectangle.

227) Release the **Ctrl** key.

228) Click **Horizontal** ━ from the Add Relations box.

229) Click **OK** ✔ from the Properties PropertyManager.

The right arc is horizontally aligned to the left arc due to symmetry from the Sketch Mirror tool.

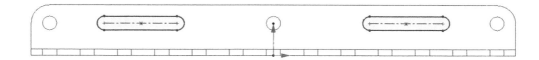

Add dimensions. Dimension the distance between the two slots.

230) Click the **Smart Dimension** Sketch tool.

231) Click the **right arc center point** of the left slot.

232) Click the **left arc center point** of the right slot.

233) Click a **position** above the top horizontal line.

234) Enter **1.000**in, **[25.40]** in the Modify dialog box.

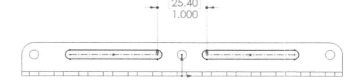

235) Click the **Green Check mark** .

236) Click the **left center point** of the left arc.

237) Click the **right center point** of the left arc.

238) Click a **position** above the top horizontal line.

239) Enter **2.000**in, **[50.80]** in the Modify dialog box.

240) Click the **Green Check mark** . The black sketch is fully defined.

Display an Isometric view.

241) Click **Isometric view** .

☀ Origin and Tangent edges are displayed for educational purposes.

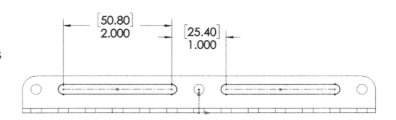

Insert an Extruded Cut Feature.

242) Click **Extruded Cut** 🔲 from the Features toolbar. The Cut-Extrude PropertyManager is displayed.

243) Select **Through All** for End Condition in Direction1. The direction arrow points to the back.

244) Click **OK** ✔ from the Extrude PropertyManager. Cut-Extrude3 is displayed in the FeatureManager.

Display Shaded With Edges.

245) Click **Shaded With Edges** 🔲 from the Heads-up View toolbar.

Save the ANGLE-13HOLE part.

246) Click **Save** 💾. The ANGLE-13HOLE is complete. All sketches in the FeatureManager should be fully defined. Add material later.

 Model about the origin; it provides a point of reference.

The dimension between the two slots is over-defined if the arc center points are aligned to the center points of the LPattern1 feature. An over-defined sketch is displayed in red, in the FeatureManager.

The mouse pointer displays a blue dashed line when horizontal and vertical sketch references are inferred.

Add relations, then dimensions. This will keep the user from having too many unnecessary dimensions. This helps to show the design intent of the model.

Right-click and drag in the Graphics area to display the mouse gesture wheel. Customize the default commands for a sketch, part, assembly or drawing document.

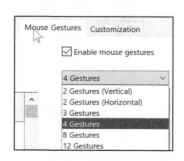

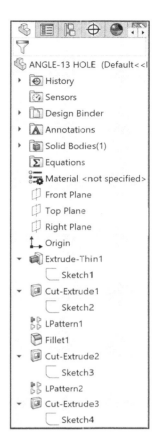

Review the ANGLE-13HOLE Part

The ANGLE-13HOLE part utilized an open L-Shaped profile sketched on the Right Plane. The Extruded Thin feature with the Mid Plane option located the part symmetrical to the Right Plane. The first Extruded Cut feature created the first hole sketched on the top face of the Extruded Thin feature.

The first Linear Pattern feature created an array of 13 holes along the bottom horizontal edge. The Fillet feature rounded the four corners. The second Extruded Cut feature created a hole on the Front face. The second Linear Pattern feature created an array of 3 holes along the top horizontal edge. The third Extruded Cut feature created two slot cuts using the Straight Slot Sketch tool.

CommandManager and FeatureManager tabs and file folders will vary depending on system setup and Add-ins.

TRIANGLE Part

The TRIANGLE part is a multipurpose supporting plane.

The TRIANGLE is manufactured from .090in, [2.3] aluminum. The TRIANGLE contains numerous features.

Utilize symmetry and Sketch tools to simplify the geometry creation.

The center points of the slots and holes locate key geometry for the TRIANGLE.

Utilize sketched construction geometry to locate the center points.

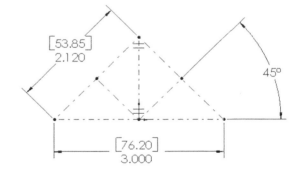

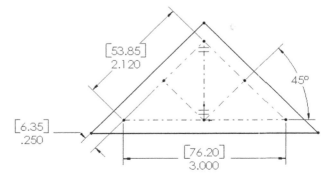

Construction geometry is not calculated in the extruded profile.

Utilize the Sketch Offset tool and Sketch Fillet to create the sketch profile for the Extruded Boss/Base 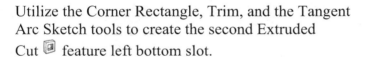 (Boss-Extrude1) feature.

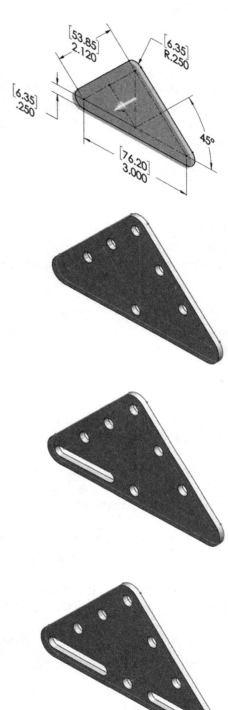

Utilize the Dynamic Mirror Sketch tool and the Circle Sketch tool to create the first Extruded Cut feature.

Utilize the Corner Rectangle, Trim, and the Tangent Arc Sketch tools to create the second Extruded Cut feature left bottom slot.

Note: A goal of this book is to expose the new user to different tools and methods. You can also apply the Straight Slot Sketch tool and eliminate steps.

Utilize the Mirror feature to create the right bottom slot.

Utilize the Straight Slot Sketch tool to create the third Extruded Cut feature.

Utilize the Circular Pattern  feature to create the three radial slot cuts.

Activity: TRIANGLE Part - Mirror, Offset and Fillet Sketch Tools

Create a New part.

247) Click **New** ⬜ from the Menu bar.

248) Click the **Templates** tab.

249) Double-click **Part**. The Part FeatureManager is displayed.

Save the part. Enter name. Enter description.

250) Click **Save As** 🖫 .

251) Select **SW-TUTORIAL-2020** for the Save in file folder.

252) Enter **TRIANGLE** for File name. Enter **TRIANGLE** for Description. Click **Save**. The TRIANGLE FeatureManager is displayed.

Set the Document Properties. Set drafting standard, units, and precision.

253) Click **Options** ⚙ from the Main menu.

254) Click the **Document Properties** tab from the dialog box.

255) Select **ANSI** from the Overall drafting standard box.

256) Click **Units**. Click **IPS**, [**MMGS**] for Unit system.

257) Select **.123**, [**.12**] for Length units Decimal places.

258) Select **None** for Angular units Decimal places.

259) Click **OK** to set the document units.

Insert a new sketch for the Extruded Base feature.

260) Right-click **Front Plane** from the FeatureManager.

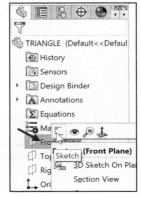

261) Click **Sketch** ⌐ from the Context toolbar. The Sketch toolbar is displayed.

Sketch a vertical centerline.

262) Click the **Centerline** ⌀ Sketch tool.

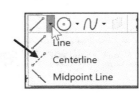

263) Click the **Origin** ↳.

264) Click a vertical **position** above the Origin as illustrated.

Deselect the Centerline Sketch tool.

265) Right-click **Select**.

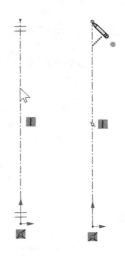

Sketch a Mirrored profile.

266) Click **Tools**, **Sketch Tools**, **Dynamic Mirror** from the Menu bar.

267) Click the **centerline** in the Graphics window.

268) Click the **Centerline** ⌀ Sketch tool.

269) Click the **Origin** ↳.

270) Click a **position** to the left of the Origin to create a horizontal line.

271) Click the **top end point** of the vertical centerline to complete the triangle as illustrated.

272) Right-click **End Chain** to end the line segment. The Centerline tool is still active.

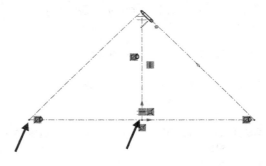

273) Click the **Origin** ↳.

274) Click a **position** coincident with the right-angled centerline. *Do not select the Midpoint.*

275) Right-click **End Chain** to end the line segment.

Deactivate the Dynamic Mirror Sketch tool.

276) Click **Tools**, **Sketch Tools**, **Dynamic Mirror** from the Menu bar.

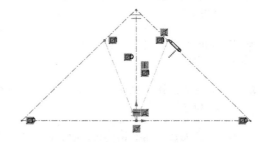

Add a dimension.

277) Click **Smart Dimension** 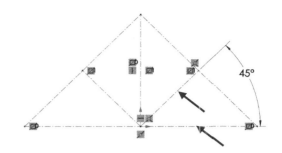 from the Sketch toolbar.

278) Click the **right horizontal** centerline.

279) Click the **inside right** centerline.

280) Click a **position** between the two lines.

281) Enter **45**deg in the Modify dialog box for the angular dimension.

282) Click the **Green Check mark** in the Modify dialog box.

283) Click the **horizontal centerline**.

284) Click a position **below** the centerline.

285) Enter **3.000**in, [**76.20**].

286) Click the **Green Check mark** in the Modify dialog box.

287) Click the **left angled** centerline.

288) Click **position** aligned to the left angled centerline.

289) Enter **2.120**in, [**53.85**].

290) Click the **Green Check mark** in the Modify dialog box.

Offset the sketch.

291) Right-click **Select** in the Graphics window.

292) Hold the **Ctrl** key down.

293) Click the **three outside centerlines**.

294) Release the **Ctrl** key.

295) Click the **Offset Entities** 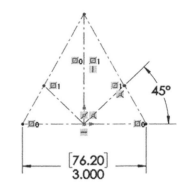 Sketch tool.

296) Enter **.250**in, [**6.35**] for Offset Distance. The yellow Offset direction is outward.

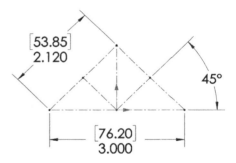

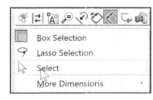

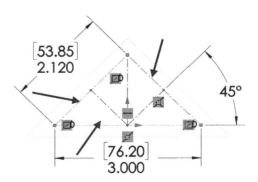

297) Click **OK** ✔ from the Offset Entities PropertyManager. If needed, deactivate the Sketch relations in the Graphics window.

Three profile lines are displayed. The centerlines are on the inside.

Insert the Sketch Fillet.

298) Click **Sketch Fillet** ⌐ from the Sketch toolbar. The Sketch Fillet PropertyManager is displayed.

299) Enter **.250**in, **[6.35]** for Radius.

300) Click the **three outside corner** points.

301) Click **OK** ✔ from the Sketch Fillet PropertyManager.

302) Click **OK** ✔ from the Sketch Fillet PropertyManager.

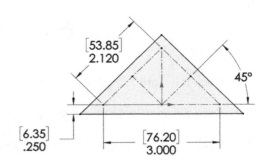

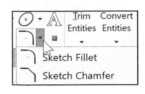

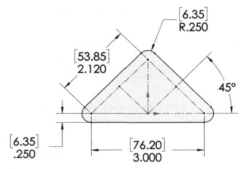

Activity: TRIANGLE Part - Extruded Boss/Base Feature

Extrude the sketch. Create the Boss-Extrude1 feature.

303) Click **Extruded Boss/Base** 🗔 from the Features toolbar. Blind is the default End Condition in Direction1.

304) Enter **.090**in, **[2.3]** for Depth in Direction 1. The direction arrow points to the front.

305) Click **OK** ✔ from the Boss-Extrude PropertyManager. Boss-Extrude1 is displayed in the FeatureManager.

Display an Isometric view. Save the TRIANGLE part.

306) Click **Isometric view** 🗗 from the Heads-up View toolbar.

307) Click **Save** 💾.

🔅 Insert centerlines and relations to build sketches that are referenced by multiple features.

Display Sketch1.

308) **Expand** Boss-Extrude1 in the FeatureManager.

309) Right-click **Sketch1**.

310) Click **Show**.

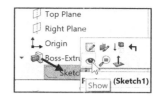

Activity: TRIANGLE Part - First Extruded Cut Feature

Insert a new sketch for the first Extruded Cut.

311) Right-click the **front face** of Boss-Extrude1. This is your Sketch plane.

312) Click **Sketch** ⌐ from the Context toolbar. The Sketch toolbar is displayed.

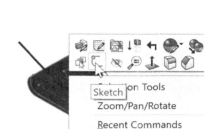

313) Click **Front view** 🔲 from the Heads-up View toolbar.

314) **Display** Hidden Lines Removed.

315) Click the **Circle** ⊙ Sketch tool. The Circle PropertyManager is displayed.

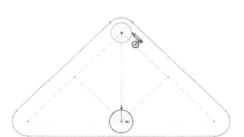

316) Sketch a **circle** centered at the Origin.

317) Sketch a **circle** centered at the endpoint of the vertical centerline as illustrated.

Sketch a vertical centerline.

318) Click the **Centerline** ⸛ Sketch tool.

319) Click the **Origin** ↳.

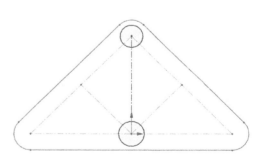

320) Click the **center point** of the top circle.

Deselect the Centerline Sketch tool.

321) Right-click **Select**.

Sketch a Mirrored profile.

322) Click **Tools**, **Sketch Tools**, **Dynamic Mirror** from the Menu bar.

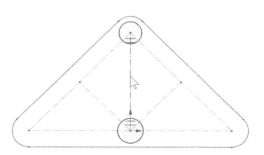

323) Click the **centerline** in the Graphics window.

324) Click the **Circle** ⊙ Sketch tool. The Circle PropertyManager is displayed.

325) Sketch a **circle** on the left side of the centerline, coincident with the left centerline, in the lower half of the triangle.

326) Sketch a **circle** on the left side of the centerline, coincident with the left centerline, in the upper half of the triangle. Right-click **Select.**

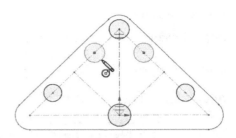

327) Deactivate the Dynamic Mirror tool. Click **Tools, Sketch Tools, Dynamic Mirror** from the Menu bar.

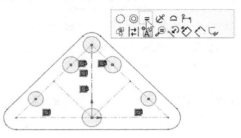

Add an Equal relation.
328) Click a **circle**. Hold the **Ctrl** key down.

329) Click the five other **circles**. Release the **Ctrl** key.

330) Click **Make Equal** ═ from the Pop-up menu.

Add a dimension.

331) Click the **Smart Dimension** ↰ Sketch tool.

332) Click the circumference of the **top circle**.

333) Click a **position** off the TRIANGLE.

334) Enter **.190**in, [**4.83**].

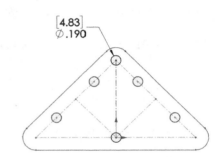

335) Click the **Green Check mark** ✔ in the Modify dialog box.

Create the aligned dimensions.

336) Click the bottom **left point**. Click the **center point** of the bottom left circle.

337) Click a **position** aligned to the angled centerline.

338) Enter **.710**in, [**18.03**]. Click the **Green Check mark** ✔ in the Modify dialog box.

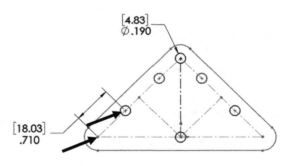

339) Click the **bottom left point**.

340) Click the **center point** of the top left circle.

341) Click a **position** aligned to the angled centerline. Enter **1.410**in, [**35.81**].

342) Click the **Green Check mark** ✔ in the Modify dialog box. The sketch is fully defined.

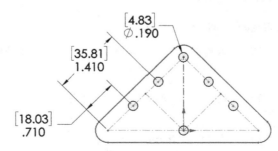

Insert an Extruded Cut Feature.

343) Click **Extruded Cut** 🔲 from the Features toolbar. The Cut-Extrude PropertyManager is displayed.

344) Select **Through All** for the End Condition in Direction1.

345) Click **OK** ✔ from the Cut-Extrude PropertyManager. Cut-Extrude1 is displayed in the FeatureManager.

Display an Isometric view. Save the model.

346) Click **Isometric view** 🔲 from the Heads-up View toolbar.

347) Click **Save** 🔲.

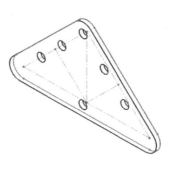

Activity: TRIANGLE Part - Second Extruded Cut Feature

Insert a new slot sketch for the second Extruded Cut.

348) Right-click the **front face** of Boss-Extrude1 for the Sketch plane. Click **Sketch** 🔲 from the Context toolbar. The Sketch toolbar is displayed.

349) Click **Front view** 🔲 from the Heads-up View toolbar.

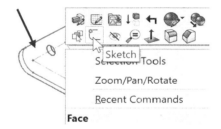

350) Click the **Corner Rectangle** 🔲 tool from the Consolidated Sketch toolbar. Note: The purpose of this book is to teach you different tools and methods. You can also apply the Straight Slot Sketch tool and eliminate many of the next steps.

351) Sketch a **rectangle** to the left of the Origin as illustrated.

Trim the vertical lines.

352) Click the **Trim Entities** 🔲 Sketch tool. The Trim PropertyManager is displayed. Click **Trim to closest** in the Options box.

The Trim to closest 🔲 icon is displayed.

353) Click the **left vertical** line of the rectangle.

354) Click the **right vertical** line of the rectangle.

355) Click **OK** ✔ from the Trim PropertyManager.

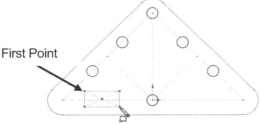

First Point

Sketch the Tangent Arcs.

356) Click the **Tangent Arc**
Sketch tool from the
Consolidated Sketch toolbar.
The Arc PropertyManager is
displayed.

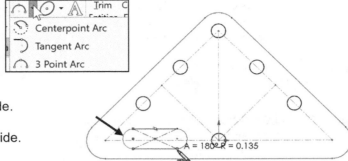

357) Sketch a **180° arc** on the left side.

358) Sketch a **180° arc** on the right side.

359) Click **OK** ✔ from the Arc
PropertyManager.

Add an Equal relation.

360) Click the **right arc**. Hold the **Ctrl** key down.

361) Click the **left arc**. Click the **bottom center
circle**. Release the **Ctrl** key.

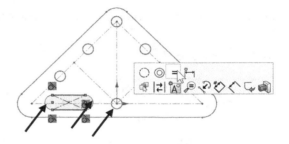

362) Click **Make Equal** ᐀ from the Pop-up
menu. Click **OK** ✔ from the Properties
PropertyManager.

Add a Coincident relation.

363) Press the **f** key to fit the model to the Graphics
window. Click the **center point** of the left arc.

364) Hold the **Ctrl** key down.

365) Click the **left lower point**.

366) Release the **Ctrl** key.

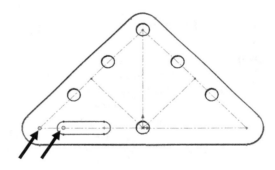

367) Click **Coincident** from the Add Relations box.

Add a Tangent relation.

368) Click the **bottom horizontal line**.

369) Hold the **Ctrl** key down.

370) Click the **first arc tangent**.

371) Release the **Ctrl** key.

372) Click **Tangent** from the Add Relations box.

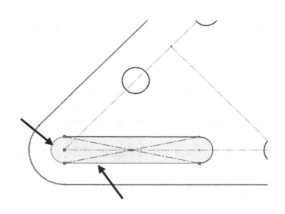

373) Click **OK** ✔ from the Properties
PropertyManager.

Add a dimension.

374) Click the **Smart Dimension** 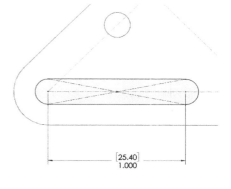 Sketch tool.

375) Click the **left center point** of the left arc.

376) Click the **right center point** of the right arc.

377) Click a **position** below the horizontal line. Enter **1.000**in, **[25.40]** in the Modify dialog box.

378) Click the **Green Check mark** .

Insert an Extruded Cut feature.

379) Click **Isometric view** .

380) Click **Extruded Cut** from the Features toolbar. The Cut-Extrude PropertyManager is displayed.

381) Select **Through All** for End Condition in Direction 1.

382) Click **OK** from the Cut-Extrude PropertyManager.

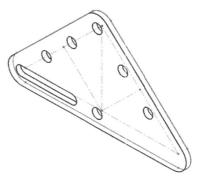

Save the model.

383) Click **Save** . Cut-Extrude2 is highlighted in the FeatureManager.

Activity: TRIANGLE Part - Mirror Feature

Mirror the Cut-Extrude2 feature.

384) Click **Mirror** from the Features toolbar. The Mirror PropertyManager is displayed. Cut-Extrude2 is displayed in the Feature to Mirror box.

385) Click **Right Plane** from the fly-out TRIANGLE FeatureManager. Right Plane is displayed in the Mirror Face/Plane box.

386) Check the **Geometry Pattern** box.

387) Click **OK** from the Mirror PropertyManager. Mirror1 is displayed in the FeatureManager.

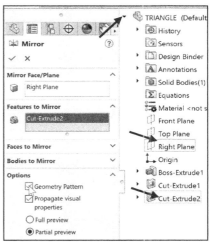

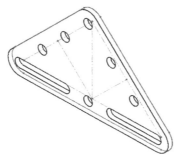

Activity: TRIANGLE Part - Third Extruded Cut Feature

Insert a new sketch for the third Extruded Cut feature.

388) Right-click the **front face** of Boss-Extrude1 for the Sketch plane.

389) Click **Sketch** from the Context toolbar. The Sketch toolbar is displayed.

390) Click **Front view** .

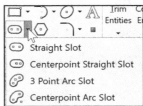

Sketch a Straight Slot.

391) Click the **Straight Slot** Sketch tool from the Consolidated Rectangle toolbar. The Straight Slot icon is displayed.

392) Click a **position** coincident with the left angled centerline as illustrated.

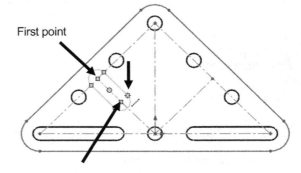

Sketch the second point.

393) Click a **position** aligned to the centerline. A dashed line is displayed.

Sketch the third point.

394) Click a **position** above the inside left centerline. The Straight Slot sketch is displayed.

 Click the Question mark ⑦ in the PropertyManager to obtain additional information on the tool.

395) Click **OK** from the Slot PropertyManager.

Add an Equal relation.

396) Click the **left arc**. Hold the **Ctrl** key down.

397) Click the **bottom circle**.

398) Release the **Ctrl** key.

399) Click **Equal** ＝ from the Add Relations box or the Pop-up menu.

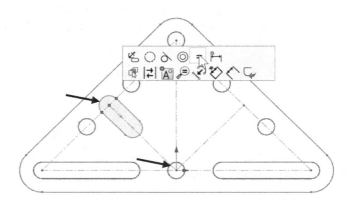

Add a dimension.

400) Click the **Smart Dimension** 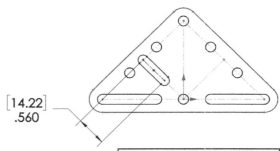 Sketch tool.

401) Click the **left center point** of the left arc.

402) Click the **right center point** of the right arc.

403) Click a **position** below the horizontal line.

404) Enter **.560**in, [**14.22**] in the Modify dialog box.

405) Click the **Green Check mark** . The sketch is fully defined.

Display an Isometric view. Insert an Extruded Cut Feature.

406) Click **Isometric view** .

407) Click **Extruded Cut** from the Features toolbar. The Cut-Extrude PropertyManager is displayed.

408) Select **Through All** for the End Condition in Direction 1.

409) Click **OK** from the Cut-Extrude PropertyManager. Cut-Extrude3 is displayed in the FeatureManager.

Save the model.

410) Click **Save** .

Display the Temporary Axis.
411) Click **View**, **Hide/Show**, check **Temporary Axes** from the Menu bar.

The book is designed to expose the new SOLIDWORKS user to many different tools, techniques and procedures. It may not always use the most direct tool or process.

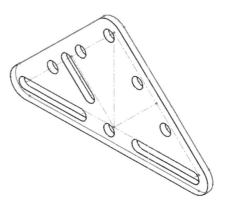

Activity: TRIANGLE Part - Circular Pattern Feature

Insert a Circular Pattern feature.

412) Click **Circular Pattern** 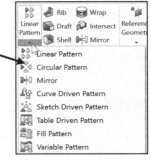 from the Features Consolidated toolbar. The Circular Pattern PropertyManager is displayed. Cut-Extrude3 is displayed in the Features to Pattern box.

413) Click the **Temporary Axis** displayed through the center hole located at the Origin. The Temporary Axis is displayed as Axis <1> in the Pattern Axis box.

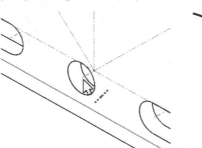

414) Enter **90**deg for Angle.

415) Enter **3** for Number of Instances.

416) Check the **Equal spacing** box. If required, click **Reverse Direction**.

417) Click the **Geometry pattern** box.

418) Click **OK** ✓ from the Circular Pattern PropertyManager.

Hide Sketch1.

419) **Expand** Boss-Extrude1 from the FeatureManager.

420) Right-click **Sketch1**.

421) Click **Hide**.

Display an Isometric view. Deactivate the Temporary Axes. Save the TRIANGLE part.

422) Click **Isometric view** .

423) Click **View**, **Hide/Show**, un-check **Temporary Axes** from the Menu bar.

424) Display **Shaded with Edges**.

425) Click **Save** . The TRIANGLE part is complete. Add material later.

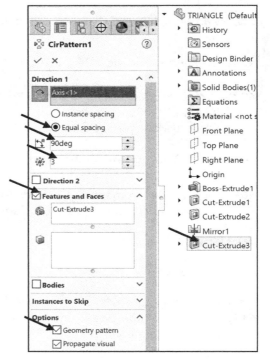

Review the TRIANGLE Part

The TRIANGLE part utilized a Boss-Extrude1 feature. A triangular shape profile was sketched on the Front Plane. Symmetry and construction geometry sketch tools located center points for slots and holes.

The Sketch Fillet tool created rounded corners for the profile. The Sketch Mirror and Circle Sketch tools were utilized to create the Extruded Cut features.

The Corner Rectangle, Sketch Trim, and Tangent Arc tools were utilized to create the second Extruded Cut feature, left bottom slot. The Mirror feature was utilized to create the right bottom slot.

The Parallelogram and Tangent Arc Sketch tools were utilized to create the third Extruded Cut feature. The Circular Pattern feature created the three radial slot cuts. The following Geometric relations were utilized: *Equal, Parallel, Coincident* and *Tangent*.

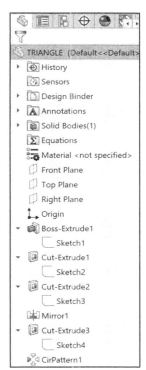

Additional details on Rectangle, Circle, Tangent Arc, Parallelogram, Mirror Entities, Sketch Fillet, Offset Entities, Extruded Boss/Base, Extruded Cut, Mirror and Circular Pattern are available in SOLIDWORKS Help.

SCREW Part

The SCREW part is a simplified model of a 10-24 x 3/8 Machine screw. Screws, nuts and washers are classified as fasteners. An assembly contains hundreds of fasteners. Utilize simplified versions to conserve model and rebuild time.

Simplified version

Machine screws are described in terms of the following:

- Nominal diameter - Size 10.

- Threads per inch - 24.

- Length - 3/8.

Screw diameter, less than ¼ inch, is represented by a size number. Size 10 refers to a diameter of .190 inch. Utilize the SCREW part to fasten components in the FRONT-SUPPORT assembly.

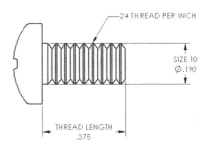

The SCREW part utilizes a Revolved Base ![screw icon] feature to add material. The Revolved Boss/Base feature requires a centerline and sketch on a Sketch plane. A Revolved feature requires an angle of revolution. The sketch is revolved around the centerline.

Sketch a centerline on the Front Sketch plane.

Sketch a closed profile.

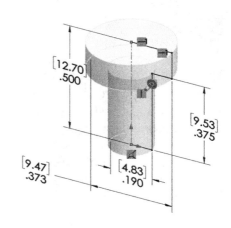

Revolve the sketch 360 degrees.

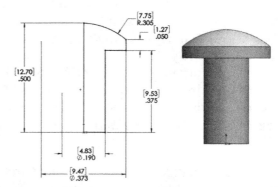

Utilize the Edit Sketch tool to modify the sketch. Utilize the Sketch Trim and Tangent Arc tool to create a new profile.

Utilize an Extruded Cut ![icon] feature sketched on the Front Plane. This is the seed feature for the Circular Pattern.

Utilize the Circular Pattern ![icon] feature to create four instances.

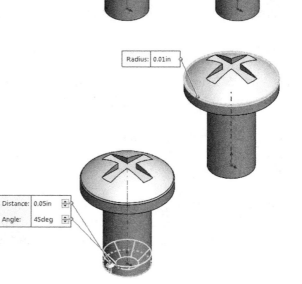

Apply the Fillet ![icon] feature to round edges and faces. Utilize the Fillet feature to round the top edge.

Apply the Chamfer ![icon] feature to bevel edges and faces. Utilize a Chamfer feature to bevel the bottom face.

Note: Utilize an M5 Machine screw for metric units.

Activity: SCREW Part-Documents Properties

Create a New part.

426) Click **New** ☐ from the Menu bar.

427) Click the **Templates** tab. Double-click **Part**.

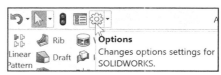

Save the part. Enter name. Enter description.

428) Click **Save As** 🖫.

429) Select **SW-TUTORIAL-2020** for the Save in file folder.

430) Enter **SCREW** for File name.

431) Enter **MACHINE SCREW 10-24x3/8** for Description.

432) Click **Save**. The SCREW FeatureManager is displayed.

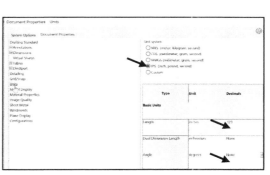

Set Document Properties. Set drafting standard, units, and precision.

433) Click **Options** ⚙ from the Main menu.

434) Click the **Documents Properties** tab from the dialog box. Select **ANSI** from the Overall drafting standard box.

435) Click **Units**. Select **IPS**, **[MMGS]** for Unit system. Select **.123**, **[.12]** for Length units Decimal places. Select **None** for Angular units Decimal places. Click **OK**.

Activity: SCREW Part - Revolved Feature

Insert a Revolved feature sketched on the Front Plane. The Front Plane is the default Sketch plane.

Insert a new sketch.

436) Right-click **Front Plane** from the FeatureManager.

437) Click **Sketch** └ from the Context toolbar. The Sketch toolbar is displayed. Click the **Centerline** ┅ Sketch tool. The Insert Line PropertyManager is displayed.

Sketch a vertical centerline.

438) Click the **Origin** ⊥.

439) Click a **position** directly above the Origin as illustrated.

Origin

440) Right-click **End Chain** to end the centerline.

Add a dimension.

441) Click the **Smart Dimension** Sketch tool.

442) Click the **centerline**. Click a **position** to the left.

443) Enter **.500**in, **[12.70]**. Click the **Green Check mark** .

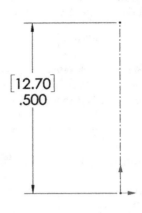

Fit the sketch to the Graphics Window.

444) Press the **f** key.

Sketch the profile.

445) Click the **Line** Sketch tool.

Sketch the first horizontal line.

446) Click the **Origin** . Click a **position** to the right of the Origin.

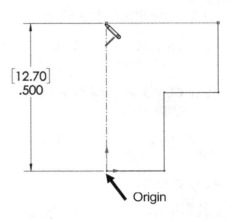

Sketch the first vertical line.

447) Click a position **above** the horizontal line endpoint.

448) Sketch the **second horizontal line**.

449) Sketch the **second vertical line**. The top point of the vertical line is collinear with the top point of the centerline.

450) Sketch the **third horizontal line**. The left endpoint of the horizontal line is coincident with the top point of the centerline.

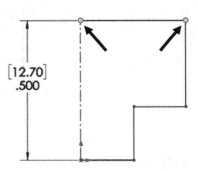

451) Right-click **Select** to deselect the Line Sketch tool.

Add a Horizontal relation.

452) Click the **top** most right point. Hold the **Ctrl** key down.

453) Click the **top** most left point. Release the **Ctrl** key.

454) Click **Horizontal** from the Add Relations box.

455) Click **OK** from the Properties PropertyManager.

Add a dimension.

456) Click the **Smart Dimension** Sketch tool. Create the first diameter dimension.

457) Click the **centerline** in the Graphics window. Click the **first vertical line**.

458) Click a **position** to the left of the Origin to create a diameter dimension.

459) Enter **.190**in, [**4.83**].

460) Click the **Green Check mark**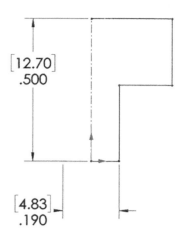

A diameter dimension for a Revolved sketch requires a centerline, profile line, and a dimension position to the left of the centerline.

A dimension position directly below the bottom horizontal line creates a radial dimension.

Create the second diameter dimension.

461) Click the **centerline** in the Graphics window.

462) Click the **second vertical line**.

463) Click a **position** to the left of the Origin to create a diameter dimension.

464) Enter **.373**in, [**9.47**].

Create a vertical dimension.

465) Click the **first vertical line**. Click a **position** to the right of the line.

466) Enter **.375**in, [**9.53**].

Center the dimension text.

467) Click the **.190**in, [**4.83**] dimension.

468) Drag the **text** between the two extension lines.

469) Click the **.373**in, [**9.47**] dimension.

470) Drag the **text** between the two extension lines. If required, click the **blue arrow dots** to flip the arrows inside the extension lines.

471) Right-click **Select** to deselect the Smart Dimension Sketch tool.

Select the Centerline for axis of revolution.

472) Click the **vertical centerline** as illustrated.

Revolve the sketch.

473) Click **Revolved Boss/Base** 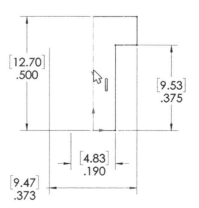 from the Features toolbar.

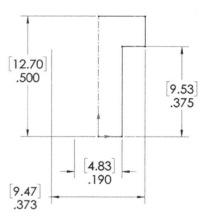

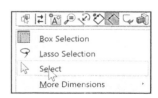

474) Click **Yes**. The Revolve PropertyManager is displayed.

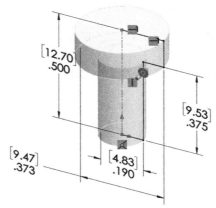

The "Yes" button causes a vertical line to be automatically sketched from the top left point to the Origin. The Graphics window displays the Isometric view and a preview of the Revolved Base feature.

The Revolve PropertyManager displays 360 degrees for the Angle of Revolution.

475) Click **OK** ✅ from the Revolve PropertyManager.

The FeatureManager displays the Revolve1 name for the first feature. The Revolved Boss/Base feature requires a centerline, sketch, and an angle of revolution. A solid Revolved Boss/Base feature requires a closed sketch. Draw the sketch on one side of the centerline.

The SCREW requires a rounded profile. Edit the Revolved Base sketch. Insert a Tangent Arc.

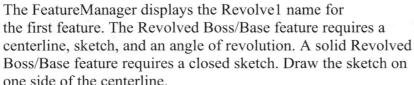

Edit the Revolved Base sketch.

476) Right-click **Revolve1** in the FeatureManager.

477) Click **Edit Sketch** 📝 from the Context toolbar.

478) Click **Front view** 🔲.

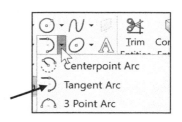

479) Click the **Tangent Arc** 🔗 Sketch tool.

480) Click the **top centerline** point as illustrated.

481) Drag the **mouse pointer** to the right and downward.

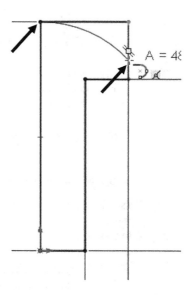

482) Click a **position** collinear with the right vertical line, below the midpoint. The arc is displayed tangent to the top horizontal line.

Deselect the Tangent Arc Sketch tool.

483) Right-click **Select**.

Delete unwanted geometry.

484) Click the **Trim Entities** ✂️ Sketch tool. The Trim PropertyManager is displayed.

485) Click **Trim to closest** from the Options box. The Trim to closest icon is displayed.

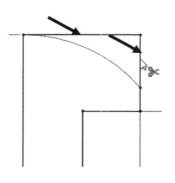

486) Click the **right top vertical line** as illustrated.

487) Click the **top horizontal line** as illustrated. The two lines are removed.

488) Click **OK** ✔ from the Trim PropertyManager. Note: You may still view lines until you exit the sketch.

Add a dimension.

489) Click the **Smart Dimension** ✛ Sketch tool.

490) Click the **arc**.

491) Click a **position** above the profile.

492) Enter **.304**in, **[7.72]**.

493) Click the **Green Check mark** ✔ . The sketch should be fully defined; if not, add a .05 inch dimension as illustrated.

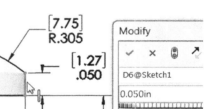

Exit the Sketch. Save the model.

494) Click **Exit Sketch** from the Sketch toolbar.

495) Click **Save** 💾 .

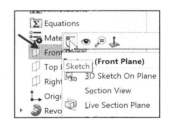

The SCREW requires an Extruded Cut feature on the Front Plane. Utilize the Convert Entities Sketch tool to extract the Revolved Base top arc edge for the profile of the Extruded Cut.

Activity: SCREW Part - Extruded Cut Feature

Insert a new sketch for the Extruded Cut feature.

496) Right-click **Front Plane** from the FeatureManager.

497) Click **Sketch** ⌐ from the Context toolbar. The Sketch toolbar is displayed.

498) Click the **top arc** as illustrated. The mouse pointer displays the silhouette edge icon for feedback.

499) Click the **Convert Entities** ▢ Sketch tool.

500) Click the **Line** ✎ Sketch tool.

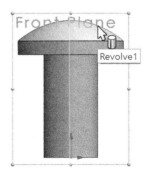

501) Sketch a **vertical line**. The top endpoint of the line is coincident with the arc, vertically aligned to the Origin.

502) Sketch a **horizontal line**. The right end point of the line is coincident with the arc. Do not select the arc midpoint. Right-click **Select**.

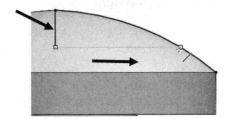

503) Click **Isometric view** .

504) Click the **Trim Entities** Sketch tool. The Trim PropertyManager is displayed.

505) Click **Power trim** from the Options box.

506) Click and drag the mouse pointer to **intersect** the **right** arc line as illustrated.

507) If needed, click and drag the mouse pointer to **intersect** the **left** arc line as illustrated.

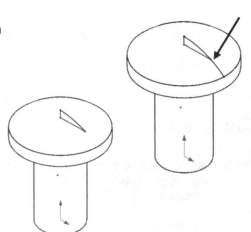

508) Click **OK** from the Trim PropertyManager.

Add a dimension.

509) Click the **Smart Dimension** Sketch tool.

510) Click the **vertical line**.

511) Click a **position** to the right of the profile.

512) Enter **.030**in, **[0.76]**.

513) Click the **Green Check mark** .

Insert an Extruded Cut Feature.

514) Click **Isometric view** .

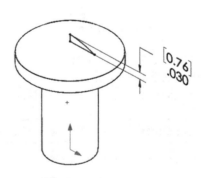

515) Click **Extruded Cut** from the Features toolbar. The Cut-Extrude PropertyManager is displayed.

516) Select **Mid Plane** for the End Condition in Direction 1.

517) Enter **.050**in, **[1.27]** for Depth.

518) Click **OK** from the Cut-Extrude PropertyManager. Cut-Extrude1 is displayed in the FeatureManager.

Activity: SCREW Part - Circular Pattern Feature

Insert the Circular Pattern feature.

519) Click **View, Hide/Show,** check **Temporary Axes** from the Menu bar. The Temporary Axis is required for the Circular Pattern feature.

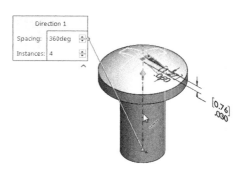

520) Click **Circular Pattern** from the Features Consolidated toolbar. Cut-Extrude1 is displayed in the Features to Pattern box.

521) Click the **Temporary Axis** in the Graphics window. Axis<1> is displayed in the Pattern Axis box.

522) Check the **Equal spacing** box.

523) Enter **360**deg for Angle.

524) Enter **4** for Number of Instances. Check the **Geometry pattern** box.

525) Click **OK** from the Circular Pattern PropertyManager. CirPattern1 is displayed in the FeatureManager.

Save the model.

526) Click **Save**.

Activity: SCREW Part - Fillet Feature

Insert a Constant Radius Fillet Type feature.

527) Click the **top circular edge** as illustrated.

528) Click **Fillet** from the Features toolbar. The Fillet PropertyManager is displayed.

529) Click the **Manual** tab. Edge<1> is displayed in the Items to Fillet box.

530) Click the **Constant Radius Fillet** icon for Fillet Type.

531) Enter **.010**in, [**.25**] for Radius.

532) Click **OK** from the Fillet PropertyManager. Fillet1 is displayed in the FeatureManager.

Activity: SCREW Part - Chamfer Feature

Insert the Chamfer feature.

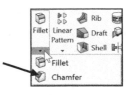

533) Click the **bottom circular edge**.

534) Click **Chamfer** 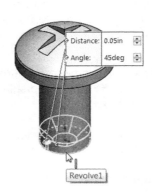 from the Features
Consolidated toolbar. Edge<1> is
displayed in the Items to Chamfer box.

535) Click **Angle distance** for type.

536) Enter **.050**in, **[1.27]** for Distance. View
the direction of the arrow.

537) Click **OK** ✔ from the Chamfer
PropertyManager. Chamfer1 is
displayed in the FeatureManager.

💡 Simplify the part. Save rebuild time. Suppress features that
are not required in the assembly.

A suppressed feature is not displayed in the Graphics window. A
suppressed feature is removed from any rebuild calculations.

Suppress the Fillet and Chamfer feature.

538) Hold the **Ctrl** key down.

539) Click **Fillet1** and **Chamfer1** from the FeatureManager.

540) Release the **Ctrl** key.

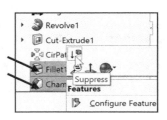

541) Right-click **Suppress** ↓▫ from the Context toolbar. Note:
Suppressed features are displayed in light gray in the
FeatureManager.

Deactivate the Temporary Axes in the Graphics window.

542) Click **View**, **Hide/Show**, un-check **Temporary Axes** from the
Menu bar.

Display an Isometric view. Save the SCREW part.

543) Click **Isometric view** 📦 from the Heads-up View toolbar.

544) Click **Save** 💾.

Close all open documents.

545) Click **Window**, **Close All** from the Menu bar.

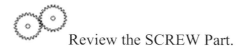 Review the SCREW Part.

The Revolved Boss/Base feature was utilized to create the simplified SCREW part. The Revolved Boss/Base feature required a centerline sketched on the Front Sketch plane and a closed profile. The sketch was revolved 360 degrees to create the Base feature for the SCREW part.

Edit Sketch was utilized to modify the sketch. The Sketch Trim and Tangent Arc tools created a new profile.

The Extruded Cut feature was sketched on the Front Plane. The Circular Pattern feature created four instances. The Fillet feature rounded the top edges.

The Chamfer feature beveled the bottom edge. The Fillet and Chamfer are suppressed to save rebuild time in the assembly. Note: suppressed features are displayed in light gray in the FeatureManager.

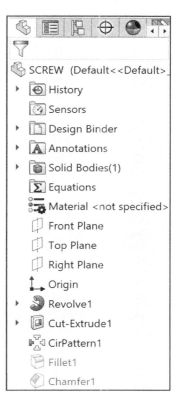

Additional details on Convert Entities, Silhouette Edge, Revolved Boss/Base, Circular Pattern, Fillet, and Chamfer are available in SOLIDWORKS Help. Keywords: Sketch tools, silhouette, features, Pattern, Revolve, Fillet and Chamfer.

The Origin and Tangent edges are displayed for educational purposes.

Think design intent. When do you use the various End Conditions and Geometric sketch relations? What are you trying to do with the design? How does the component fit into an assembly?

Mate to the first part added to the assembly. If you mate to the first part or base part of the assembly and decide to change its orientation later, all the parts will move with it.

FRONT-SUPPORT Assembly

The FRONT-SUPPORT assembly consists of the following parts:

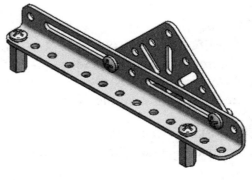

- ANGLE-13HOLE

- TRIANGLE

- HEX-STANDOFF

- SCREW

Create the FRONT-SUPPORT assembly. Insert the ANGLE-13HOLE part. The ANGLE-13HOLE part is fixed to the FRONT-SUPPORT Origin. Insert the first HEX-STANDOFF part.

Utilize Concentric and Coincident mates to assemble the HEX-STANDOFF to the left hole of the ANGLE-13HOLE part. Insert the second HEX-STANDOFF part.

Utilize Concentric and Coincident mates to assemble the HEX-STANDOFF to the third hole from the right side. Insert the TRIANGLE part. Utilize Concentric, Distance, and Parallel mates to assemble the TRIANGLE. Utilize Concentric/Coincident SmartMates to assemble the four SCREWS. Note: Other mate types can be used.

Determine the static and dynamic behavior of mates in each sub-assembly before creating the top-level assembly.

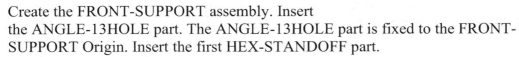

Activity: FRONT-SUPPORT Assembly-Insert ANGLE-13HOLE

Create a new assembly.

546) Click **New** ☐ from the Menu bar.

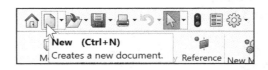

547) Double-click **Assembly** from the Templates tab. The Begin Assembly PropertyManager is displayed. Open models are displayed in the Open documents box. If there are no Open SOLIDWORKS models, the Window Open dialog is also displayed.

548) Double-click the **ANGLE-13HOLE** part SW-TUTORIAL-2020.

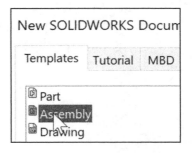

Fix the first component to the Origin.

549) Click **OK** ✔ from the Begin Assembly PropertyManager. The first component is fixed to the Origin (f).

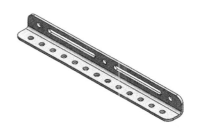

Save the assembly. Enter name. Enter description.

550) Click **Save As** 🖫.

551) Select **SW-TUTORIAL-2020** for the Save in file folder.

552) Enter **FRONT-SUPPORT** for File name.

553) Enter **FRONT SUPPORT ASSEMBLY** for Description.

554) Click **Save**. The FRONT-SUPPORT assembly FeatureManager is displayed.

Set Document Properties. Set drafting standard.

555) Click **Options** ⚙ from the Main menu.

556) Click the **Document Properties** tab from the dialog box.

557) Select **ANSI** from the Overall drafting standard box.

Set document units. Set precision.

558) Click **Units**.

559) Select **IPS**, [**MMGS**] for Unit system.

560) Select **.123**, [**.12**] for Length units Decimal places.

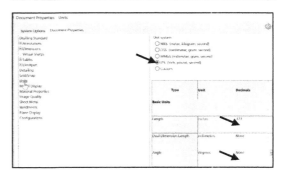

561) Select **None** for Angular units Decimal places.

562) Click **OK**.

The ANGLE-13HOLE name in the FeatureManager displays an (f) symbol. The (f) symbol indicates that the ANGLE-13HOLE component is fixed to the FRONT-SUPPORT assembly Origin. The component cannot move or rotate.

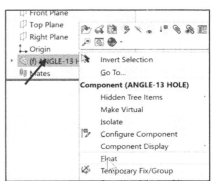

💡 To remove the fixed state, Right-click a component name in the FeatureManager. Click **Float**. The component is free to move.

Display an Isometric view.

563) Click **Isometric view** from the Heads-up View toolbar.

Save the assembly.

564) Click **Save** .

Activity: FRONT-SUPPORT Assembly - Insert HEX-STANDOFF

Insert the HEX-STANDOFF part.

565) Click the **Insert Components** Assembly tool.

566) Double-click **HEX-STANDOFF** from the SW-TUTORIAL 2020 folder.

567) Click a **position** near the left top front hole as illustrated.

Enlarge the view.

568) **Zoom in** on the front left side of the assembly.

Move the component.

569) Click and drag the **HEX-STANDOFF** component below the ANGLE-13HOLE left hole.

The HEX-STANDOFF name in the FeatureManager displays a (-) minus sign. The minus sign indicates that the HEX-STANDOFF part is free to move.

Insert a Concentric mate.

570) Click the **left inside cylindrical hole face** of the ANGLE-13HOLE component.

571) Hold the **Ctrl** key down.

572) Click inside the **cylindrical hole face** of the HEX-STANDOFF component.

573) Release the **Ctrl** key. The Mate pop-up menu is displayed.

574) Click **Concentric** from the Mate pop-up menu.

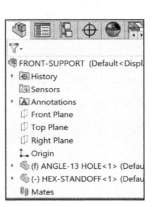

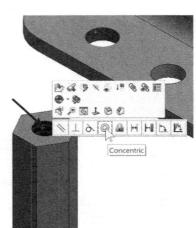

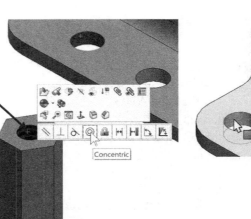

575) Click and drag the **HEX-STANDOFF** component below the ANGLE-13HOLE component until the top face is displayed.

Boss-Extrude1

Insert a Coincident mate.

576) Click the **HEX-STANDOFF top** face.

577) Rotate the assembly to view the bottom face of the ANGLE-13HOLE component.

578) Hold the **Ctrl** key down.

579) Click the **ANGLE-13HOLE bottom face**.

580) Release the **Ctrl** key. The Mate pop-up menu is displayed.

581) Click **Coincident** from the Mate pop-up menu.

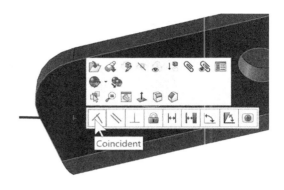

Display an Isometric view.

582) Click **Isometric view** .

583) Click and drag the **HEX-STANDOFF** component. The HEX-STANDOFF rotates about its axis.

Insert a Parallel mate.

584) Click **Front view** .

585) Click the **HEX-STANDOFF front face**.

586) Hold the **Ctrl** key down.

587) Click the **ANGLE-13HOLE front face**.

588) Release the **Ctrl** key. The Mate pop-up menu is displayed.

589) Click **Parallel** from the Mate pop-up menu.

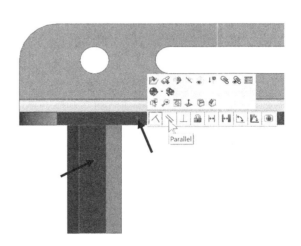

Display the created mates.

590) Expand the Mates folder in the FRONT-SUPPORT FeatureManager. Three mates are displayed between the ANGLE-13HOLE component and the HEX-STANDOFF component. The HEX-STANDOFF component is fully defined.

> (f) ANGLE-13 HOLE<1> (Default<<Default>_Display State 1>
> HEX-STANDOFF<1> (Default<<Default>_Display State 1>)
▼ Mates
  ◎ Concentric1 (ANGLE-13 HOLE<1>,HEX-STANDOFF<1>)
  ⅄ Coincident1 (ANGLE-13 HOLE<1>,HEX-STANDOFF<1>)
  ⟍ Parallel1 (ANGLE-13 HOLE<1>,HEX-STANDOFF<1>)

Save the FRONT-SUPPORT assembly.

591) Click **Save** 🖫.

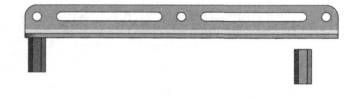

Insert the second HEX-STANDOFF part.

592) Hold the **Ctrl** key down.

593) Click and drag the **HEX-STANDOFF<1>**
🔩 HEX-STANDOFF<1> name from the FeatureManager into
the FRONT-SUPPORT assembly Graphics window.

594) Release the **mouse pointer**
below the far right hole of the
ANGLE-13HOLE component.

595) Release the **Ctrl** key. HEX-
STANDOFF<2> is displayed in
the Graphics window and listed in
the FeatureManager.

💡 The number <2> indicates the second instance or
copy of the same component. The instance number
increments every time you insert the same component. If
you delete a component and then reinsert the component in
the same SOLIDWORKS session, the instance number
increments by one.

Enlarge the view.

596) **Zoom in** on the right side of the assembly.

Insert a Concentric mate.

597) Click inside the **cylindrical hole face** of the
second HEX-STANDOFF component.

598) Hold the **Ctrl** key down.

599) Click the **third hole cylindrical face** from the
right ANGLE-13HOLE component.

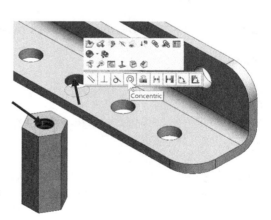

600) Release the **Ctrl** key. The Mate pop-up menu is
displayed.

601) Click **Concentric** from the Mate pop-up menu.

Move the second HEX-STANDOFF part.

602) Click and drag the **HEX-STANDOFF** component below the
ANGLE-13HOLE component until its top face is displayed.

Insert a Coincident mate.

603) Click the **second HEX-STANDOFF** top face.

604) **Rotate** the assembly to view the bottom face of the ANGLE-13HOLE component.

605) Hold the **Ctrl** key down.

606) Click the **ANGLE-13HOLE bottom face**.

607) Release the **Ctrl** key.

608) Click **Coincident** from the Mate pop-up menu.

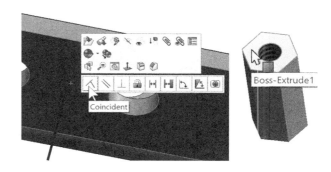

Insert a Parallel mate.

609) Click **Front view** 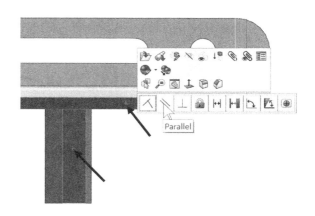.

610) Click the **front face** of the second HEX-STANDOFF.

611) Hold the **Ctrl** key down.

612) Click the **front face** of the ANGLE-13HOLE.

613) Release the **Ctrl** key. The Mate pop-up menu is displayed.

614) Click **Parallel** from the Mate pop-up menu.

Display the created mates.

615) **Expand** the Mates folder in the FRONT-SUPPORT FeatureManager. Three mates are displayed between the ANGLE-13HOLE component and the second HEX-STANDOFF component. The second HEX-STANDOFF is fully defined.

Activity: FRONT-SUPPORT Assembly - Insert the TRIANGLE

Insert the TRIANGLE part.

616) Click **Isometric view** .

617) Click the **Insert Components** Assembly tool.

618) Double-click **TRIANGLE** from the SW-TUTORIAL-2020 folder.

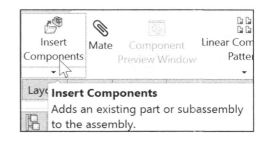

619) Click a **position** in back of the ANGLE-13HOLE component as illustrated.

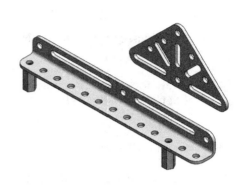

Enlarge the view.
620) Zoom in on the right side of the TRIANGLE and the ANGLE-13HOLE.

Insert a Concentric mate.

621) Click the **inside right arc face** of the TRIANGLE.

622) Hold the **Ctrl** key down.

623) Click the **inside right arc face** of the ANGLE-13HOLE slot.

624) Release the **Ctrl** key. The Mate pop-up menu is displayed.

625) Click **Concentric** from the Mate pop-up menu.

Fit the model to the Graphics window.

626) Press the **f** key.

Insert a Distance mate.
627) Click the **front face** of the TRIANGLE.

628) Rotate the model to view the back face of the ANGLE-13HOLE component.

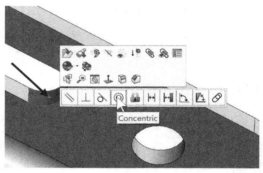

629) Hold the **Ctrl** key down.

630) Click the **back face** of the ANGLE-13HOLE component.

631) Release the **Ctrl** key. The Mate pop-up menu is displayed.

632) Click Standard **Distance** |↔| from the Mate pop-up menu.

633) Enter **0**.

634) Click the **Green Check mark** ✅.

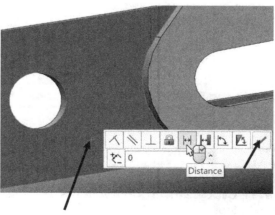

A Distance Mate of 0 provides additional flexibility compared to a Coincident mate. A Distance mate value can be modified.

Insert a Parallel mate.

635) Click **Bottom view** .

636) Click the **narrow bottom face** of the TRIANGLE.

637) Hold the **Ctrl** key down.

638) Click the **bottom face** of the ANGLE-13HOLE.

639) Release the **Ctrl** key.

640) Click **Parallel** from the Mate pop-up menu.

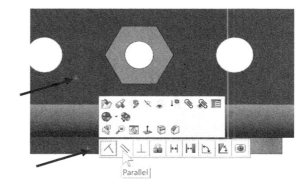

Display the Isometric view.

641) Click **Isometric view** .

View the created mates.

642) **Expand** the Mates folder. View the created mates.

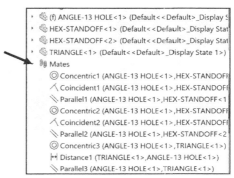

Save the FRONT-SUPPORT assembly.

643) Click **Save** .

Assemble the four SCREW parts with SmartMates.

A SmartMate is a mate that automatically occurs when a component is placed into an assembly.

The mouse pointer displays a SmartMate feedback symbol when common geometry and relationships exist between the component and the assembly.

SmartMates are Concentric, Coincident, or Concentric and Coincident.

A Concentric SmartMate assumes that the geometry on the component has the same center as the geometry on an assembled reference.

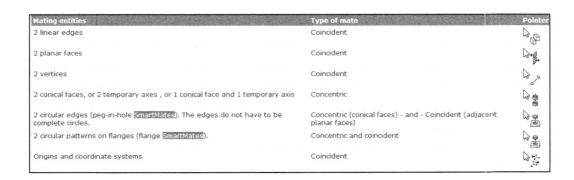

Mating entities	Type of mate	Pointer
2 linear edges	Coincident	
2 planar faces	Coincident	
2 vertices	Coincident	
2 conical faces, or 2 temporary axes , or 1 conical face and 1 temporary axis	Concentric	
2 circular edges (peg-in-hole SmartMates). The edges do not have to be complete circles.	Concentric (conical faces) - and - Coincident (adjacent planar faces)	
2 circular patterns on flanges (flange SmartMates).	Concentric and coincident	
Origins and coordinate systems	Coincident	

As the component is dragged into place, the mouse pointer provides various feedback icons. The SCREW utilizes a Concentric and Coincident SmartMate. Assemble the first SCREW. The circular edge of the SCREW mates Concentric and Coincident with the circular edge of the right slot of the TRIANGLE.

Activity: FRONT-SUPPORT Assembly - Insert the SCREW

Insert the SCREW part.

644) Click **Open** 📂 from the Menu bar.

645) Double-click **SCREW** from the SW-TUTORIAL-2020 folder. The SCREW PropertyManager is displayed.

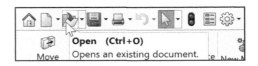

Display the SCREW part and the FRONT-SUPPORT assembly.

646) Click **Window**, **Tile Horizontally** from the Menu bar.

647) **Zoom in** on the right side of the FRONT-SUPPORT assembly. Note: Work between the two tile windows.

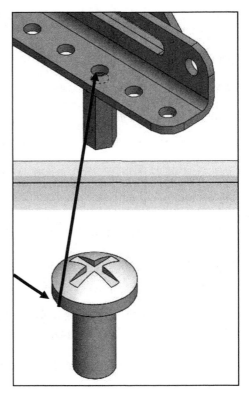

Insert the first SCREW.

648) Click and drag the **circular edge** of the SCREW part into the FRONT-SUPPORT assembly Graphic window.

649) Release the mouse pointer on the **top 3ʳᵈ circular hole edge** of the ANGLE-13HOLE. The mouse pointer displays the Coincident/Concentric circular edges icon.

💡 Use the Pack and Go option to save an assembly or drawing with references. The Pack and Go tool saves either to a folder or creates a zip file to email. View SOLIDWORKS help for additional information.

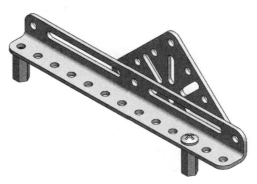

Insert the second SCREW.

650) **Zoom in** on the right side of the FRONT-SUPPORT assembly.

651) Click and drag the **circular edge** of the SCREW part into the FRONT-SUPPORT assembly Graphic window.

652) Release the mouse pointer on the **right arc edge** of the ANGLE-13HOLE. The mouse pointer displays the Coincident/Concentric circular edges icon.

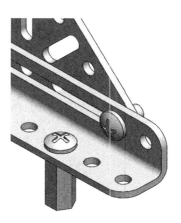

Insert the third SCREW part.

653) **Zoom in** on the left side of the FRONT-SUPPORT assembly.

654) Click and drag the **circular edge** of the SCREW part into the FRONT-SUPPORT Assembly Graphic window.

655) Release the mouse pointer on the **left arc edge** of the ANGLE-13HOLE. The mouse pointer displays the Coincident/Concentric circular edges icon.

Insert the fourth SCREW part.

656) **Zoom in** on the bottom circular edge of the ANGLE-13HOLE.

657) Click and drag the **circular edge** of the SCREW part into the FRONT-SUPPORT Assembly Graphic window.

658) Release the mouse pointer on the **bottom circular edge** of the ANGLE-13HOLE. The mouse pointer displays the Coincident/Concentric circular edges icon.

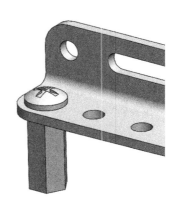

659) **Close** the SCREW part window.

660) **Maximize** the FRONT-SUPPORT assembly window.

Display an Isometric view.

661) Click **Isometric view** . View the results in the Graphics window.

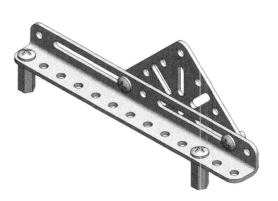

Save the FRONT-SUPPORT assembly.

662) Click **Save** 💾.

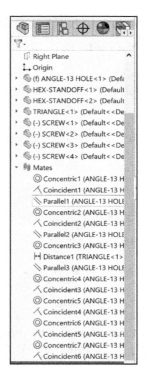

Select the Ctrl-Tab keys to quickly alternate between open SOLIDWORKS documents.

Close all open parts and assemblies.

663) Click **Windows**, **Close All** from the Menu bar.

The FRONT-SUPPORT assembly is complete.

Display the Mates in the FeatureManager to check that the components and the mate types correspond to the design intent.

When a component is in the Lightweight 🪶 state, only a subset of its model data is loaded in memory. The remaining model data is loaded on an as-needed basis.

 Review the FRONT-SUPPORT Assembly.

The ANGLE-13HOLE part was the first part inserted into the FRONT-SUPPORT assembly. The ANGLE-13HOLE part was fixed to the FRONT-SUPPORT Origin.

Concentric, Coincident, and Parallel mates were utilized to assemble the HEX-STANDOFF to the ANGLE-13HOLE. *Concentric*, *Distance*, and *Parallel* mates were utilized to assemble the TRIANGLE to the ANGLE-13HOLE. The *Concentric/Coincident* SmartMate was utilized to mate the four SCREW parts to the FRONT-SUPPORT assembly.

Chapter Summary

In this chapter you created four parts, HEX-STANDOFF, ANGLE-13HOLE, TRIANGLE and SCREW, utilizing the Top, Front, Right Planes and the FRONT-SUPPORT assembly.

You obtained the knowledge of the following SOLIDWORKS features: Extruded Boss/Base, Extruded Thin, Extruded Cut, Revolved Boss/Base, Hole Wizard, Linear Pattern, Circular Pattern, Fillet, and Chamfer. You also applied sketch techniques with various Sketch tools: Line, Circle, Corner Rectangle, Centerline, Dynamic Mirror, Straight Slot, Trim Entities, Polygon, Tangent Arc, Sketch Fillet, Offset Entities, and Convert Entities.

You utilized centerlines as construction geometry to reference dimensions and relationships. You incorporated the four new parts to create the FRONT-SUPPORT assembly. Concentric, Distance, and Parallel mates were utilized to assemble the TRIANGLE to the ANGLE-13HOLE.

🔅 You can define the end condition of a hole to the depth of the drill tip or to the depth of the shoulder. The options are available for all Hole Wizard features (including the Hole Wizard Assembly feature) and Advanced Hole types with the following end conditions: Blind, Up to Vertex, Up to Surface, Offset to Surface. Previously, the end condition of a hole was calculated only up to the full diameter of the shoulder.

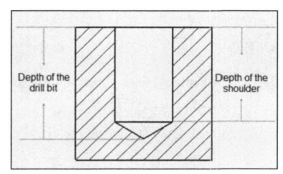

You applied the Quick mate procedure in the FRONT-SUPPORT assembly. Quick Mate is a procedure to mate components together. No command (click Mate from the Assembly CommandManager) is required. Hold the Ctrl key down, and make your selections. Release the Ctrl key; a Quick Mate pop-up is displayed below the context toolbar. Select your mate and you are finished.

Utilize the Quick Mate procedure for Standard mates, Cam mate, Slot mate, Symmetric mate, Profile Center, Advance Distance and Angle mates. To activate the Quick Mate functionality, click Tools, Customize. On the toolbars tab, under Context toolbar settings, select Show Quick Mates. Quick Mate is selected by default.

The Concentric/Coincident SmartMate was utilized to mate the four SCREW parts to the FRONT-SUPPORT assembly.

🔅 CommandManager and FeatureManager tabs and file folders will vary depending on system setup and Add-ins.

🔅 Use Selection filters to select difficult individual features such as *faces*, *edges* and *points* when applying mates in an assembly.

Questions

1. Identify the three default Reference planes used in SOLIDWORKS.

2. True or False. Sketches are created only on the Front Plane.

3. Identify the sketch tool required to create a hexagon.

4. Describe the profile required for an Extruded Thin feature.

5. Mirror Entities, Offset Entities, Sketch Fillet, and Trim Entities are located in the _____ toolbar.

6. List the six standard principle orthographic views in a drawing: _____, _____, _____, _____, _____, _____,

7. Identify the type of Geometric relations that can be added to a sketch.

8. Describe the difference between a Circular Pattern feature and a Linear Pattern feature.

9. Describe the difference between a Fillet feature and a Chamfer feature.

10. Identify the function of the Hole Wizard feature.

11. Four 10-24x3/8 Machine Screws are required for an assembly. The diameter is _____. The threads per inch is _____. The length is _____.

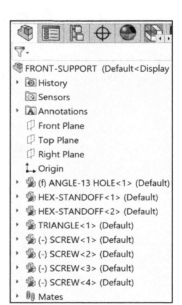

12. Describe the difference between a Distance mate and a Coincident mate.

13. True or False. A fixed component cannot move in an assembly.

14. Describe the procedure to remove the fix state (f) of a component in an assembly.

15. Determine the procedure to rotate a component in an assembly.

16. Describe the procedure to resolve a Lightweight component in an assembly.

17. Describe the procedure to resolve a Lightweight assembly.

Exercises

Exercise 3.1: HEX-NUT Part

Create an ANSI, IPS HEX-NUT part. Apply 6061 Alloy material. Apply the following dimensions:

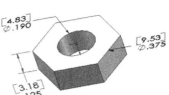

- Depth: .125 in, [3.18].

- Inside hole diameter: .190in, [4.83].

- Outside diameter: .375in, [9.53].

Use the Top Plane as the Sketch plane.

Exercise 3.2: FRONT-SUPPORT-2 Assembly

Create an ANSI, IPS FRONT-SUPPORT-2 assembly.

- Name the new assembly FRONT-SUPPORT-2.

- Insert the FRONT-SUPPORT assembly. The FRONT-SUPPORT assembly was created in this Chapter. Note: The FRONT-SUPPORT assembly is provided in the Chapter 3 Homework folder.

- Fix the FRONT-SUPPORT assembly to the Origin.

- Insert the first HEX-NUT (Exercise 3.1) into the FRONT-SUPPORT-2 assembly.

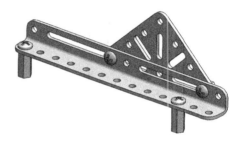

- Insert a Concentric mate and Coincident mate.

- Insert the second HEX-NUT part.

- Insert a Concentric mate and Coincident mate.

You can also insert a Parallel mate between the HEX-NUT parts and the FRONT-SUPPORT assembly.

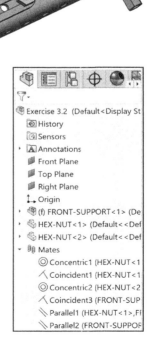

Exercise 3.3: Weight-Hook Assembly

Create an ANSI, IPS Weight-Hook assembly. The Weight-Hook assembly has two components: WEIGHT and HOOK.

- Create a new assembly document.

- Copy and insert the WEIGHT part from the Chapter 3 Homework folder.

- Fix the WEIGHT to the Origin in the Assem1 FeatureManager.

- Insert the HOOK part from the Chapter 3 Homework folder into the assembly.

- Insert a Concentric mate between the inside top cylindrical face of the WEIGHT and the cylindrical face of the thread. Concentric is the default mate.

- Insert the first Coincident mate between the top edge of the circular hole of the WEIGHT and the top circular edge of Sweep1, above the thread.

- Coincident is the default mate. The HOOK can rotate in the WEIGHT.

- Fix the position of the HOOK. Insert the second Coincident mate between the Right Plane of the WEIGHT and the Right Plane of the HOOK. Coincident is the default mate.

- Expand the Mates folder and view the created mates.

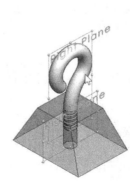

- Calculate the Mass and Volume of the assembly.

- Identify the Center of Mass for the assembly.

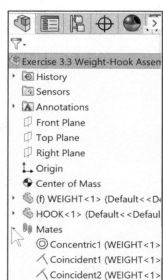

Exercise 3.4: Weight-Link Assembly

Create an ANSI, IPS Weight-Link assembly. The Weight-Link assembly has two components and a sub-assembly: Axle component, FLATBAR component, and the Weight-Hook sub-assembly that you created in a previous exercise.

- Create a new assembly document. Copy and insert the Axle part from the Chapter 3 Homework folder.

- Fix the Axle component to the Origin.

- Copy and insert the FLATBAR part from the Chapter 3 Homework folder. Rotate as needed.

- Insert a Concentric mate between the Axle cylindrical face and the FLATBAR inside face of the top circle.

- Insert a Coincident mate between the Front Plane of the Axle and the Front Plane of the FLATBAR.

- Insert a Coincident mate between the Right Plane of the Axle and the Top Plane of the FLATBAR. Position the FLATBAR as illustrated.

- Insert the Weight-Hook sub-assembly that you created in Exercise 3.3.

- Insert a Tangent mate between the inside bottom cylindrical face of the FLATBAR and the top circular face of the HOOK, in the Weight-Hook assembly. Tangent mate is selected by default. Click Flip Mate Alignment if needed.

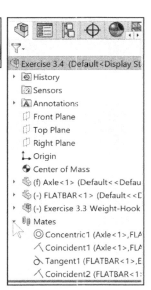

- Insert a Coincident mate between the Front Plane of the FLATBAR and the Front Plane of the Weight-Hook sub-assembly. Coincident mate is selected by default. The Weight-Hook sub-assembly is free to move in the bottom circular hole of the FLATBAR.

- Calculate the Mass and Volume and the Center of Mass for the assembly.

Exercise 3.5: Binder Clip

Create a simple Gem® binder
clip.

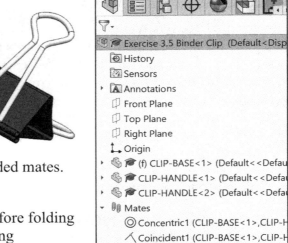

Create an ANSI - IPS assembly.
Create two components - BASE
and HANDLE.

Apply material (plain carbon
steel) to each component and address all needed mates.
Think about where you would start.

Note: The cuts in the main body are made before folding
the metal and strengthening the material during
manufacturing. You do not need to model this exactly.
The key to this exercise is to apply Advanced distance
mates (Limit mates) to limit the movement of the handle.

What is the Base Sketch for each component?

What are the dimensions? Measure all dimensions
(approximately) from a small or large Gem binder clip.

View SOLIDWORKS Help or Chapter 5 on the Swept
Base feature to create the Binder Clip handle.

🔅 Determine the static and dynamic behavior of mates
in each sub-assembly before creating the top level
assembly.

🔅 Use the Quick mate procedure to apply Advanced
distance angle (LimitAngle) mates.

Exercise 3.6: Limit Mate (Advanced Distance Mate)

- Copy all files from the Chapter 3 Homework\Limit Mate folder to your hard drive.

- Create the final assembly with all needed mates for proper movement.

- Insert a Distance (Advanced Limit Mate) to restrict the movement of the Slide Component - lower and upper movement.

- Use the Measure tool to obtain maximum and minimum distances.

- Use SOLIDWORKS Help for additional information.

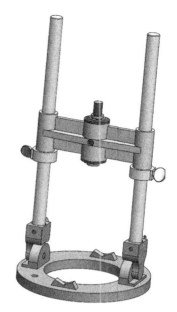

A Distance (Limit Mate) is an Advanced Mate type. Limit mates allow components to move within a range of values for distance and angle. You specify a starting distance or angle as well as a maximum and minimum value.

- Save the model and move the slide to view the results in the Graphics window.

- Think about how you would use this mate type in other assemblies.

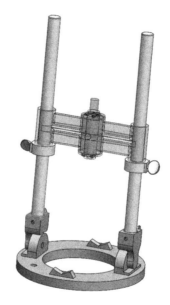

Use the Quick mate procedure to apply Advanced distance (Limit) mates.

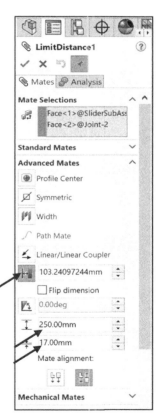

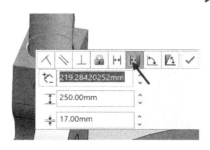

Exercise 3.7: Screw Mate (Mechanical Mate)

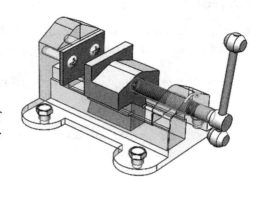

- Copy the Chapter 3 Homework\Screw Mate folder to your hard drive.

- Open the Screw Mate assembly.

- Insert a Screw mate between the inside Face of the Base and the Face of the vice and any other mates that are required. A Screw is a Mechanical Mate type.

- Calculate the total mass and volume of the assembly.

A Screw mate constrains two components to be concentric and also adds a pitch relationship between the rotation of one component and the translation of the other. Translation of one component along the axis causes rotation of the other component according to the pitch relationship. Likewise, rotation of one component causes translation of the other component. Use SOLIDWORKS Help if needed.

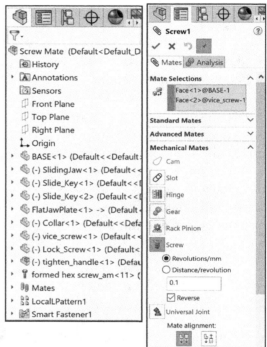

- Rotate the handle and view the results. Think about how you would use this mate type in other assemblies?

Use the Select Other tool (See SOLIDWORKS Help) to select faces and edges that are hidden in an assembly.

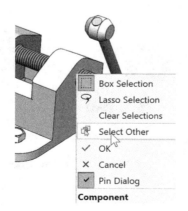

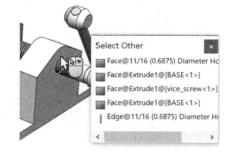

Exercise 3.8: Angle Mate

- Copy the Chapter 3 Homework\Angle Mate folder to your hard drive.

- Open the Angle Mate assembly.

- Move the Handle in the assembly. The Handle is free to rotate.

- Set the angle of the Handle.

- Insert an Angle mate (165 degree) between the Handle and the Side of the valve using Planes. An Angle mate places the selected items at the specified angle to each other. The Handle has a 165 degree Angle mate to restrict flow through the valve.

- Think about how you would use this mate type in other assemblies?

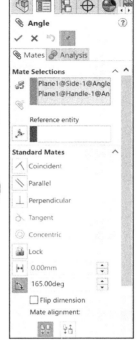

Exercise 3.8A: Angle Mate (Cont:)

Create two end caps (lids) for the ball valve using the Top-down Assembly method. Note: The Reference - In-Content symbols in the FeatureManager.

- Modify the Appearance of the body to observe the change, and enhance visualization.

- Apply the Select-other tool to obtain access to hidden faces and edges.

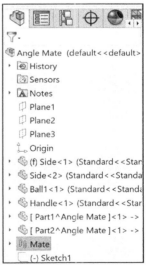

Exercise 3.9: Symmetric Mate (Advanced Mate)

- Copy the Chapter 3 Homework\Symmetric Mate folder to your hard drive.

- Open the Symmetric Mate assembly.

- Insert a Symmetric Mate between the two Guide Rollers.

- Apply LimitMates (Advanced Distance mate) to limit movement of the two Guide Rollers.

- Think about how you would use this mate type in other assemblies?

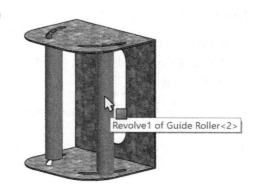

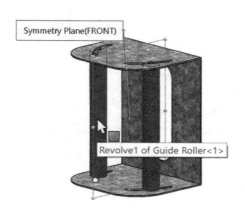

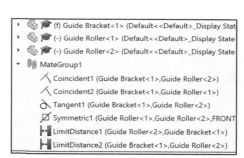

Exercise 3.10: Gear Mate (Mechanical mate)

View the ppt presentation and avi file on Gears located in the Chapter 3 Homework\Gears folder.

Create the Gear Mate assembly as illustrated.

The Gear assembly consists of a **Base Plate**, two **Shafts** for the gears, and two different gears from the **SOLIDWORKS toolbox**.

Use the SOLIDWORKS toolbox to insert the needed gear components. You create the **Base Plate**, and **Shafts**.

Insert all needed mates for proper movement.

Gear mate: Forces two components to rotate relative to one another around selected axes. The Gear mate provides the ability to establish gear type relations between components without making the components physically mesh.

SOLIDWORKS provides the ability to modify the gear ratio without changing the size of the gears. Align the components before adding the Mechanical gear mate.

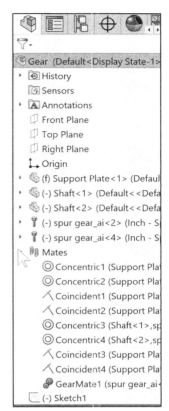

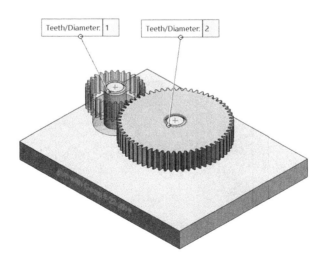

Exercise 3.10A: Slot Mate (Center in Slot option)

Copy the Chapter 3 Homework\Slot Mate folder to your hard drive.

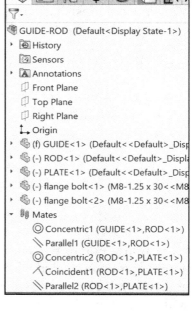

Open the GUIDE-ROD assembly. Create two Slot mates for the two flange bolts. You can mate bolts to straight or arced slots and you can mate slots to slots. Use the Quick Mate procedure.

Insert a Slot mate (Center in Slot option) between the first flange bolt and the inside face of the right slot.

Insert the second slot mate (Center in Slot option) between the second flange bolt and the inside face of the left slot.

Insert a Coincident mate between the top face of the GUIDE and the bottom face of the second flange bolt cap.

Insert a Coincident mate between the top face of the GUIDE and the bottom face of the first flange bolt cap.

Insert a parallel mate between the front face of the first hex head flange bolt and the front of the GUIDE.

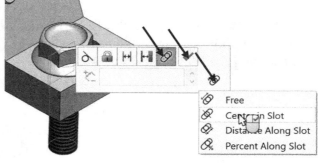

Inset a parallel mate between the front face of the second hex head flange bolt and the front face of the GUIDE.

Expand the Mates folder. View the created mates.

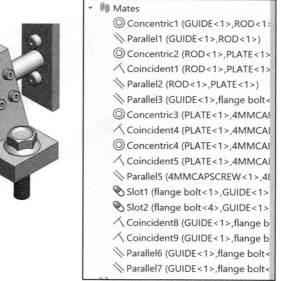

Exercise 3.11: Slider Part

Create the part from the illustrated A-ANSI - MMGS Third Angle Projection drawing below: Front, Top, Right and Isometric view.

Note: The location of the Origin (Shown in an Isometric view).

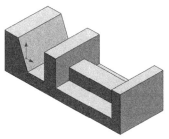

- Apply Cast Alloy steel for material.

- The part is symmetric about the Front Plane.

- Apply Mid Plane for End Condition in Boss-Extrude1.

- Apply Through All for End Condition in Cut-Extrude1.

- Apply Through All for End Condition in Cut-Extrude2.

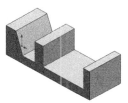

- Apply Up to Surface for End Condition in Boss Extrude2

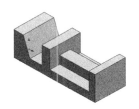

- Calculate the Volume of the part and locate the Center of mass.

Think about the steps that you would take to build the model. Do you need the Right view for manufacturing? Does it add any important information?

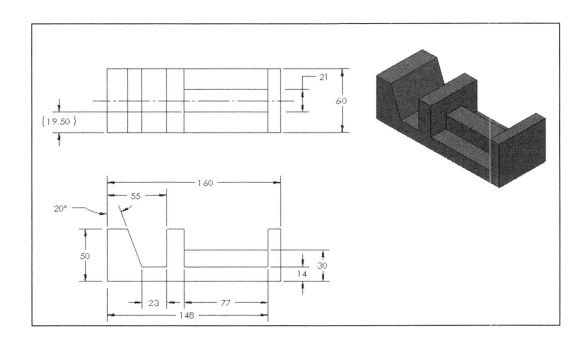

Exercise 3.12: Cosmetic Thread Part

Apply a Cosmetic thread: 1/4-20-1 UNC 2A. A cosmetic thread represents the inner diameter of a thread on a boss or the outer diameter of a thread.

Copy the Cosmetic thread part from the Chapter 3 Homework folder to your hard drive.

Open the part. Create a Cosmetic thread. Produce the geometry of the thread. Click the bottom edge of the part as illustrated.

Click Insert, Annotations, Cosmetic Thread from the Menu bar menu. View the Cosmetic Thread PropertyManager. Edge<1> is displayed.

Select ANSI Inch for Standard.

Select Machine Threads for Type.

Select ¼-20 for Size.

End-Condition - Blind.

Enter 1.00 for depth.

Select 2A for Thread class. Click OK ✔ from the Cosmetic Thread FeatureManager.

Expand the FeatureManager. View the Cosmetic Thread feature. If needed, right-click the Annotations folder, click Details.

Check the Cosmetic threads and Shaded cosmetic threads box.

Click OK. View the cosmetic thread on the model.

☼ ¼-20-1 UNC 2A: ¼ inch major diameter - 20 threads / inch, 1 inch long, Unified National Coarse thread series, Class 2 (General Thread), A - External threads.

Exercise 3.13: Hole - Block

Create the Hole-Block ANSI - IPS part using the Hole Wizard feature as illustrated. Create the Hole-Block part on the Front Plane.

The Hole Wizard 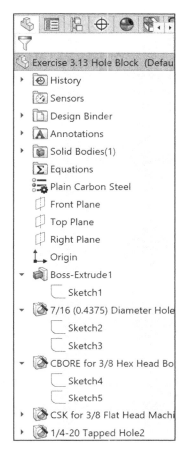 feature creates either a 2D or 3D sketch for the placement of the hole in the FeatureManager.

You can consecutively place multiple holes of the same type. The Hole Wizard creates 2D sketches for holes unless you select a non-planar face or click the 3D Sketch button in the Hole Position PropertyManager.

Hole Wizard creates two sketches. A sketch of the revolved cut profile of the selected hole type and the other sketch of the center placement for the profile. Both sketches should be fully defined.

Create a rectangular prism 2 inches wide by 5 inches long by 2 inches high. On the top surface of the prism, place four holes, 1 inch apart.

- Hole #1: Simple Hole Type: Fractional Drill Size, 7/16 diameter, End Condition: Blind, 0.75 inch deep.
- Hole #2: Counterbore hole Type: for 3/8 inch diameter Hex bolt, End Condition: Through All.
- Hole #3: Countersink hole Type: for 3/8 inch diameter Flat head screw, 1.5 inch deep.
- Hole #4: Tapped hole Type, Size ¼-20, End Condition: Blind -1.0 inch deep.

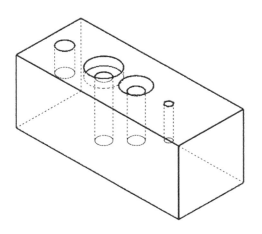

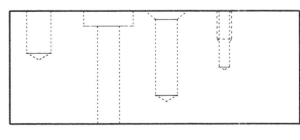

Exercise 3.14: Hole Wizard Part

Apply the 3D sketch placement method as illustrated in the FeatureManager. Insert and dimension a hole on a cylindrical face.

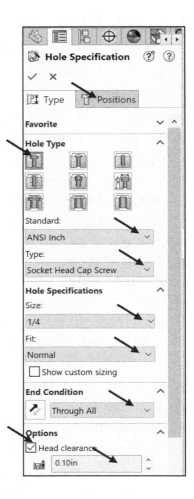

Copy the Hole Wizard 3-14 part from the Chapter 3 Homework folder to your hard drive.

Open the Hole Wizard 3-14 part.

Note: With a 3D sketch, press the Tab key to move between planes.

Click the Hole Wizard 🔧 Features tool. The Hole Specification PropertyManager is displayed.

Select the Counterbore Hole Type. Select ANSI Inch for Standard. Select Socket Head Cap Screw for fastener Type.

Select 1/4 for Size. Select Normal for Fit.

Select Through All for End Condition.

Enter .100 for Head clearance in the Options box.

Click the Positions Tab. The Hole Position PropertyManager is displayed.

Click the 3D Sketch button. SOLIDWORKS displays a 3D interface with the Point ✏️ tool active.

💡 When the Point tool is active, wherever you click, you will create a point.

Click the cylindrical face of the model as illustrated. The selected face is displayed in blue.

Insert a dimension between the top face and the Sketch point. Click the Smart Dimension 🖊️ Sketch tool.

Click the top flat face of the model and the sketch point.

Enter .25in.

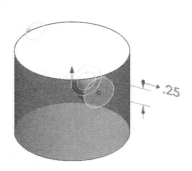

Locate the point angularly around the cylinder. Apply construction geometry.

Activate the Temporary Axes. Click View, Hide/show, check the Temporary Axes box from the Menu bar toolbar.

Click the Line ✏ Sketch tool. Note: 3D sketch is still activated.

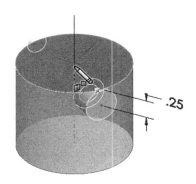

Ctrl+click the top flat face of the model. This moves the red space handle origin to the selected face. This also constrains any new sketch entities to the top flat face. Note the mouse pointer

⊹⊡✗ icon.

Move the mouse pointer near the center of the activated top flat face as illustrated. View the small black circle. The circle indicates that the end point of the line will pick up a Coincident relation.

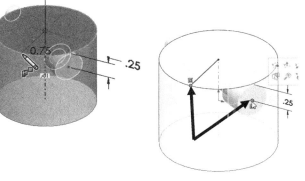

Click the center point of the circle.

Sketch a line so it picks up the **AlongZ sketch relation**. The cursor displays the relation to be applied. **This is a very important step**. Insert an AlongZ relation is needed.

Create an AlongY sketch relation between the centerpoint of the hole on the cylindrical face and the endpoint of the sketched line as illustrated. The sketch is fully defined.

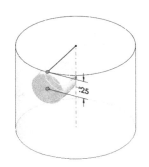

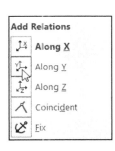

Add Relations	
⌐ˣ	Along X
⌐ʸ	Along Y
⌐ᶻ	Along Z
✗	Coincident
✗	Fix

Click OK ✔ from the Properties PropertyManager. Click OK ✔ from the Hole Position PropertyManager.

Expand the FeatureManager and view the results. The two sketches are fully defined. One sketch is the hole profile, the other sketch is to define the position of the feature.

Close the model.

Exercise 3.15: Counter Weight Assembly

Create the Counter Weight assembly as illustrated using SmartMates and Standard mates.

Copy the Chapter 3 Homework\Counter-Weight folder to your local hard drive.

Create a new assembly. The Counter Weight assembly consists of the following items:

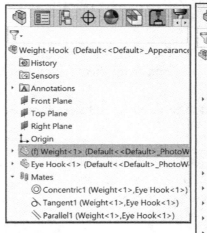

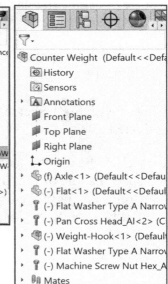

- Weight-Hook sub-assembly.

- Weight.

- Eye Hook.

- Axle component (f). Fixed to the origin.

- Flat component.

- Flat Washer Type A (from the SOLIDWORKS Toolbox).

- Pan Cross Head Screw (from the SOLIDWORKS Toolbox).

- Flat Washer Type A (from the SOLIDWORKS toolbox).

- Machine Screw Nut Hex (from the SOLIDWORKS Toolbox).

Apply SmartMates with the Flat Washer Type A Narrow_AI, Machine Screw Nut Hex_AI and the Pan Cross Head_AI components.

Use a Distance mate to fit the Axle in the middle of the Flat. Note a Symmetric mate could replace the Distance mate. Think about the design of the assembly. Apply all needed Lock mates.

The symbol (f) represents a fixed component. A fixed component cannot move and is locked to the assembly Origin.

Exercise 3.16: Gear Mate

Insert a Mechanical Gear mate between two gears. Gear mates force two components to rotate relative to one another about selected axes. Valid selections for the axis of rotation for gear mates include cylindrical and conical faces, axes, and linear edges.

1. **Copy** the Chapter 3 Homework folder to your hard drive.

2. Open **Mechanical Gear Mate** from the Chapter 3 Homework\Gear Mate folder.

3. Click and drag the **right gear**.

4. Click and drag the **left gear**. The gears move independently.

5. Click the **Mate** ✎ tool from the Assembly tab. The Mate PropertyManager is displayed.

6. **Clear** all selections.

7. **Align the components** before adding the Mechanical mate. To align the two components, select a Front view and move them manually.

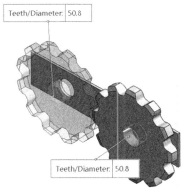

8. Select the **inside cylindrical face** of the left hole.

9. Select the **inside cylindrical face** of the right hole.

10. **Expand** the Mechanical Mates box from the Mate PropertyManager.

11. Click **Gear** ⚙ mate from the Mechanical Mates box. The GearMate1 PropertyManager is displayed. Accept the default settings.

12. Click **OK** ✔ from the GearMate1 PropertyManager.

13. **Expand** the Mates folder. View the created GearMate1.

14. **Rotate** the Gears in the Graphics window.

15. **View** the results.

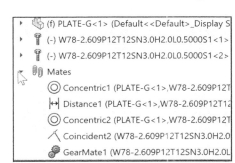

Exercise 3.17: Cam Mate

Create a Tangent Cam follower mate and a Coincident Cam follower mate. A cam-follower mate is a type of Tangent or Coincident mate. It allows you to mate a cylinder, plane or point to a series of tangent extruded faces, such as you would find on a cam.

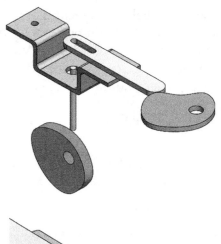

Create the profile of the cam from lines, arcs and splines as long as they are tangent and form a closed loop.

1. **Copy** the Chapter 3 Homework folder to your hard drive.

2. Open **Cam Mate** from the Chapter 3 Homework\Cam Mate folder. View the model. The model contains two cams.

3. **Expand** the Mates folder to view the existing mates.

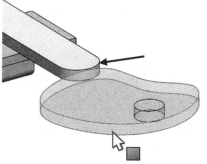

Insert a tangent Cam follower mate.

4. Click the **Mate** ✎ tool from the Assembly tab.

5. **Pin** the Mate PropertyManager.

6. **Expand** the Mechanical Mates box from the Mate PropertyManager.

7. Click **Cam Follower** ⬭ mate.

8. Click inside of the **Cam Path** box.

9. Click the **outside face** of cam-s<1> as illustrated.

10. Click the **face of Link-c<1>** as illustrated. Face<2>@Link-c-1 is displayed.

11. Click **OK** ✔ from the CamMateTangent1 PropertyManager.

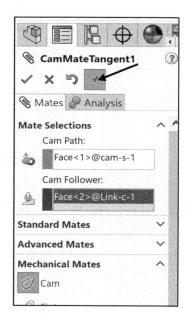

Insert a Coincident Cam follower mate.

12. Click **Cam Follower** ⌒ mate from the Mechanical Mates box.

13. Click the **outside face** of cam-s2<1>.

14. Click the vertex of **riser<1>** as illustrated. Vertex<1>@riser-1 is displayed.

15. Click **OK** ✔ from the CamMateCoincident1 PropertyManager.

16. **Un-pin** the Mate PropertyManager.

17. Click **OK** ✔ from the Mate PropertyManager.

18. **Expand** the Mates folder. View the new created mates.

19. Display an **Isometric** view.

20. **Rotate** the cams.

21. **View** the results in the Graphics window.

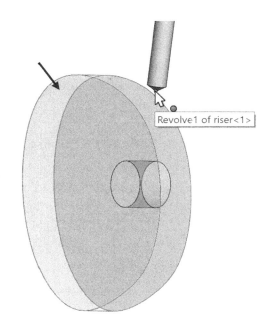

When creating a cam follower mate, make sure that your Spline or Extruded Boss/Base feature, which you use to form the cam contact face, does nothing but form the face.

In the next exercise of the spur gear and rack, the gear's pitch circle must be tangent to the rack's pitch line, which is the construction line in the middle of the tooth cut.

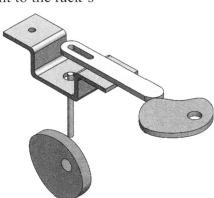

Exercise 3.18: Rack Pinion Mate

The Spur gear and rack components for this exercise were
obtained from the SOLIDWORKS Toolbox folder marked Power
Transmission. Insert a Mechanical Rack Pinion mate.

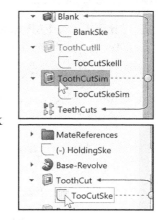

1. **Copy** the Chapter 3 Homework folder to your hard drive.

2. Open **Rack Pinion Mate** from the Chapter 3 Homework\Rack
   Pinion Mate folder.

3. **View** the displayed sketches; (TooCutSkeSim and
   ToothCutSim). If needed, display the
   illustrated sketches to mate the assembly.

The Diametral Pitch, Pressure Angle and Face
Width values should be the same for the spur gear
as the rack.

🔆 Tangency is required between the spur gear
and rack.

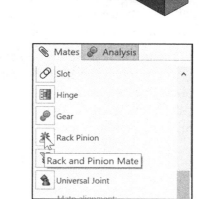

4. **Expand** the Mates folder to view the existing
   mates. Four Coincident mates and a Distance
   mate were created in the assembly.

Insert a Rack Pinion mate.

5. Click the **Mate** 🖉 tool from the Assembly
   tab.

6. **Expand** the Mechanical Mates box from the
   Mate PropertyManager.

7. Click **Rack Pinion** 🦷 mate from the
   Mechanical Mates box.

8. Mate the rack. Make sure the teeth of the spur
   gear and rack are **meshing properly**; they are
   not interfering with each
   other. It is important to
   create a starting point
   where the gear and rack
   have no interference.

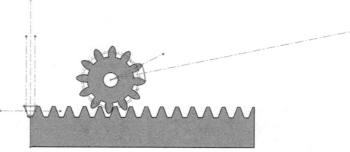

9. Click the **bottom linear edge of the rack** in the direction of travel. Any linear edge that runs in the direction of travel is acceptable.

10. Click the **spur gear pitch circle** as illustrated. The Pinion pitch diameter, 1in is taken from the sketch geometry. Accept the default settings.

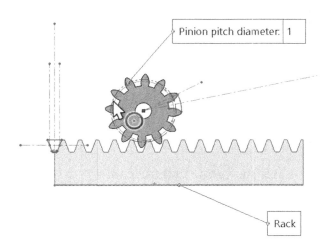

11. Click **OK** ✔ from the RackPinionMate1 PropertyManager.

12. **Rotate** the spur gear on the rack. View the results.

13. **Hide** the sketches in the model. **Close** the model.

The Pitch Diameter is the diameter of an imaginary pitch circle on which a gear tooth is designed. Pitch circles of the two spur gears are required to be tangent for the gears to move and work properly.

The distance between the center point of the gear and the pitch line of the rack represents the theoretical distance required.

🔆 Diametral Pitch is the ratio equal to the number of teeth on a gear per inch of pitch diameter.

🔆 Before applying the Rack Pinion mate, make sure the teeth of the spur gear and rack are meshing properly.

Pressure Angle is the angle of direction of pressure between contacting teeth. Pressure angle determines the size of the base circle and the shape of the involute teeth. It is common for the pressure angle to be 20 or 14 1/2 degrees.

Exercise 3.19: Hinge Mate

Create a Mechanical Hinge mate. The two components in this tutorial were obtained from the SOLIDWORKS What's New section.

1. **Copy** the Chapter 3 Homework folder to your hard drive.

2. Open **Hinge Mate** from the Chapter 3 Homework\Hinge Mate folder. Two components are displayed.

A Hinge mate has the same effect as adding a Concentric mate plus a Coincident mate. You can also limit the angular movement between the two components.

3. Click the **Mate** ✎ tool from the Assembly tab.

4. **Expand** the Mechanical Mates box from the Mate PropertyManager.

5. Click **Hinge** ▦ mate from the Mechanical Mates box.

Set two Concentric faces.

6. Click the **inside cylindrical face** of the first component as illustrated.

7. Click the **outside cylindrical face** of the second component as illustrated. The two selected faces are displayed.

Set two Coincident faces.

8. Click the **front flat face** of the first component as illustrated.

9. Click the **bottom flat face** of the second component as illustrated. View the results in the Graphics window.

You can specify a limit angle for rotation by checking the **Specify angle limits** box and selecting the required faces.

10. Click **OK** ✔ from the Hinge1 PropertyManager.

11. **Rotate** the flap component about the pin.

12. **Close** the model.

A Hinge mate limits the movement between two components to one rotational degree of freedom.

Similar to other mate types, Hinge mates do not prevent interference or collisions between components. To prevent interference, use Collision Detection or Interference Detection.

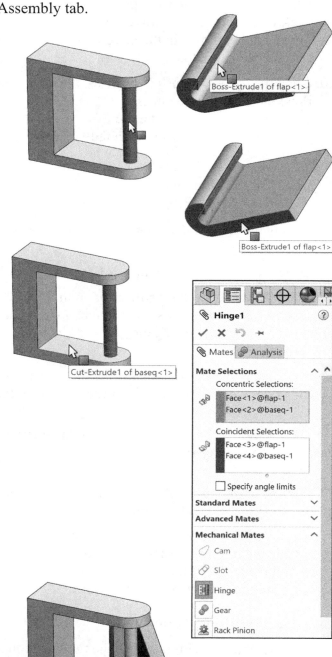

Exercise 3.20: AIR RESERVOIR SUPPORT AND PLATE Assembly

The project team developed a concept sketch of the PNEUMATIC TEST MODULE assembly. Develop the AIR RESERVOIR SUPPORT AND PLATE assembly.

Create three new parts:

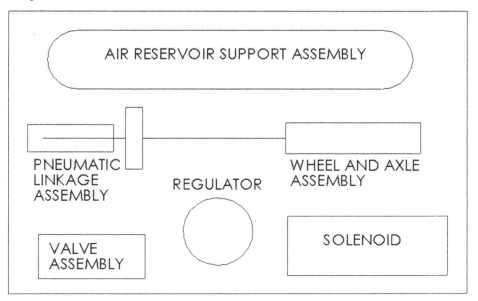

PNEUMATIC TEST MODULE Assembly Layout

- FLAT-PLATE

- IM15-MOUNT

- ANGLE-BRACKET

The Reservoir is a purchased part. The assembly file is located in the Chapter 3 Homework folder.

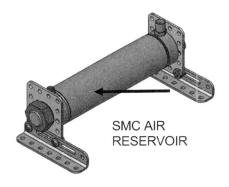

AIR RESERVOIR SUPPORT Assembly
Courtesy of Gears Educational Systems & SMC
Corporation of America

- Create a new assembly named AIR RESERVOIR SUPPORT AND PLATE.

- Two M15-MOUNT parts and two ANGLE-BRACKET parts hold the SMC AIR RESERVOIR.

- The ANGLE-BRACKET parts are fastened to the FLAT-PLATE.

Exercise 3.21a: FLAT-PLATE Part

Create the FLAT-PLATE Part on the Top Plane. The FLAT-PLATE is machined from .090, [2.3] 6061 Alloy flat stock. The default units are inches.

Utilize the Top Plane for the Sketch plane.

Locate the Origin at the Midpoint of the left vertical line.

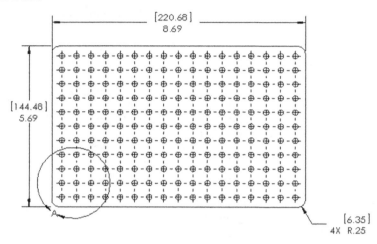

The 8.690, [220.68mm] x 5.688, [144.48mm] FLAT PLATE contains a Linear Pattern of ⌀.190, [4.83mm] Thru holes.

The Holes are equally spaced, .500, [12.7mm] apart.

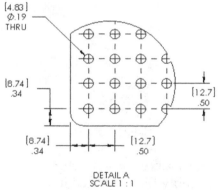

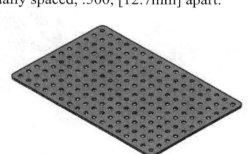

Determine the maximum
number of holes contained in
the FLAT-PLATE.

Maximum # of
holes_____.

- Utilize a Linear Pattern in
  two Directions to create the
  holes.

- Utilize the Geometric
  Pattern Option.

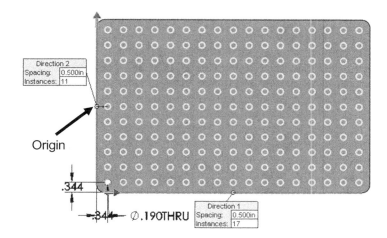

Exercise 3.21b: IM15-MOUNT Part

- Create the IM15-MOUNT part on the Right plane.

- Center the part on the Origin. Utilize the features in the FeatureManager.

- The IM15-MOUNT Part is machined from 0.060, [1.5mm] 6061 Alloy flat stock. The
  default units are inches.

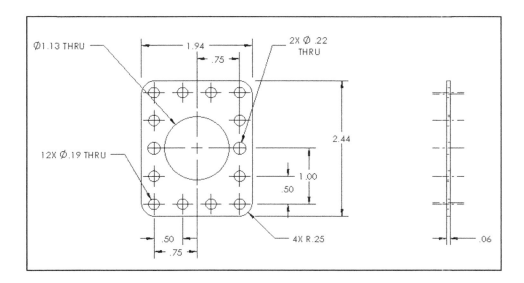

Exercise 3.21c: **ANGLE BRACKET Part**

- Create the ANGLE BRACKET part.

- The Extruded Base (Boss-Extrude1) feature is sketched with an L-Shaped profile on the Right Plane. The ANGLE BRACKET Part is machined from 0.060, [1.5mm] 6061 Alloy flat stock. The default units are inches.

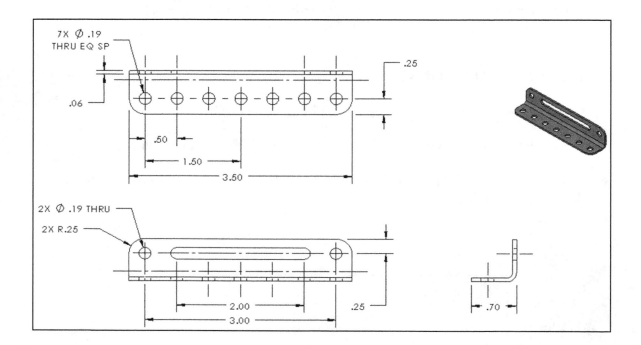

Exercise 3.21d: **Reservoir Assembly**

The Reservoir stores compressed air. Air is pumped through a Schrader Valve into the Reservoir.

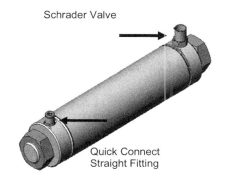

A Quick Connect Straight Fitting is utilized to supply air to the Pneumatic Test Module Assembly. Quick Connect Fittings allow air tubing to be assembled and disassembled without removing the fitting.

Copy the Pneumatic Components folder to your hard drive.

Open the part Reservoir from the Pneumatics Components folder. The Reservoir default units are in millimeters (MMGS).

Reservoir and Fittings
Courtesy of SMC Corporation of America and
Gears Educational Systems

Engineers and designers work in metric and english units. Always verify your units for parts and other engineering data. In pneumatic systems, common units for volume, pressure and temperature are defined in the below table.

Magnitude	Metric Unit (m)	English (e)
Mass	kg	pound
	g	ounce
Length	m	foot
	m	yard
	mm	inch
Temperature	°C	°F
Area, Section	m ^{2}	sq.ft
	cm ^{2}	sq.inch
Volume	m ^{3}	cu.yard
	cm ^{3}	cu.inch
	dm ^{3}	cu.ft.
Volume Flow	m ^{3}n / min	scfm
	dm ^{3}n /min (ℓ/min)	scfm
Force	N	pound force (ℓbf.)
Pressure	bar	ℓbf./sq.inch (psi)

Common Metric and English Units

The ISO unit of pressure is the Pa (Pascal). 1Pa = 1N/m.

Exercise 3.21e: AIR RESERVOIR SUPPORT AND PLATE Assembly

Create the AIR RESERVOIR SUPPORT AND PLATE assembly. Note: There is more than one solution for the mate types illustrated below.

The FLAT-PLATE is the first component in the AIR RESERVOIR SUPPORT AND PLATE assembly. Insert the FLAT-PLATE. The FLAT-PLATE is fixed to the Origin.

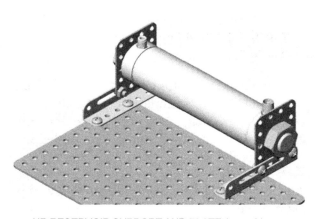

AIR RESERVOIR SUPPORT AND PLATE Assembly
Courtesy of SMC Corporation of America

- Insert the ANGLE BRACKET.

- Mate the ANGLE BRACKET to the FLAT-PLATE. The bottom flat face of the ANGLE BRACKET is coincident to the top face of the FLAT-PLATE.

- The center hole of the ANGLE BRACKET is concentric to the upper left hole of the FLAT-PLATE.

- The first hole of the ANGLE bracket is concentric with the hole in the 8th row, 1st column of the FLAT-PLATE.

- Insert the IM15-MOUNT.

- Mate the IM15-MOUNT. The IM15-MOUNT flat back face is coincident to the flat inside front face of the ANGLE BRACKET.

- The bottom right hole of the IM15-MOUNT is concentric with the right hole of the ANGLE BRACKET.

- The bottom edge of the IM15-MOUNT is parallel to the bottom edge of the ANGLE BRACKET.

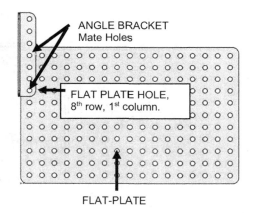

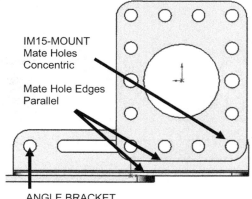

- Insert the Reservoir Assembly.

- Mate the Reservoir Assembly. The conical face of the Reservoir is concentric to the IM15-MOUNT center hole.

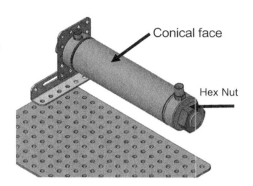

- The left end cap of the Reservoir Assembly is coincident to the front face of the IM15-MOUNT.

- The Hex Nut flat face is parallel to the top face of the FLAT-PLATE.

- Insert the second ANGLE BRACKET.

- Mate the ANGLE BRACKET to the FLAT-PLATE. The bottom flat face of the ANGLE BRACKET is coincident to the top face of the FLAT-PLATE.

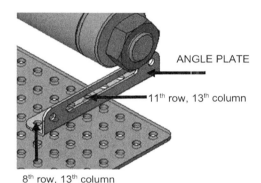

- The center hole of the ANGLE BRACKET is concentric with the hole in the 11th row, 13th column of the FLAT-PLATE.

- The first hole of the ANGLE bracket is concentric with the hole in the 8th row, 13th column of the FLAT-PLATE.

- Insert the second IM15-MOUNT.

- Mate the IM15-MOUNT to the outside face of the ANGLE BRACKET. The bottom right hole of the IM15-MOUNT is concentric with the right hole of the ANGLE BRACKET.

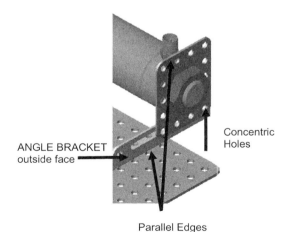

- The top edge of the IM15-MOUNT is parallel to the top edge of the ANGLE BRACKET.

- Save the assembly. Insert the required SCREWS. The AIR RESERVOIR SUPPORT AND PLATE assembly is complete.

Notes:

Chapter 4

Fundamentals of Drawing

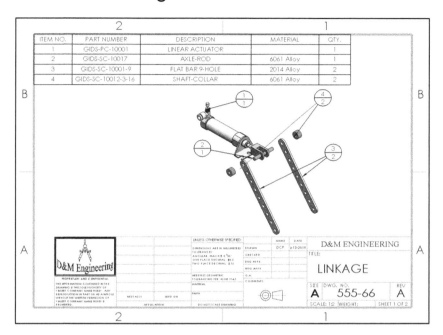

Below are the desired outcomes and usage competencies based on the completion of Chapter 4.

Desired Outcomes:	**Usage Competencies:**
• CUSTOM-A Sheet Format. • A-ANSI-MM Drawing Template.	• Ability to create a Custom Sheet Format, Drawing Template, Company logo and Title block.
• FLATBAR Configurations. • FLATBAR Part Drawing. • LINKAGE Assembly Drawing. • FLATBAR-SHAFTCOLLAR Assembly.	• Understand Standard, Isometric, Detail, Section and Exploded views. • Knowledge of the View Palette. • Ability to incorporate a Bill of Materials with Custom Properties.
	• Proficiency to create and edit drawing dimensions and annotations. • Aptitude to create a Design Table.

Notes:

Chapter 4 - Fundamentals of Drawing

Chapter Objective

Create a FLATBAR drawing with a customized Sheet Format and a Drawing Template containing a Company logo and Title block.

Obtain an understanding to display the following views with the ability to insert, add, and edit dimensions and annotations:

- Standard: Top, Front, and Right.

- Isometric, Detail, Section, and Exploded.

Create a LINKAGE assembly drawing with a Bill of Materials. Obtain knowledge to develop and incorporate a Bill of Materials with Custom Properties. Create a FLATBAR-SHAFTCOLLAR assembly.

On the completion of this chapter, you will be able to:

- Create a customized Sheet Format.

- Generate a custom Drawing Template.

- Produce a Bill of Materials with Custom Properties.

- Develop various drawing views.

- Reposition views on a drawing.

- Move dimensions in the same view.

- Apply Edit Sheet Format mode and Edit Sheet mode.

- Modify the dimension scheme.

- Create a Parametric drawing note.

- Link notes in the Title block to SOLIDWORKS properties.

- Generate an Exploded view.

- Create and edit a Design Table.

- Insert a Center of Mass Point.

Chapter Overview

Generate the LINKAGE assembly drawing with two sheets and the FLATBAR part drawing using configurations from a design table.

The LINKAGE assembly drawing contains two sheets:

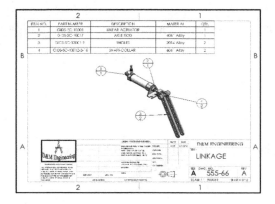 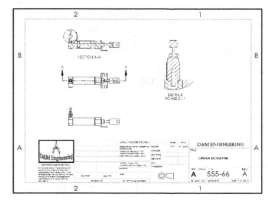

- Sheet1 contains the LINKAGE assembly in an Exploded view with a Bill of Materials.

- Sheet2 contains the AirCylinder assembly with a Section view, Detail view and a Scale view.

The FLATBAR drawing utilizes a custom Drawing Template and a custom Sheet Format. The FLATBAR drawing contains two sheets:

- Sheet1 contains a Front, Top and Isometric view with dimensions and a linked Parametric note.

- Sheet2 contains the 3HOLE configuration of the FLATBAR. Configurations are created with a Design Table.

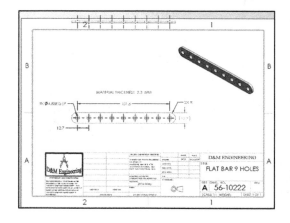 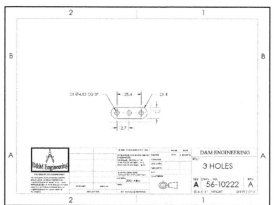

Create the FLATBAR-SHAFTCOLLAR assembly. Utilize a Design Table to create four new configurations of the assembly.

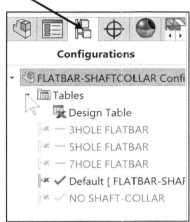

There are two major design modes used to develop a drawing:

- Edit Sheet Format and Edit Sheet.

The Edit Sheet Format mode provides the ability to:

- Change the Title block size and text headings.

- Incorporate a Company logo.

- Define the Zone Tag Editor.

- Add Custom Properties, text and more.

The Edit Sheet mode provides the ability to:

- Add or modify views.

- Add or modify dimensions.

- Add or modify notes and more.

Drawing Template and Sheet Format

The foundation of a SOLIDWORKS drawing is the Drawing Template. Drawing size, drawing standards, company information, manufacturing and/or assembly requirements, units and other properties are defined in the Drawing Template.

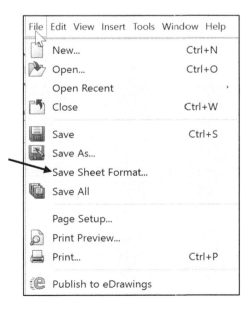

The Sheet Format is incorporated into the Drawing Template. The Sheet Format contains the border, Title block information, Revision block information, Company name and or Logo information, Custom Properties, and SOLIDWORKS Properties.

Custom Properties and SOLIDWORKS Properties are shared values between documents. Utilize an A-size Drawing Template with Sheet format for the FLATBAR drawing and LINKAGE assembly drawing.

During the initial SOLIDWORKS installation, you are requested to select either the ISO or ANSI drafting standard. ISO is typically a European drafting standard and uses First Angle Projection. The book is written using the ANSI (US) overall drafting standard and Third Angle Projection for drawings.

Views from the part or assembly are inserted into the SOLIDWORKS drawing.

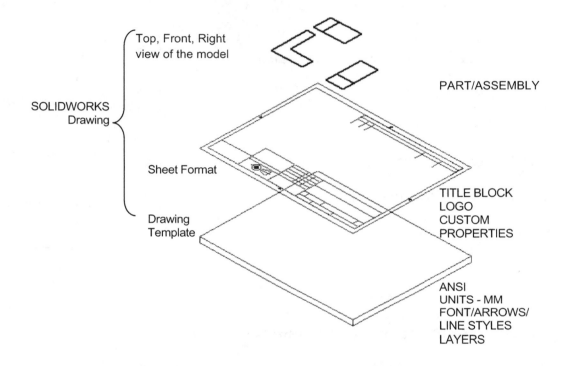

Create Sheet Formats for different parts types. Example: sheet metal parts, plastic parts, and high precision machined parts.

Create Sheet Formats for each category of parts that are manufactured with unique sets of title block notes.

Note: The Third Angle Projection scheme is illustrated in this chapter.

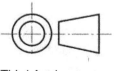

Third Angle Projection icon

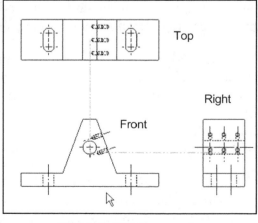

Third Angle Projection

For non-ANSI dimension standards, the dimensioning techniques are the same, even if the displayed arrows and text size are different. For printers supporting millimeter paper sizes, select A4 (ANSI) Landscape (297mm × 210mm).

The default Drawing Templates contain predefined Title block Notes linked to Custom Properties and SOLIDWORKS Properties.

Activity: Create a New Drawing

Create a New drawing. Close all parts and drawings.
1) Click **Windows**, **Close All** from the Menu bar.

2) Click **New** ⬜ from the Menu bar.

3) Double-click **Drawing** from the Templates tab.

4) Select **A (ANSI) Landscape**. If needed uncheck the Only show standard formats box.

5) Click **OK** from the Sheet Format/Size box. The Model View PropertyManager is displayed.

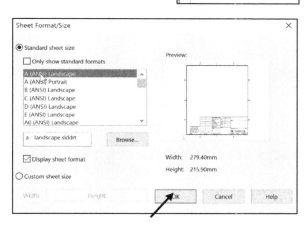

6) Click **Cancel** ✕ from the Model View PropertyManager. The Draw1 FeatureManager is displayed.

The A (ANSI) Landscape paper is displayed in a new Graphics window. The sheet border defines the drawing size, 11″ × 8.5″ or (279.4mm × 215.9mm).

The Model View PropertyManager is displayed if the Start command when creating new drawing is checked.

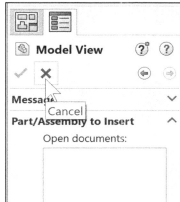

💡 To view the Sheet properties, right-click Properties in the drawing sheet. Expand the drop-down menu. View the Sheet Properties dialog box. Click OK to return to the drawing sheet.

Draw1 is the default drawing name. Sheet1 is the default first Sheet name. CommandManager tabs will vary depending on system setup and Add-ins.

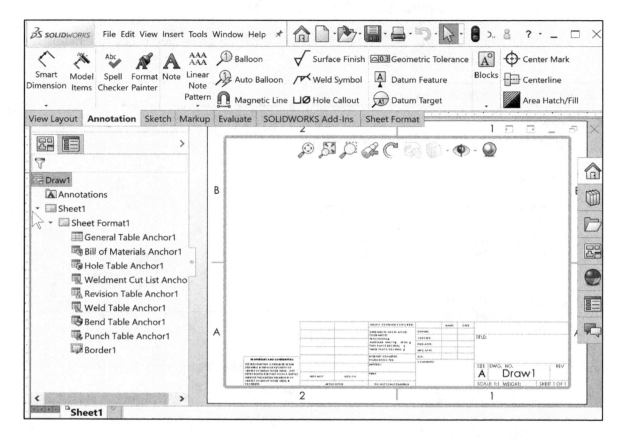

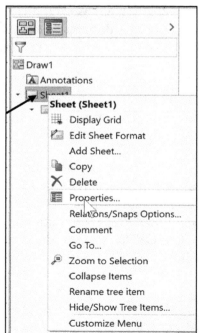

Utilize the CommandManager tabs or individual toolbars to access the needed tools in this chapter.

Set Sheet Properties.

7)   Right-click **Sheet1** in the Draw FeatureManager.

8)   Click **Properties**. The Sheet Properties are displayed.

9)   **View** your options.

10)   Enter Sheet Scale **1:1**.

11)   Check the **Third angle** box for Type of projection.

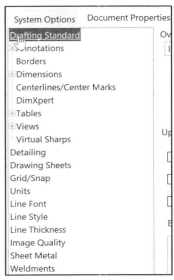

12)   Click **Apply Changes** or **Cancel** from the Sheet Properties dialog box. Apply Changes is active if you changed any of the default settings.

Activity: Drawing-Document Properties

Set Document properties. Set drafting standard, units, and precision.

13)   Click **Options** ⚙ from the Main menu.

14)   Click the **Document Properties** tab from the dialog box.

15)   Select **ANSI** for Overall drafting standard from the drop-down menu.

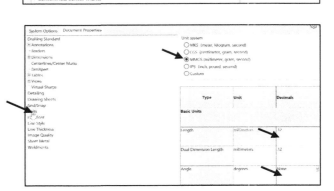

16)   Click **Units**. Click **MMGS** (millimeter, gram, second) for Unit system.

17)   Select **.12** for Length units Decimal places.

18)   Select **None** for Angular units Decimal places.

🔆 Available Document Properties are document dependent.

Companies develop drawing format standards and use specific text height for Metric and English drawings. Numerous engineering drawings use the following format:

- *Font*: Century Gothic - All capitals.

- *Text height*: 3mm for drawings up to B Size, 17in. × 22in.

- *Text height*: 5mm for drawings larger than B Size, 17in × 22in.

- *Arrowheads*: Solid filled with a 1:3 ratio - width to height.

Set the dimension font.
19)   Click **Annotations** folder as illustrated.

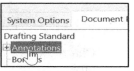

20)   Click the **Font** button. The Choose Font dialog box is displayed.

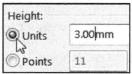

Set the dimension text height.
21)   Click the **Units** box from the Choose Font dialog box.

22)   Enter **3.0**mm for Height.

23)   Click **OK** from the Choose Font dialog box.

Set the arrow size.
24)   Click the **Dimensions** folder as illustrated.

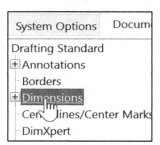

25)   Enter **1**mm for arrow Height. Enter **3**mm for arrow Width.

26)   Enter **6**mm for arrow Length.

27)   Click **OK** from the Document Properties dialog box.

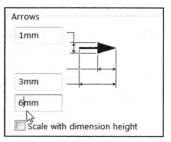

There are three dimension style type buttons: Outside, Inside, and Smart. Smart is the default option.

Title Block

The Title block contains text fields linked to System Properties and Custom Properties. System Properties are determined from the SOLIDWORKS documents. Custom Property values are assigned to named variables.

The Automatic Border command provides the ability to select columns and then select annotation notes that represent column labels for the sheet zones. Click rows and then select annotation notes that represent row labels for the sheet zones.

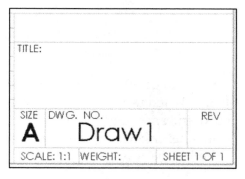

Save time. Utilize System Properties and define Custom Properties in your Sheet Formats.

System Properties Linked to fields in default Sheet Formats:	Custom Properties of drawings linked to fields in default Sheet Formats:		Custom Properties of parts and assemblies linked to fields in default Sheet Formats:
SW-File Name (in DWG. NO. field):	CompanyName:	EngineeringApproval:	Description (in TITLE field):
SW-Sheet Scale:	CheckedBy:	EngAppDate:	Weight:
SW-Current Sheet:	CheckedDate:	ManufacturingApproval:	Material:
SW-Total Sheets:	DrawnBy:	MfgAppDate:	Finish:
	DrawnDate:	QAApproval:	Revision:
	EngineeringApproval:	QAAppDate:	

The Title block is located in the lower right hand corner of Sheet1. The Drawing contains two modes:

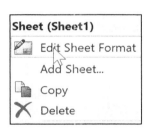

1. Edit Sheet Format.

2. Edit Sheet.

Insert views and dimensions in the Edit Sheet mode. Modify the Sheet Format text, lines or title block information in the Edit Sheet Format mode.

The CompanyName Custom Property is located in the title block above the TITLE box. There is no value defined for CompanyName. A small text box indicates an empty field.

Define a value for the Custom Property CompanyName. Example: D&M ENGINEERING.

Activity: Create a Title Box

Activate the Edit Sheet Format Mode.
28)   Right-click in **Sheet1**. Do not click in a drawing view.

29)   Click **Edit Sheet Format**. The Title block lines turn blue.

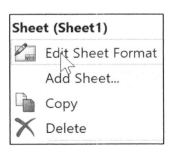

View the right side of the Title block.

30)   Click the **Zoom to Area** tool from the Heads-up View toolbar.

31)   **Zoom in** on the Title block.

32) Click the **Zoom to Area** tool to deactivate.

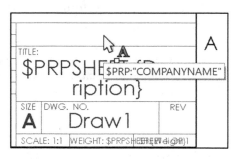

Define CompanyName Custom Property.
33) Position the **mouse pointer** in the middle of the box above the TITLE box as illustrated.

34) Click **File**, **Properties** from the Menu bar. The Summary Information dialog box is displayed.

35) Click the **Custom** tab.

36) Click inside the **Property Name** box.

37) Click the **drop down arrow** in the Property Name box.

38) Select **CompanyName** from the Property menu.

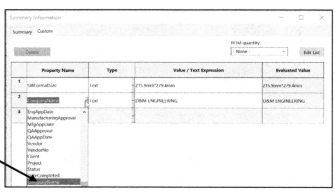

39) Enter **D&M ENGINEERING** (or your company/school name) in the Value/Text Expression box.

40) Click inside the **Evaluated Value** box. The CompanyName is displayed in the Evaluated Value box.

41) Click **OK** from the dialog box.

42) Move your **mouse pointer** in the center of the block as illustrated. The Custom Property, $PRP: "COMPANYNAME", is displayed in the Title block.

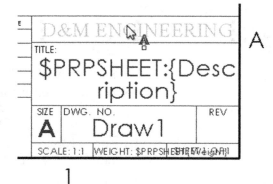

Modify the font size.
43) Double-click the **D&M ENGINEERING** (or your company/school name) text. The Formatting dialog box is displayed.

44) Click the **drop down arrows** to set the Text Font and Height from the Formatting toolbar.

45) Click the **Style buttons** and **Justification buttons** to modify the selected text.

46) **Close** the Formatting dialog box.

47) Click **OK** from the Note PropertyManager.

Click a position outside the selected text box to save and exit the text.

The Tolerance block is located in the Title block. The Tolerance block provides information to the manufacturer on the minimum and maximum variation for each dimension on the drawing. If a specific tolerance or note is provided on the drawing, the specific tolerance or note will override the information in the Tolerance block.

General tolerance values are based on the design requirements and the manufacturing process.

Create Sheet Formats for different part types; examples: sheet metal parts, plastic parts and high precision machined parts. Create Sheet Formats for each category of parts that are manufactured with unique sets of Title block notes.

Modify the Tolerance block in the Sheet Format for ASME Y14.5 machined millimeter parts. Delete unnecessary text. The FRACTIONAL text refers to inches. The BEND text refers to sheet metal parts. The Three Decimal Place text is not required for this millimeter part in the chapter.

Modify the Tolerance Note.
48) Double-click the text **INTERPRET GEOMETRIC TOLERANCING PER:**

49) Enter **ASME Y14.5**.

50) Click **OK** ✅ from the Note PropertyManager.

51) Double-click inside the **Tolerance block** text. The Formatting dialog box and the Note PropertyManager is displayed.

52) Delete the text **INCHES**.

53) Enter **MILLIMETERS**.

54) Delete the line **FRACTIONAL ±**.

55) Delete the text **BEND ±**.

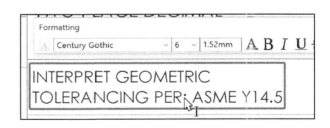

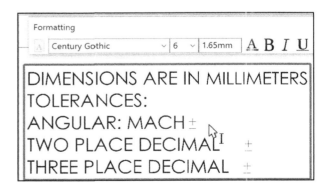

56) Click a **position** at the end of the
ANGULAR: MACH ± line.

57) Enter **0**.

58) Click the **Add Symbol** button from the Text
Format box. The Symbols dialog box is displayed.

59) Select **Degree** ° from the Symbols dialog box.

60) Enter **30'** for minutes of a degree.

Modify the TWO and THREE PLACE
DECIMAL LINES.

61) Delete the **TWO** and **THREE
PLACE DECIMAL** lines.

62) Enter **ONE PLACE DECIMAL ± 0.5**.

63) Enter **TWO PLACE DECIMAL ±
0.15**.

64) Click **OK** from the Note
PropertyManager.

Fit the drawing to the Graphics window.
65) Press the **f** key.

Save Draw1.

66) Click **Save** . Accept the default name. Draw1 is the first default drawing file name.

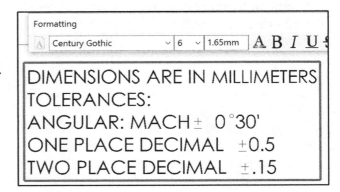

Various symbols are available through the Symbol dialog box. The ± symbol is located in the Modify Symbols list. The ± symbol is sometimes displayed as <MOD-PM>. The degree symbol ° is sometimes displayed as <MOD-DEG>.

Interpretation of tolerances is as follows:

- The angular dimension 110° is machined between 109.5° and 110.5°.

- The dimension 2.5 is machined between 2.0 and 3.0.

- The dimension 2.05 is machined between 1.90 and 2.20.

Company Logo

A Company logo is normally located in the Title block of the drawing. Create your own Company logo or copy and paste an existing picture.

The Logo.jpeg file is provided. Copy the LOGO folder to your hard drive. Insert the Company logo in the Edit Sheet Format mode.

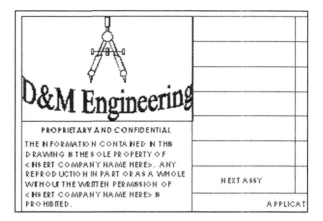

Activity: Insert a Company or School Logo

Insert a Company or School Logo.

67) Copy the **LOGO** folder to your hard drive. Note: If you have your own Logo, use it for the drawing.

68) Click **Insert**, **Picture** from the Menu bar. The Open dialog box is displayed.

69) Double-click the **Logo.jpg**. The Sketch Picture PropertyManager is displayed.

70) Uncheck the **Lock aspect ratio** box.

71) Drag the picture handles to size the **picture** to the left side of the Title block. Note: Text was added to the picture.

72) Click **OK** ✔ from the Sketch Picture PropertyManager.

🔆 Text can be added to create a custom logo. You can insert a picture or an object.

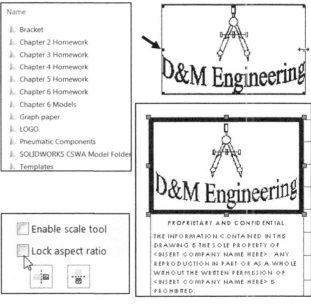

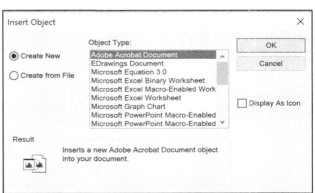

Return to the Edit Sheet mode.

73) Right-click in the **Graphics window**.

74) Click **Edit Sheet**. The Title block is displayed in black.

Fit the Sheet Format to the Graphics window.

75) Press the **f** key.

Draw1 displays Editing Sheet1 in the Status bar. The Title block is displayed in black when in Edit Sheet mode.

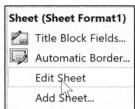

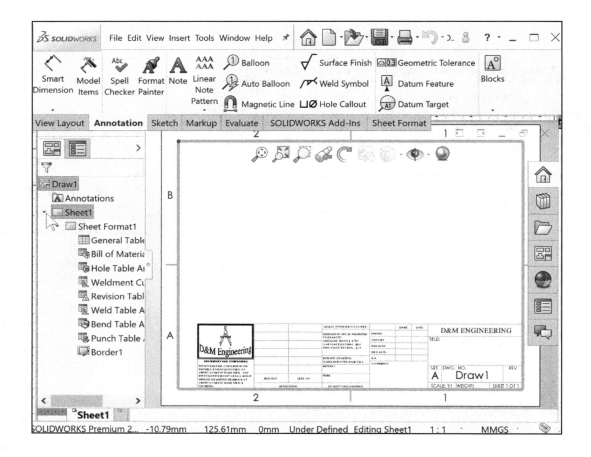

Save the Sheet Format. Save As a Drawing Template. Create the MY-TEMPLATES Tab.

Save the drawing document in the Graphics window in two forms: Sheet Format and Drawing Template. Save the Sheet Format as a custom Sheet Format named CUSTOM-A. Use the CUSTOM-A Sheet Format for the drawings in this chapter. The Sheet Format file extension is .slddrt.

The Drawing Template can be displayed with or without the Sheet Format. Combine the Sheet Format with the Drawing Template to create a custom Drawing Template named A-ANSI-MM. Utilize the Save As option to save a Drawing Template. The Drawing Template file extension is .drwdot.

Always select the Save as type option first, then select the Save in folder to avoid saving in default SOLIDWORKS installation directories.

Set the File Locations option in order to view the SW-TUTORIAL-2020\MY-TEMPLATE tab in the New Document dialog box.

If you did not create the MY-TEMPLATE tab or the drawing template, use the standard SOLIDWORKS drawing template and apply all of the needed document properties and custom properties.

Activity: Save Sheet Format. Save As Drawing Template. Create the MY-TEMPLATES Tab.

Save the Sheet Format.

76) Click **File**, **Save Sheet Format** from the Menu bar. The Save Sheet Format dialog box is displayed.

77) Select **SW-TUTORIAL-2020\MY-TEMPLATES** for the Save in folder. If needed, create the MY-TEMPLATES sub-folder under the SW-TUTORIAL-2020 folder.

78) Enter **CUSTOM-A** for File name.

79) Click **Save** from the Save Sheet Format dialog box.

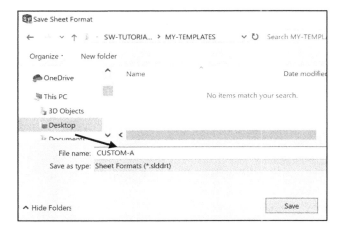

Save the Drawing Template.

80) Click **Save As** .

81) Click **Drawing Templates (*.drwdot)** from the Save as type box.

82) Select **SW-TUTORIAL-2020\MY-TEMPLATES** for the Save in folder.

83) Enter **A-ANSI-MM** for File name.

84) Click **Save**.

The A-ANSI-MM drawing template is displayed in the Graphics window. Create and add the MY-TEMPLATES folder, under the SW-TUTORIAL-2020 folder, to the File Locations Document Template System Option.

Add the MY-TEMPLATES tab. Set System Options - File Locations. Only perform this procedure if you are working on a non-network, private system.

85) Click **Options** ⚙ from the Main menu. The System Options General dialog box is displayed.

86) Click **File Locations** under the System Options tab.

87) Select **Document Templates** from Show folders for.

Add a new tab.

88) Click the **Add** button.

89) Select the **MY-TEMPLATES** folder. If needed create the folder under the **SW-TUTORIAL-2020** folder.

90) Click **Select Folder**. You should see the new folder added to the Document Templates. If needed, click the Move Down button.

91) Click **OK** from the System Options, File Locations dialog box.

92) Click **Yes** to the SOLIDWORKS dialog box. The SOLIDWORKS dialog box displays the new Folder location.

93) Click **Yes**. Click **Yes**. Return back to the drawing.

Close all files.

94) Click **Windows, Close All** from the Menu bar.

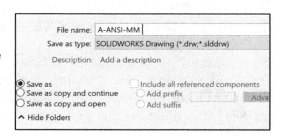

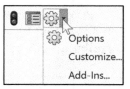

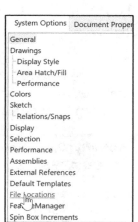

Create a New drawing.

95) Click **New** ☐ from the Menu bar.

96) Click the **MY-TEMPLATES** tab. View the created drawing template. Note: Additional templates are displayed.

97) Double-click the **A-ANSI-MM** Drawing Template.

98) Click **Cancel** ✕ from the Model View PropertyManager. The Draw2 FeatureManager is displayed in the Graphics window.

New SOLIDWORKS Document

Templates   MBD   Tutorial   MY-TEMPLATES

- PART-IN-ANSI
- PART-IN-ISO
- PART-MM-ANSI-AL6061
- PART-MM-ANSI
- PART-MM-ISO
- ASM-IN-ANSI
- ASM-IN-ISO
- ASM-MM-ANSI
- ASM-MM-ISO
- A-ANSI-IN
- A-ANSI-MM
- A-ISO-MM
- B-ANSI-MM
- C-ANSI-MM

🔅 The Draw2-Sheet1 drawing is displayed in the Graphics window. You have successfully created a new drawing Template with a Custom sheet format.

Close all files.
99) Click **Windows**, **Close All** from the Menu bar.

🔅 Combine customized Drawing Templates and Sheet Formats to match your company's drawing standards. Save the empty Drawing Template and Sheet Format separately to reuse information.

🔍 Additional details on Drawing Templates, Sheet Format and Custom Properties are available in SOLIDWORKS Help Topics. Keywords: Documents (templates, properties); Sheet Formats (new, new drawings, note text); Properties (drawing sheets); Customize Drawing Sheet Formats.

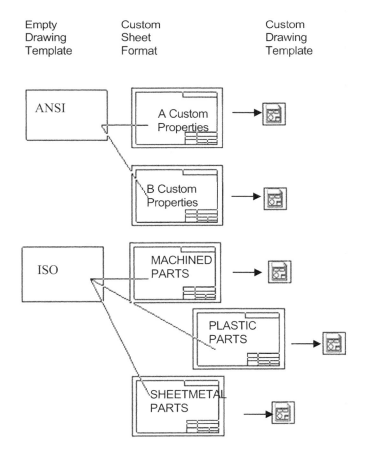

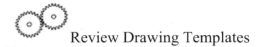 Review Drawing Templates

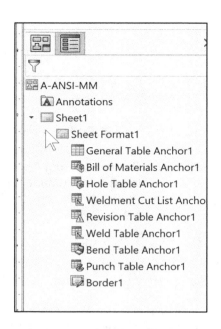

The Custom Drawing Template was created from the default Drawing Template. You modified Sheet Properties and Document Properties to control the Sheet size, Scale, Annotations and Dimensions.

The Sheet Format contained a Title block and Custom Property information. You inserted a Company Logo and modified the Title block.

The Save Sheet Format option was utilized to save the CUSTOM-A.slddrt Sheet Format. The Save As option was utilized to save the A-ANSI-MM.drwdot template.

The Sheet Format and Drawing Template were saved in the MY-TEMPLATES folder.

FLATBAR Drawing

A drawing contains part views, geometric dimensioning and tolerances, notes and other related design information. When a part is modified, the drawing automatically updates. When a dimension in the drawing is modified, the part is automatically updated.

Create the FLATBAR drawing from the FLATBAR part. Display the Front, Top, Right and Isometric views. Utilize the Model View tool from the View Layout toolbar.

Insert dimensions from the part. Utilize the Insert Model Items tool from the Annotation toolbar. Insert and modify dimensions and notes.

Insert a Parametric note that links the dimension text to the part depth. Utilize a user defined Part Number. Define the part material with the Material Editor. Add Custom Properties for Material and Number.

Activity: FLATBAR Drawing. Open the FLATBAR Part

Open the FLATBAR part. The FLATBAR part was created in Chapter 2.

100) Click **Open** 📂 from the Menu bar.

101) Select the **SW-TUTORIAL-2020** folder.

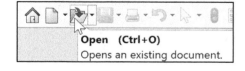

102) Double-click **FLATBAR**. The FLATBAR FeatureManager is displayed.

Create a New drawing.

103) Click **New** ⬜ from the Menu bar.

104) Click the **MY-TEMPLATES** tab.

105) Double-click **A-ANSI-MM**. The Model View PropertyManager is displayed. Additional templates are displayed. Note: if you did not create the MY-TEMPLATES tab, create a drawing document, A-(ANSI) Landscape, Third Angle projection using the MMGS unit system. Add all needed document properties and custom properties.

The Model View PropertyManager is displayed if the Start command when creating new drawing box is checked. If the Model View PropertyManager is not displayed, click the Model View tool from the View Layout toolbar.

The FLATBAR 🗏 FLATBAR part icon is displayed in the Open documents box. Drawing view names are based on the part view orientation. The Front view is the first view inserted into the drawing. The Top view and Right view are projected from the Front view.

Insert the Front, Top and Right view.

106) Click **Next** 🔄 from the Model View PropertyManager.

107) Check the **Create multiple views** box.

108) Click ***Front**, ***Top** and ***Right** view from the Standard views box.

De-activate the Isometric view.

109) Click the ***Isometric** icon from the Standard views box.

110) Click **OK** ✅ from the Model View PropertyManager. View the three views on the drawing.

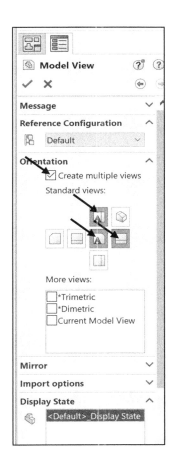

A part cannot be inserted into a drawing when the *Edit Sheet Format* mode is selected. You are required to be in the *Edit Sheet* mode.

Insert an Isometric view using the View Palette.

111) Click the **View Palette** tab on the right side of the Graphics window.

112) Select **FLATBAR** from the View Palette drop-down menu as illustrated. View the available views.

113) Click and drag the **Isometric view** in the top right corner as illustrated.

114) Click **OK** from the Drawing View PropertyManager. Your Title block may look different if you did not use the **A-ANSI-MM** Drawing Template.

*Isometric

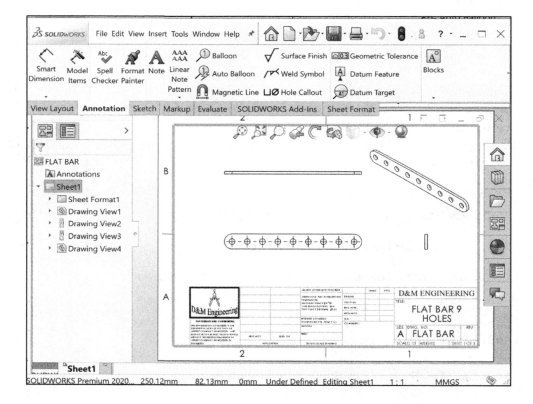

Click the View Palette tab in the Task Pane. Click the drop-down arrow to view any active documents or click the Browse button to locate a document. Click and drag the desired view/views into the active drawing sheet.

Modify the Sheet Scale.
115) Right-click Sheet1 in the FLATBAR Drawing
FeatureManager as illustrated.

116) Click **Properties**. The Sheet Properties dialog box is
displayed.

117) If needed, enter **1:1** for Sheet Scale.

118) Click **Apply Changes** or **Cancel** from the Sheet Properties
dialog box. Apply Changes is active if you changed any of
the default settings.

If needed, hide the Origins.
119) Click **View**, **Hide/Show**, un-check **Origins** from the
Menu bar.

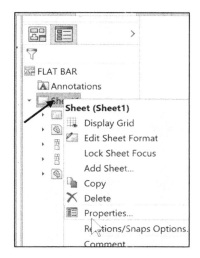

Save the drawing.

120) Click **Save As** .

121) Enter **FLATBAR** in the SW-TUTORIAL-2020 folder.
Click **Save**. Click **Save** .

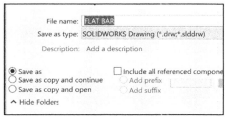

Text in the Title block is linked to the Filename and
Description created in the part. The DWG. NO. text
box utilizes the Property, $PRP:"SW-File Name"
passed from the FLATBAR part to the FLATBAR drawing.

The Title text box utilizes the Property, $PRPSHEET:
"Description."

The filename FLATBAR is displayed in the DWG. NO. box.
The Description FLATBAR 9 HOLE is displayed in the Title
box. The Description was added in Chapter 2.

The FLATBAR drawing contains three Principle views
(Standard views): Front, Top, Right and Isometric view.

Insert drawing views as follows:

- Utilize the Model View tool.

    o Drag a part into the drawing to create three Standard
      views.

    o Predefine views in a custom Drawing Template.

    o Drag a hyperlink through Internet Explorer.

- Drag an active part view from the View Palette. The View Palette is located in the Task Pane. With an open part, drag the selected view into the active drawing sheet.

The View Palette populates when you:

- Click Make Drawing from Part/Assembly.

- Browse to a document from the View Palette.

- Select from a list of open documents in the View Palette.

Move Views and Properties of the Sheet

Move Views on Sheet1 to create space for additional Drawing View placement.

The mouse pointer provides feedback in both the Drawing Sheet and Drawing View modes.

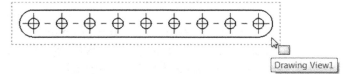

The mouse pointer displays the Drawing Sheet icon when the Sheet properties and commands are executed.

The mouse pointer displays the Drawing View icon when the View properties and commands are executed.

View the mouse pointer icon for feedback to select Sheet, View, and Component and Edge properties in the Drawing.

Use the Pack and Go option to save an assembly or drawing with references. The Pack and Go tool saves either to a folder or creates a zip file to e-mail. View SOLIDWORKS help for additional information.

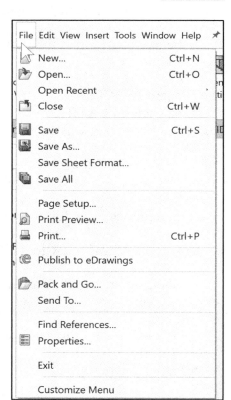

Sheet Properties

- Sheet Properties display properties of the selected sheet. Right-click in the sheet boundary 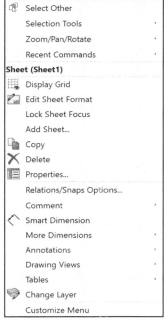 to view the available commands.

View Properties

- View Properties display properties of the selected view. Right-click inside the view boundary. Modify the View Properties in the Display Style box or the View Toolbar.

FLAT BAR 9 HOLES

Component Properties

- Component Properties display properties of the selected component. Right-click on the face of the component . View the available options.

Edge Properties

- Edge Properties display properties of the selected geometry. Right-click on an edge inside the view boundary. View the available options.

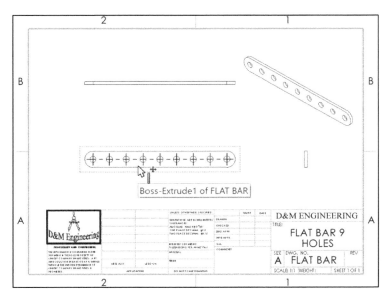

Reposition the views on the drawing. Provide approximately 25mm - 50mm between each view for dimension placement.

Activity: FLATBAR Drawing-Position Views

Position the views.

122) Click inside the view boundary of **Drawing View1** (Front). The mouse pointer displays the Drawing View icon.

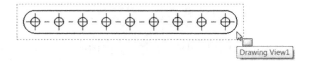

123) Position the **mouse pointer** on the edge of the view boundary until the Drawing View icon is displayed.

124) Drag **Drawing View1** in an upward vertical direction. The Top and Right views move aligned to Drawing View1 (Front).

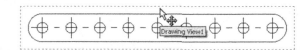

125) Press **Shift + Z** key to Zoom in on Sheet1.

126) Click the **Right view** boundary.

127) Position the **mouse pointer** on the edge of the view until the Drawing Move View icon is displayed.

128) Drag the **Right view** in a right to left direction towards the Front view.

💡 Tangent edges are displayed for educational purposes.

Move the Top view in a downward vertical direction.

129) Click the **Top view**, "Drawing View3" boundary.

130) Drag the **Top view** in a downward direction towards Drawing View1.

Fit Sheet1 to the Graphics window.

131) Press the **f** key.

Detail Drawing

The design intent of this project is to work with dimensions inserted from parts and to incorporate them into the drawings. Explore methods to move, hide and recreate dimensions to adhere to a drawing standard.

There are other solutions to the dimensioning schemes illustrated in this project. Detail drawings require dimensions, annotations, tolerance, materials, Engineering Change Orders, authorization, etc. to release the part to manufacturing and other notes prior to production.

Review a hypothetical "worst case" drawing situation. You just inserted dimensions from a part into a drawing. The dimensions, extension lines and arrows are not in the correct locations. How can you address the position of these details? Answer: Dimension to an ASME Y14.5 standard.

No.	Situation:
1	Extension line crosses dimension line. Dimensions not evenly spaced.
2	Largest dimension placed closest to profile.
3	Leader lines overlapping.
4	Extension line crossing arrowhead.
5	Arrow gap too large.
6	Dimension pointing to feature in another view. Missing dimension – inserted into Detail view (not shown).
7	Dimension text over centerline, too close to profile.
8	Dimension from other view – leader line too long.
9	Dimension inside section lines.
10	No visible gap.
11	Arrows overlapping text.
12	Incorrect decimal display with whole number (millimeter), no specified tolerance.

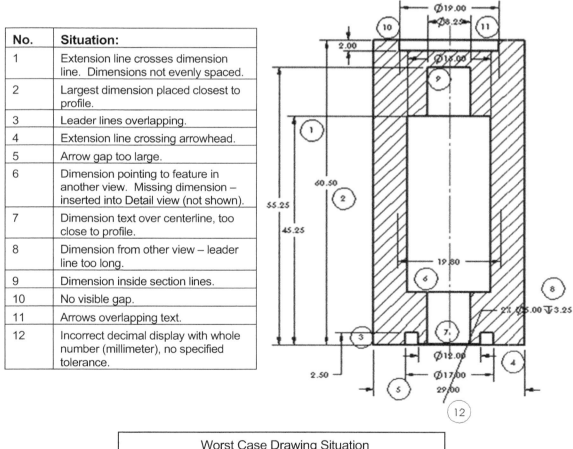

Worst Case Drawing Situation

The ASME Y14.5 standard defines an engineering drawing standard.

Review the twelve changes made to the drawing to meet the standard.

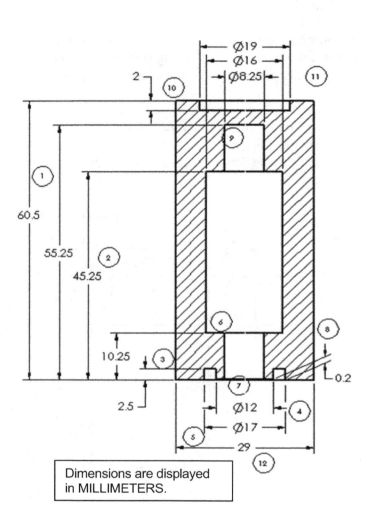

No.	Preferred Application of the Dimensions:
1	Extension lines do not cross unless situation is unavoidable. Stagger dimension text.
2	Largest dimension placed farthest from profile. Dimensions are evenly spaced and grouped.
3	Arrowheads do not overlap.
4	Break extension lines that cross close to arrowhead.
5	Flip arrows to the inside.
6	Move dimensions to the view that displays the outline of the feature. Ensure that all dimensions are accounted for.
7	Move text off of reference geometry (centerline).
8	Drag dimensions into their correct view boundary. Create reference dimensions if required. Slant extension lines to clearly illustrate feature.
9	Locate dimensions outside off section lines.
10	Create a visible gap between extension lines and profile lines.
11	Arrows do not overlap the text.
12	Whole numbers displayed with no zero and no decimal point (millimeter).

Apply these dimension practices to the FLATBAR and other drawings in this project.

A Detailed drawing is used to manufacture a part. A mistake on a drawing can cost your company substantial loss in revenue. The mistake could result in a customer liability lawsuit.

Dimension and annotate your parts clearly to avoid common problems and mistakes.

Dimensions and Annotations

Dimensions and annotations are inserted from the part. The annotations are not in the correct location. Additional dimensions and annotations are required.
Dimensions and annotations are inserted by selecting individual features, views or the entire sheet. Select the entire sheet. Insert Model Items command from the Annotations toolbar.

Activity: FLATBAR Drawing - Dimensions and Annotations

Insert dimensions.

132) Click **Sheet1** in the center of the drawing. The mouse pointer displays the Sheet icon.

133) Click the **Annotation** tab from the CommandManager.

134) Click the **Model Items** tool from the Annotation toolbar. The Model Items PropertyManager is displayed. The Import items into all views option is checked.

135) Select **Entire model** from the Source box.

136) Click **OK** from the Model Items PropertyManager. Dimensions are inserted into the drawing. View the results.

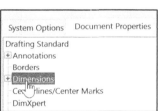

Dimensions are inserted into the drawing. The dimensions MAY NOT BE in the correct location with respect to the feature lines per the ANSI Y-14.5 standard. Move them later in the chapter and address extension line gaps (5mm).

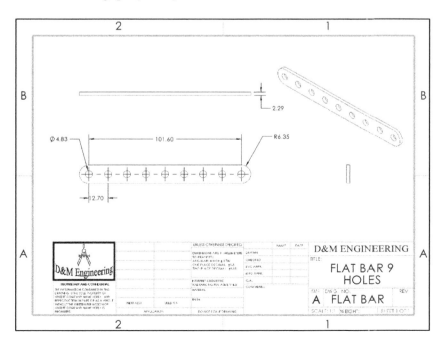

The dimensions and text in the next section have been enlarged for visibility. Drawing dimension location is dependent on *Feature dimension creation* and *Selected drawing views*.

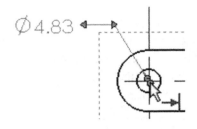

Move dimensions within the same view. Use the mouse pointer to drag dimensions and leader lines to a new location. Leader lines reference the size of the profile. A gap must exist between the profile lines and the leader lines. Shorten the leader lines to maintain a drawing standard. Use the blue Arrow buttons to flip the dimension arrows.

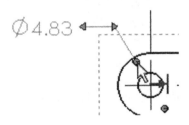

Plan ahead for general drawing notes. Notes provide relative part or assembly information. Example: Material type, material finish, special manufacturing procedure or considerations, preferred supplier, etc.

Below are a few helpful guidelines to create general drawing notes:

- Use capital letters.

- Use left text justification.

- Font size should be the same as the dimension text.

Create parametric notes by selecting dimensions in the drawing. Example: Specify the material thickness of the FLATBAR as a note in the drawing. If the thickness is modified, the corresponding note is also modified.

Move the linear dimensions in Drawing View1, (Front).

137) Click the dimension text **101.60**. The dimension text turns blue. **Remove** the trailing zero. Select **Precision** of 1.

138) Drag the **dimension text** downward.

139) Click the dimension text **12.70**. The dimension text turns blue. **Remove** the trailing zero. Select **Precision** of 1.

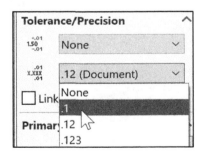

140) Drag the **text** approximately 10mm from the profile. The smallest linear dimensions are closest to the profile.

141) Click the radial dimension **R6.35**.

142) Drag the **text** diagonally off the profile if required.

Modify dimension text.

143) Click the diameter dimension **4.83**. The Dimension PropertyManager is displayed.

144) Click inside the **Dimension Text** box.

145) Enter **9X** before <MOD-DIAM>.

146) Enter **EQ SP** after <DIM>.

147) Click **OK** ✔ from the Dimension PropertyManager.

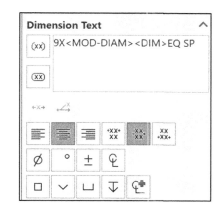

💡 Inserted dimensions can be moved from one drawing view to another. Hold the Shift key down. Click and drag the dimension text from one view into the other view boundary. Release the Shift key.

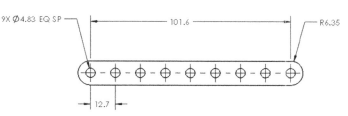

Modify the precision of the material thickness.

148) Click the **depth dimension** text in the Top view. Note: The part was created in Chapter 2.

149) Select **.1** from the Unit Precision drop-down menu.

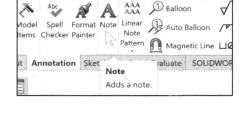

150) Click **OK** ✔ from the Dimension PropertyManager. The text displays the dimension.

Insert a Parametric note.

151) Click the **Annotation** tab from the CommandManager.

152) Click the **Note** 🅰 tool from the Annotation toolbar.

The Note icon is displayed.

153) Click a **position** above the Front view.

154) Enter **MATERIAL THICKNESS**. Click the **depth dimension text (2.3)** in the Top view. The variable name for the dimension is displayed in the text box. Note: Your material thickness may vary if you modified the FLATBAR model.

MATERIAL THICKNESS 2.3 MM

155) Enter **MM**.

156) Click **OK** ✔ from the Note PropertyManager.

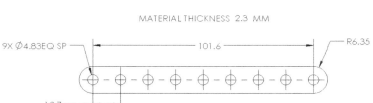

Hide superfluous dimensions.
157) Right-click the **dimension (2.3)** in the Top view. Note: Your material thickness may vary if you modified the FLATBAR model.

158) Click **Hide**.

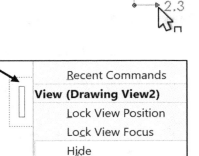

Hide the Right view.
159) Right-click the **Right view** boundary.

160) Click **Hide**. Note: If required, expand the drop-down menu. The Right view is not displayed in the Graphics window.

161) Click **OK** ✔ from the Drawing View PropertyManager.

Fit the model to the drawing. Address all needed extension line gaps.
162) Press the **f** key.

Save the drawing.
163) Click **Save** 💾.

Locate the Top view off of Sheet1.
164) Click and drag the **Top view boundary** off of the Sheet1 boundary as illustrated.

Views and notes outside the sheet boundary do not print. The Parametric note is controlled through the FLATBAR Boss-Extrude1 feature depth. Modify the depth to update the note.

🔅 Click the dimension Palette rollover button 🖱 to display the dimension palette. Use the dimension palette in the Graphics window to save mouse travel to the Dimension PropertyManager. Click on a dimension in a drawing view, and modify it directly from the dimension palette.

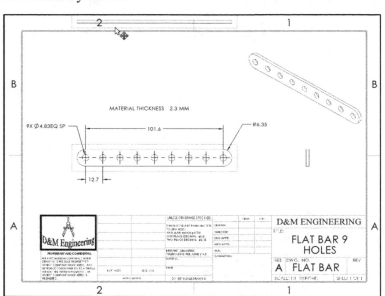

🔅 To show a hidden dimension, click **View**, **Hide/Show**, **Annotations** from the Menu bar.

Open the FLATBAR part.
165) Right-click inside the **Front view** boundary.

166) Click **Open Part**. The FLATBAR part is displayed.

Modify the Boss-Extrude1 depth dimension.
167) Click **Boss-Extrude1** from the FeatureManager.

168) Click **.090**in, [2.29] in the Graphics window.

169) Enter **2.5mm** as illustrated. Note: You need to enter mm.

170) Rebuild the model.

Save the model.

171) Click **Save** .

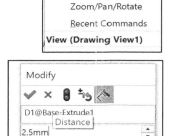

Return to the drawing.
172) Click **Window, FLATBAR - Sheet1** from the Menu bar. The Parametric note is updated to reflect the dimension change in the part.

173) Return to the FLATBAR part and **undo** the dimension change from 2.5mm to 2.3mm.

174) Return to the drawing. View the update.

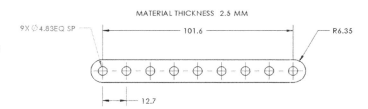

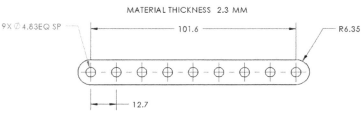

The FLATBAR drawing references the FLATBAR part. Do not delete the part or move the part location. Work between multiple documents:

- Press Ctrl-Tab to toggle between open SOLIDWORKS documents.

- Right-click inside the Drawing view boundary. Select Open Part.

- Right-click the part icon in the FeatureManager. Select Open Drawing.

Commands are accessed through the toolbars and drop-down menus. Commands are also accessed with a right-click in the Graphics window and FeatureManager.

A majority of FLATBAR drawing dimensions are inserted from the FLATBAR part. An overall dimension is required to dimension the slot shape profile. Add a dimension in the drawing.

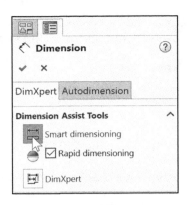

Add a dimension to Drawing View1.

175) Click the **Smart Dimension** tool from the Annotation toolbar. The Autodimension tab is selected by default.

176) Click the **Smart dimensioning** box.

177) Click the **top horizontal line** of the FLATBAR.

178) Click the **bottom horizontal line** of the FLATBAR.

179) Click a **position** to the right of the Front view as illustrated.

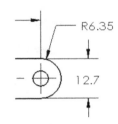

Modify the Radius text. R will be a stated reference dimension.
180) Click the **R6.36** dimension text.

181) Delete **R<DIM>** in the Dimension Text box.

182) Click **Yes** to confirm dimension override.

183) Enter **2X R** for Dimension text. Do not enter the radius value.

184) Click **OK** from the Dimension PropertyManager.

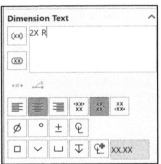

185) Insert all needed **extension line gaps (5mm)** as illustrated.

Save the FLATBAR drawing.

186) Click **Save** .

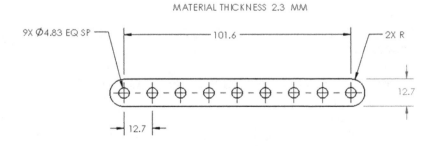

Part Number and Document Properties

Engineers manage the parts they create and modify. Each part requires a Part Number and Part Name. A part number is a numeric representation of the part. Each part has a unique number. Each drawing has a unique number. Drawings incorporate numerous part numbers or assembly numbers.

There are software applications that incorporate unique part numbers to create and perform: Bill of Materials, Manufacturing procedures, Cost analysis and Inventory control/Just in Time, JIT.

You are required to procure the part and drawing numbers from the documentation control manager. Utilize the following prefix codes to categorize created parts and drawings. The part name, part number and drawing numbers are as follows:

Category:	Prefix:	Part Name:	Part Number:	Drawing Number:
Machined Parts	56-	FLATBAR	GIDS-SC-10001-9	56-10222
		AXLE	GIDS-SC-10017	56-10223
		SHAFT-COLLAR	GIDS-SC-10012-3-16	56-10224
Purchased Parts	99-	AIRCYLINDER	99-FBM8x1.25	999-101-8
Assemblies	10-	LINKAGE ASM	GIDS-SC-1000	10-10123

Link notes in the Title block to SOLIDWORKS Properties. Properties are variables shared between documents and applications.

The machined parts are manufactured from Aluminum. Specify the Material Property in the part. Link the Material Property to the drawing title block. Create a part number that is utilized in the Bill of Materials. Create additional notes in the title block to complete the drawing.

Activity: FLATBAR Drawing - Part Number and Document Properties

Return to the FLATBAR part. Apply Material.
187) Right-click in the **Front view** boundary.

188) Click **Open Part**. Material was not added when you created the FLATBAR in Chapter 2.

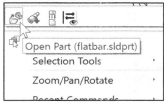

189) Right-click **Material** in the FLATBAR FeatureManager.

190) Click **Edit Material**. The Material dialog box is displayed.

191) **Expand** the Aluminum Alloys folder.

192) Select **2014 Alloy**. View the material properties.

193) Click the **Apply** button.

194) Click **Close**. The 2014 Alloy is displayed in the FeatureManager.

Save the FLATBAR.

195) Click **Save** 💾.

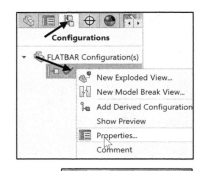

Define the part number property for the BOM.

196) Click the FLATBAR **ConfigurationManager** 🔁 tab.

197) Right-click **Default**.

198) Click **Properties**. The Configuration Properties PropertyManager is displayed.

199) Click **User Specified Name** from the drop-down box under Document Name.

200) Enter **GIDS-SC-10001-9** in the Part number text box.

Define a material property.
201) Click the **Custom Properties** button.

202) Click inside the **Property Name** box.

203) Click the **down arrow**. Select **Material** from the Property Name list.

204) Click inside the **Value/Text Expression** box.

205) Click the **down arrow**. Select **Material**.

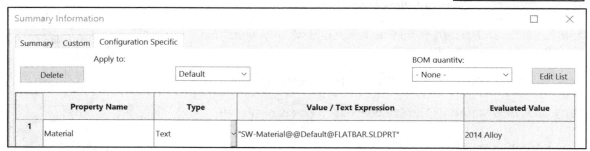

Define the Number Property.
206) Click inside the **second Property Name** box.

207) Click the **down arrow**.

208) Select **Number** from the Name list.

209) Click inside the **Value/Text Expression** box.

210) Enter **56-10222** for Drawing Number.

211) Click **OK** from the Summary Information dialog box.

212) Click **OK** ✔ from the Configuration Properties PropertyManager.

Return to the FeatureManager.

213) Click the FLATBAR **FeatureManager** tab.

Save the FLATBAR part.

214) Click **Save** .

Return to the drawing.

215) Click **Windows, FLATBAR - Sheet1** from the Main menu.

The Material Property is inserted into the Title block.

Activity: FLATBAR Drawing - Linked Note

Create a Linked Note. Set the Drawing Number.

216) Click **File, Properties** from the Main menu. Click the **Custom** tab.

217) Click inside the **Property Name** box. Note: The CompanyName was set the created drawing template.

218) Click the **down arrow**. Select **Number** from the Property Name list.

219) Click inside the **Value/Text Expression** box.

220) Enter **56-10222**.

221) Click **OK** from the Summary Information dialog box.

222) **Right-click** in Sheet1.

223) Click **Edit Sheet Format**.

224) **Zoom in** on the lower right corner of the drawing.

225) Double-click on the DWG. NO. text **FLATBAR**. The Note PropertyManager is displayed.

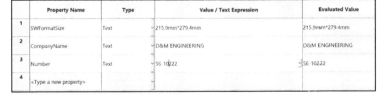

226) Delete **FLATBAR**. Note: The Formatting dialog is still active.

227) Click **Link to Property**  from the Text Format box.

228) Select **Current** document.

229) Select **Number** from the Link to Property drop-down menu.

230) Click **OK** from the Link to Property box.

231) Click **OK** ✔ from the Note PropertyManager.

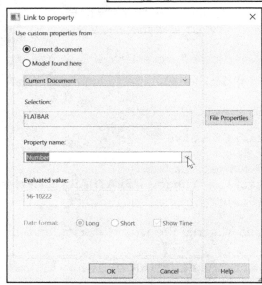

Return to the drawing sheet.
232) Right-click a **position** in the Graphics window.

233) Click **Edit Sheet**.

234) If needed, **remove all Tangent Edges** from the Top and Isometric view.

235) Address **all extension line gaps (5mm) and any additional needed information** as illustrated below.

Save the FLATBAR drawing.

236) Click **Save** 💾.

As an exercise, insert the Third Angle projection icon located in the LOGO folder.

Add drawing Custom properties: DrawnBy and Drawndate.

Insert the reference dimension symbol for the second 12.7mm dimension as illustrated.

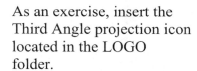

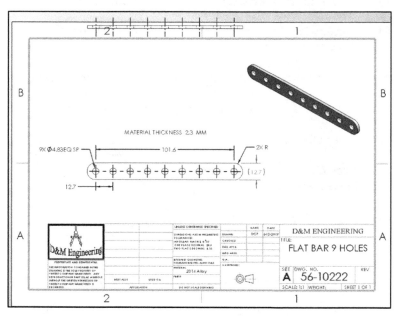

Additional details on Drawing Views, New Drawing, Details, Dimensions, Dimensions and Annotations are available in SOLIDWORKS Help.

Keywords: Drawing Views (overview); Drawing Views (model); Move (drawing views); Dimensions (circles, extension lines, inserting into drawings, move, parentheses); Annotations Hole Callout; Centerline; Center Mark; Properties and Sheet format.

Review the FLATBAR Drawing.

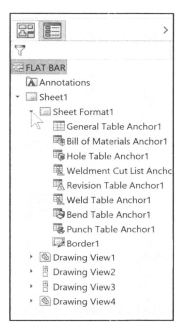

You created the FLATBAR drawing with the A-ANSI-MM Drawing Template. The FLATBAR drawing utilized the FLATBAR part with the Model View tool and the View Palette tool.

The Model View PropertyManager provided the ability to insert new views of a document. You selected the Front, Top and Right view. You applied the View Palette to insert an Isometric view.

You moved the views by dragging the blue view boundary. You inserted part dimensions and annotations into the drawing with the Insert Model Items tool. Dimensions were moved to new positions. Leader lines and dimension text were repositioned. Annotations were edited to reflect the dimension standard.

You created a Parametric note that referenced part dimensions in the drawing text. Material was assigned in the FLATBAR part. The Material Custom Property and Number Custom Property were assigned in the FLATBAR part and referenced in the drawing Title block.

Know inch/mm decimal display. The ASME Y14.5 standard states:

- For millimeter dimensions < 1, display the leading zero. Remove trailing zeros.

- For inch dimensions < 1, delete the leading zero. The dimension is displayed with the same number of decimal places as its tolerance.

Note: The FLATBAR drawing linked Title block notes to Custom Properties in the drawing and in the part. The additional drawings in this project utilize drawing numbers linked to the model file name. The Title of the drawing utilizes a Note.

LINKAGE Assembly Drawing - Sheet1

The LINKAGE assembly drawing Sheet 1 utilizes the LINKAGE assembly. Add an Exploded view and a Bill of Materials to the drawing.

Create an Exploded view in the LINKAGE assembly. The Bill of Materials reflects the components of the LINKAGE assembly. Create a drawing with a Bill of Materials. Perform the following steps:

- Create a new drawing with the custom A-ANSI-MM size Drawing Template with the CUSTOM-A sheet format.

- Create and display the Exploded view of the LINKAGE assembly.

- Insert the Exploded view of the assembly into the drawing.

- Insert a Bill of Materials.

- Label each component with Balloon text.

Activity: LINKAGE Assembly Drawing - Sheet1

Close all parts and drawings.
237) Click **Windows**, **Close All** from the Menu bar.

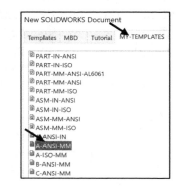

Create a New drawing.
238) Click **New** 🗋 from the Menu bar.

239) Click the **MY-TEMPLATES** tab.

240) Double-click **A-ANSI-MM**. Additional templates are displayed. Note: if you did not create the MY-TEMPLATES tab, create a drawing document, A-(ANSI) Landscape, Third Angle projection using the MMGS unit system. Add all needed document properties and custom properties.

Insert the LINKAGE assembly using Model View feature.
241) **Browse** to locate the LINKAGE assembly. There is a LINKAGE assembly located in the Chapter 4 homework section.

242) Double-click **LINKAGE**. Single view is selected by default.

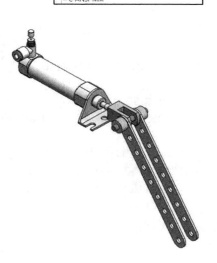

243) Click **Isometric view** from the Standard views box.

244) Select **Shaded With Edges**  from the Display Style box.

245) Check the **Use custom scale** box.

246) Enter **2:3** for Scale.

247) Click a **position** in the middle of Sheet1.

248) Click **OK** ✔ from the Drawing View1 PropertyManager.

Deactivate the Origins if needed.
249) Click **View**, **Hide/Show**, un-check **Origin** from the Menu bar.

Save the LINKAGE assembly drawing.

250) Click **Save** 💾. Accept the default file name.

251) Click **Save All**.

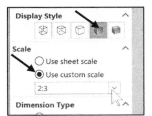

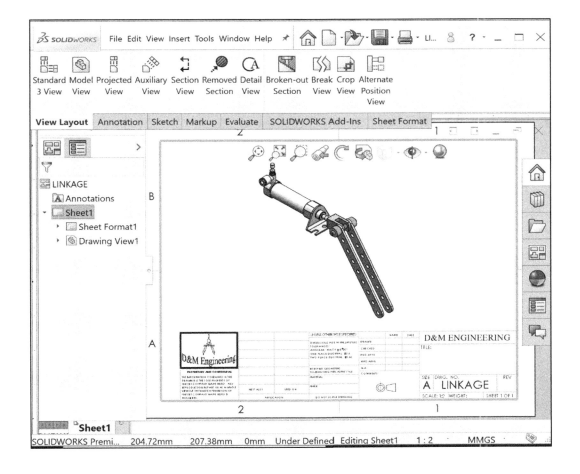

Display modes for a Drawing view are similar to a part document. The 3D Drawing View 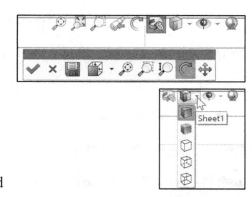 tool provides the ability to manipulate the model view in 3D, to select a difficult face, edge, or point.

Wireframe and Shaded Display modes provide the best graphic performance. Mechanical details require Hidden Lines Visible display and Hidden Lines Removed display. Select Shaded/Hidden Lines Removed to display Auxiliary Views to avoid confusion.

Tangent Edges Visible provides clarity for the start of a Fillet edge. Tangent Edges Removed provides the best graphic performance.

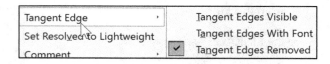

Right-click in the view boundary to access the Tangent Edge options.

Utilize the Lightweight Drawing option to improve performance for large assemblies.

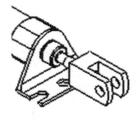

Wireframe   Hidden Lines Visible   Hidden Lines Removed   Shaded

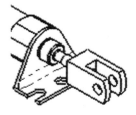

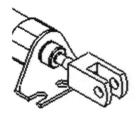

Tangent Edges Visible   Tangent Edges With Font   Tangent Edges Removed

To address Tangent lines views:

- Right-click in a Drawing view.

- Click Tangent Edge.

- Click a Tangent Edge view option.

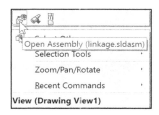

Return to the LINKAGE assembly. Set Configuration Properties.
252) Right-click inside the **Isometric view** boundary.

253) Click **Open Assembly**. The LINKAGE assembly is displayed.

254) Click the **ConfigurationManager** tab.

255) Right-click **Default [LINKAGE]**. Click **Properties**. The Configuration Properties PropertyManager is displayed.

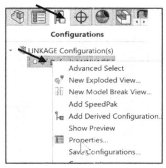

256) Select **User Specified Name** from the Part number displayed when used in Bill of Materials.

257) Enter **GIDS-SC-1000** in the Part number text box.

258) Click **OK** ✔ from the Configuration Properties PropertyManager.

Return to the FeatureManager.
259) Click the LINKAGE **FeatureManager** tab.

Save the model.

260) Click **Save**.

Exploded View

The Exploded View illustrates how to assemble the components in an assembly. Create an Exploded View with four steps. Click and drag components in the Graphics window. The Manipulator icon indicates the direction to explode. Select an alternate component edge for the Explode direction. Drag the component in the Graphics window or enter an exact value in the Explode distance box.

Manipulate the top-level components in the assembly. Access the Explode view option as follows:

- Right-click the configuration name in the ConfigurationManager.

- Select the Exploded View tool in the Assembly toolbar.

- Select Insert, Exploded View from the Menu bar.

Activity: LINKAGE Assembly Drawing - Exploded View

Insert an Exploded view.

261) Click the **ConfigurationManager** tab.

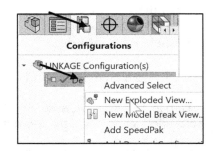

262) Right-click **Default [GIDS-SC-1000]**.

263) Click **New Exploded view** . The Explode
PropertyManager is displayed.

264) Click the **Regular step (translate and rotate)** icon.

Create Chain1. Use the distance box option.
265) Click the back **SHAFT-COLLAR** as illustrated.

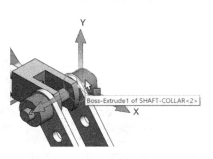

266) Enter **50**mm(1.97in) in the Explode Distance box.

267) Click the **Reverse Direction** box.

268) Click **Add Step**. The SHAFT-COLLAR moves 50mm to
the back of the model. Chain 1 is created.

Create Chain2. Use the Manipulator icon.
269) Click the front **SHAFT-COLLAR** in the Graphics window
as illustrated.

270) Drag the **Manipulator icon** to the front of the assembly
approximately 50mm (1.97in).

271) Click **Done**. Chain 2 is created.

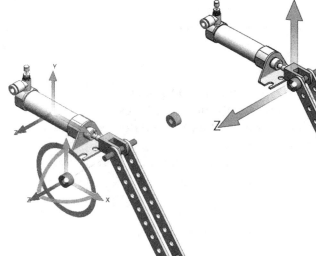

Create Chain3.

272) Click the **back FLATBAR** in the Graphics window as illustrated.

273) Drag the **Manipulator icon** approximately 32mm (1.26in) to the back of the assembly. Chain 3 is created.

274) Click **Done**.

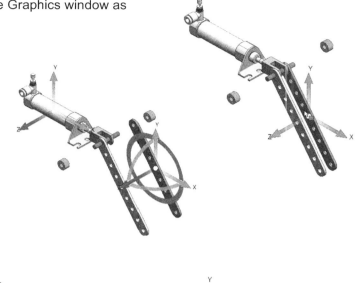

Create Chain4.

275) Click the **front FLATBAR** in the Graphics window.

276) Drag the **Manipulator icon** approximately 32mm (1.26in) to the front of the assembly. Chain 4 is created.

277) Click **Done**.

278) Click **OK** ✔ from the Explode PropertyManager.

Save the LINKAGE part in the Exploded State.

279) Click **Save** 🖫.

🔆 Create exploded view steps that rotate a component with or without linear translation. Use the rotation and translation handles of the triad. Include rotation and translation in the same explode step. You can also edit the explode step translation distance and rotation angle values in the PropertyManager.

Activity: LINKAGE Assembly Drawing - Animation

Animate the Exploded view.
280) Expand the Default [GIDS-SC-1000] folder.

281) Right-click **Exploded View1** in the ConfigurationManager.

282) Click **Animate collapse** to play the animation. View the Animation.

283) Click **Pause** 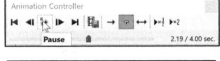.

284) Click **End** .

Return the Exploded view in its collapsed state.
285) Close the Animation Controller dialog box.

Return to the Assembly FeatureManager.
286) Click the LINKAGE **FeatureManager** tab.

Save the assembly.
287) Click **Save**.

Open the LINKAGE drawing.
288) Click **Window, LINKAGE - SHEET1** from the Menu bar.

Display the Exploded view in the drawing.
289) Right-click inside the **Isometric view.**

290) Click **Show in Exploded State.** Note: You can also click **Properties,** check the **Show in Exploded State** box and then click **OK** from the Drawing Views Properties box.

291) Click **OK** from the Drawing View1 PropertyManager. View the exploded state in the drawing on Sheet1.

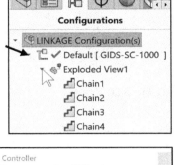

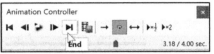

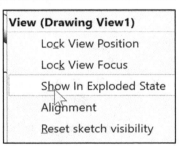

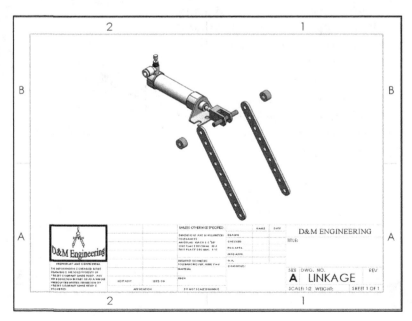

Bill of Materials

A Bill of Materials (BOM) is a table inserted into a drawing to keep a record of the parts used in an assembly. The default BOM template contains the Item Number, Quantity, Part No. and Description. The default Item number is determined by the order in which the component is inserted into the assembly. Quantity is the number of instances of a part or assembly.

Part No. is determined by the following: file name, default and the User Defined option, Part number used by the Bill of Materials. Description is determined by the description entered when the document is saved.

Activity: LINKAGE Assembly Drawing - Bill of Materials

Create a Bill of Materials.
292) Click inside the **Isometric view** boundary.

293) Click the **Annotation** tab from the CommandManager.

294) Click the **Bill of Materials** tool from the Consolidated Tables toolbar. The Bill of Materials PropertyManager is displayed.

295) Select **bom-material** for Table Template.

296) Select **Top-level only** for BOM Type.

297) Click **OK** from the Bill of Materials PropertyManager. Double-click a position in the **upper left corner** of the Sheet1.

298) Click a **position** in Sheet1. The Descriptions were created in Chapter 2.

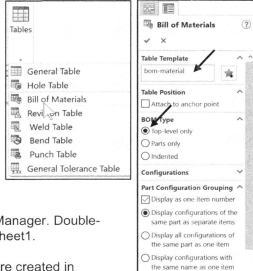

ITEM NO.	PART NUMBER	DESCRIPTION	MATERIAL	QTY.
1	GIDS-PC-10001	LINEAR ACTUATOR		1
2	AXLE	AXLE-ROD		1
3	FLATBAR	FLAT BAR 9-HOLE		2
4	SHAFT	SHAFT-COLLAR		2

The Bill of Materials requires some editing. The AXLE, FLATBAR, and SHAFT-COLLAR PART NUMBER values are not defined. The current part file name determines the PART NUMBER value. If needed, address the description for each part in the BOM.

The current part description determines the DESCRIPTION values. Redefine the PART NUMBER for the Bill of Materials. Note: You will also scale the LINKAGE assembly.

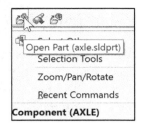

Modify the AXLE Part number.
299) Right-click the **AXLE** part in the LINKAGE drawing.

300) Click **Open Part**. The AXLE FeatureManager is displayed.

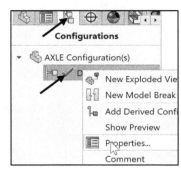

301) Click the AXLE **ConfigurationManager** tab. Right-click **Default [AXLE]** in the ConfigurationManager.

302) Click **Properties**. Select **User Specified Name** from the Configuration Properties dialog box.

303) Enter **GIDS-SC-10017** for the Part Number to be utilized in the Bill of Materials. Click **OK** ✅ from the Configuration Properties PropertyManager.

304) **Return** to the FeatureManager. Click **Save** 💾.

Return to the LINKAGE drawing. Save the drawing.
305) Click **Window**, **LINKAGE - Sheet1** from the Menu bar.

Modify the SHAFT-COLLAR PART NUMBER.
306) Right-click the front **SHAFT-COLLAR** part in the LINKAGE drawing. Click **Open Part**. The SHAFT-COLLAR FeatureManager is displayed. Click the SHAFT-COLLAR **ConfigurationManager** tab.

307) Right-click **Default [SHAFT-COLLAR]** from the ConfigurationManager. Click **Properties**.

308) Select the **User Specified Name** from the Configuration Properties box. Enter **GIDS-SC-10012-3-16** for the Part Number in the Bill of Materials. Click **OK** ✅ from the Configuration Properties PropertyManager.

309) **Return** to the FeatureManager. Click **Save** 💾.

Modify the FLATBAR Part number.

310) **Follow** the above procedure.

311) Enter **GIDS-SC-10001-9** for the Part Number in the Bill of Materials.

💡 Double-click the values in the BOM table (Keep Link) to directly edit the Custom Property of the Part/Assembly.

Return to the LINKAGE assembly drawing.
312) Click **Window**, **LINKAGE - Sheet1** from the Menu bar.

313) Rebuild the drawing and update the BOM.

Modify the LINKAGE assembly scale.
314) Click inside the **Isometric view** boundary.

315) Enter **1:2** for Scale.

316) Click **OK** ✔ from the Drawing View1 PropertyManager.

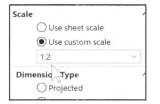

Save the drawing.
317) Click **Save** 💾 .

ITEM NO.	PART NUMBER	DESCRIPTION	MATERIAL	QTY.
1	GIDS-PC-10001	LINEAR ACTUATOR		1
2	GIDS-SC-10017	AXLE-ROD	6061 Alloy	1
3	GIDS-SC-10001-9	FLAT BAR 9-HOLE	2014 Alloy	2
4	GIDS-SC-10012-3-16	SHAFT-COLLAR	6061 Alloy	2

Note: As an exercise, complete the Bill of Materials. Add material to each component. Label each component with a unique item number. The item number is placed inside a circle. The circle is called Balloon text. List each item in a Bill of Materials table. Utilize Auto Balloon to apply Balloon text to all BOM components.

The Circle Split Line option contains the Item Number and Quantity. Item number is determined by the order listed in the assembly FeatureManager. Quantity lists the number of instances in the assembly.

Activity: LINKAGE Assembly Drawing - Automatic Balloons

Insert the Automatic Balloons.
318) Click inside the **Isometric view** boundary of the LINKAGE.

319) Click the **Auto Balloons** tool from the Annotation toolbar. The Auto Balloon PropertyManager is displayed. Accept the defaults. Note: The Insert Magnet line option inserts one or more magnetic lines.

320) Click **OK** ✔ from the Auto Balloon PropertyManager.

Reposition the Balloon text.
321) Click and drag each **Balloon** to the desired position.

322) Click and drag the **Balloon arrowhead** to reposition the arrow on a component.

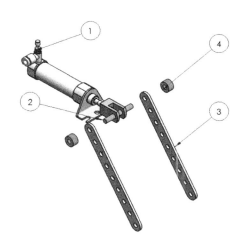

323) Click **OK** ✓ from the Balloon PropertyManager.

Display Item Number/Quantity.
324) Ctrl-Select the **four Balloon text** in the Graphics window. The Balloon PropertyManager is displayed. Select **Circular Split Line** for Style.

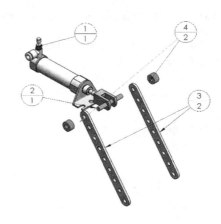

325) Click **OK** ✓ from the Balloon PropertyManager.

Split the Component lines.

326) Click the **SHAFT Circular Split Line**. The Circular Split Line is displayed in blue. **Zoom in** on the arrow point. Hold the **Ctrl** key down. **Click and drag** the arrow point to the second SHAFT-COLLAR. Release the **Ctrl** key. **Release** the mouse button.

327) **Perform** the above procedure for the FLAT BAR.

Save the LINKAGE assembly drawing.

328) Click **Save** 💾 .

LINKAGE Assembly Drawing - Sheet2

A drawing consists of one or more sheets. Utilize the Model View tool in the View Layout toolbar to insert the AirCylinder assembly. The LINKAGE drawing Sheet2 displays the Front view and Top view of the AirCylinder assembly.

Activity: LINKAGE Assembly - Drawing-Sheet2

Add Sheet2.
329) **Right-click** in the Graphics window.

330) Click **Add Sheet**. Sheet2 is displayed.

331) Right-click **Sheet2** in the Drawing FeatureManager.

332) Click **Properties**.

Select the CUSTOM-A Sheet Format.
333) Click **Browse** from the Sheet Properties box.

334) Select the **SW-TUTORIAL-2020\MY-TEMPLATES** folder.

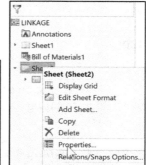

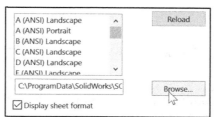

335) Double-click **CUSTOM-A**. Click **Apply Changes** from the Sheet Properties box. Sheet 2 of 2 is displayed in the Graphics window. CUSTOM-A was created at the beginning of the chapter. If you did not create CUSTOM A, accept the default sheet format and fill in all of the needed icons and Custom Properties.

Insert the AirCylinder assembly.

336) Click the **View Layout** tab from the CommandManager.

337) Click the **Model View** tool from the View Layout toolbar. The Model View PropertyManager is displayed.

338) Click the **Browse** button.

339) Double-click the **AirCylinder** assembly from the SW-TUTORIAL-2020 folder.

340) Check the **Create multiple views** box. Create a Top and Front view drawing. Click **Top view** as illustrated in the Standard views box.

341) Click **OK** from the Model View PropertyManager.

Save Sheet2.

342) Click **Save** .

Drawing sheets are ordered as they are created. The names are displayed in the FeatureManager design tree and as Excel-style tabs at the bottom of the Graphics window. Activate a sheet by right-clicking in FeatureManager design tree and clicking Activate or click the tab name.

Enter Title Name and Modify font size.

343) Right-click in the Graphics window.

344) Click **Edit Sheet Format**.

345) Double-click on the Title:

346) If needed, enter **LINEAR ACTUATOR**. Resize the text to the Title block. Enter **5**mm for text height.

347) Click inside the **Graphics window**.

348) Right-click in the Graphics window.

349) Click **Edit Sheet**.

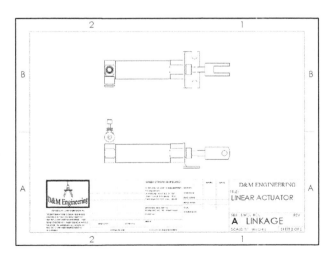

Section views display the interior features. Define a cutting plane with a sketched line in a view perpendicular to the Section view. Create a full Section view by sketching a section line in the Top view. Detailed views enlarge an area of an existing view. Specify location, shape and scale. Create a Detail view from a Section view at a 2:1 scale.

Activity: LINKAGE Assembly Drawing - Sheet2 Section View

Add a Section View to the drawing.
350) Click inside the **Drawing View3** boundary.

351) Click the **Section View** ⤢ tool from the View Layout toolbar. The Section View PropertyManager is displayed.

352) Click the **Section** tab from the PropertyManager.

353) Click the **Horizontal** Cutting Line button. Check the **Auto-start section view** box.

354) Click the **midpoint** of the model as illustrated.

Position Section View A-A and Modify the Drawing Scale.
355) Click **OK** from the Section View dialog box. Click a **location** above the Top view. The section arrows point upward.

356) If needed, check the **Flip direction** box.

357) Click **OK** ✔ from the Section View PropertyManager.

358) **Modify** the Drawing Scale to 1:2.

Fit the views to the drawing.
 359) Press the **f** key.

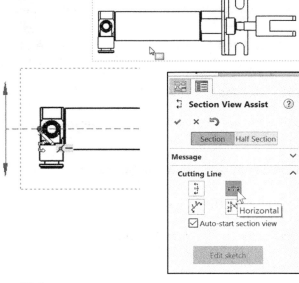

SECTION A-A

Activity: LINKAGE Assembly Drawing - Sheet2 Detail view

Add a Detail view to the drawing.
360) Click inside the **Section View** boundary. The Section View A-A PropertyManager is displayed.

361) **Zoom in** to enlarge the view.

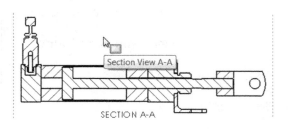

SECTION A-A

362) Click the **Detail View** 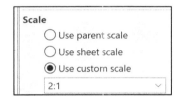 tool from the View Layout toolbar. The Circle Sketch tool is selected.

363) Click the **center** of the air fitting on the left side in the Section View as illustrated.

364) Sketch a **Circle** to encompass the air fitting. If required, enter **B** for Detail View Name in the Label text box.

Position Detail View B.
365) Press the **f** key.

366) Click a **location** on Sheet2 to the right of the SECTION View.

367) Enter **2:1** for Scale.

368) Click **OK** ✔ from the Detail View B PropertyManager.

🔅 Select a view boundary before creating Projected Views, Section Views or Detail Views.

Move views if required.
369) Click and drag the **view boundary** to allow for approximately 1 inch, [25mm] spacing between views.

Save the LINKAGE assembly drawing.
370) Click **Save** 💾.

Close all parts and assemblies.
371) Click **Window**, **Close All** from the Menu bar.

🔍 Additional details on Exploded View, Notes, Properties, Bill of Materials, Balloons, Section View and Detail View are available in SOLIDWORKS Help. Keywords: Exploded, Notes, Properties (configurations), Bill of Materials, Balloons, Auto Balloon, Section and Detail.

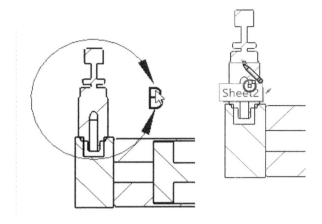

Scale
- ○ Use parent scale
- ○ Use sheet scale
- ● Use custom scale
- [2:1]

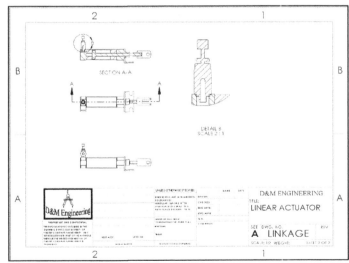

Review the LINKAGE Assembly Drawing

The LINKAGE Assembly drawing consisted of two sheets. Sheet1 contained an Exploded view. The Exploded view was created in the LINKAGE assembly.

The Bill of Materials listed the Item Number, Part Number, Description, Material and Quantity of components in the assembly. Balloons were inserted to label top level components in the LINKAGE assembly. You developed Custom Properties in the part and utilized the Properties in the drawing and Bill of Materials.

Sheet2 contained the Front view, Top view, Section view and Detail view of the AirCylinder assembly.

Design Tables

A Design Table is a spreadsheet used to create multiple configurations in a part or assembly. The Design Table controls the dimensions and parameters in the part. Utilize the Design Table to modify the overall length and number of holes in each FLATBAR.

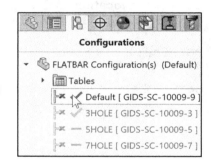

Create three configurations of the FLATBAR: 3HOLE, 5HOLE and 7HOLE

Utilize the Design Table to control the Part Number and Description in the Bill of Materials. Insert the custom parameter $PRP@DESCRIPTION into the Design Table. Insert the system parameter $PARTNUMBER into the Design Table.

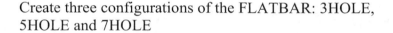

Activity: FLATBAR Part - Design Table

Open the FLATBAR part.

372) Click **Open** from the Menu bar.

373) Double-click the **FLATBAR** part from the SW-TUTORIAL-2020 folder. You created the FLATBAR in Chapter 2.

Insert a Design Table.
374) Click **Insert, Tables, Design Table** from the Menu bar. The Auto-create option is selected. Accept the default settings.

375) Click **OK** from the Design Table PropertyManager.

Select the input dimension.
376) Hold the **Ctrl key** down.

377) Click the **D1@Sketch1**, **D2@Sketch1**, **D1@Boss-Extude1**, **D1@Sketch2**, **D3@LPattern1** and **D1@LPattern1** from the Dimensions box.

378) Release the **Ctrl key**.

379) Click **OK** from the Dimensions dialog box.

The illustrated dimension variable names in the Dimensions dialog box will be different if sketches or features were deleted when creating the FLATBAR part.

The input dimension names and default values are automatically entered into the Design Table. The Design Table displays the Primary Units of the Part. Example: Inches. The value Default is entered in Cell A3.

The values for the FLATBAR are entered in Cells B3 through G9. The FLATBAR length is controlled in Column B. The Number of Holes is controlled in Column G.

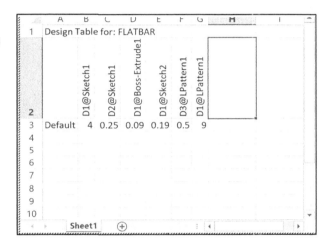

Enter the three configuration names.
380) Click **Cell A4**.

381) Enter **3HOLE**.

382) Click **Cell A5**.

383) Enter **5HOLE**.

384) Click **Cell A6**.

385) Enter **7HOLE**.

386) Click **Cell D3**.

Enter the dimension values for the 3HOLE configuration.
387) Click **Cell B4**. Enter **1**.

388) Click **Cell G4**. Enter **3**.

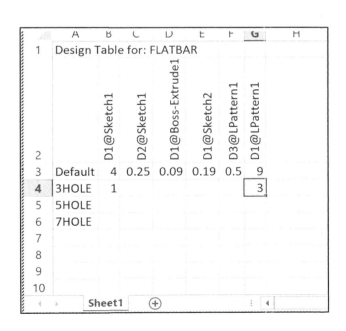

Enter the dimension values for the 5HOLE configuration.
389) Click **Cell B5**.

390) Enter **2**.

391) Click **Cell G5**.

392) Enter **5**.

Enter the dimension values for the 7HOLE configuration.
393) Click **Cell B6**.

394) Enter **3**.

395) Click **Cell G6**.

396) Enter **7**.

Build the three configurations.
397) Click a **position** outside the EXCEL Design Table in the Graphics window.

398) Click **OK** to generate the configurations. The Design Table icon is displayed in the FLATBAR FeatureManager.

Display the configurations.
399) Double-click **3HOLE**.

400) Double-click **5HOLE**.

401) Double-click **7HOLE**.

402) Double-click **Default**.

Save the model.
403) Click **Save** 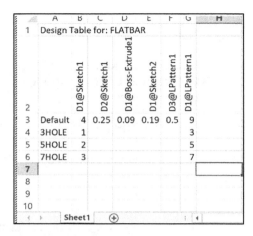.

Edit the Design Table.
404) Right-click **Design Table** in the ConfigurationManager.

405) Click **Edit Table**. The Add Rows and Columns dialog box is displayed.

406) Click **Cancel** from the Add Rows and Columns dialog box.

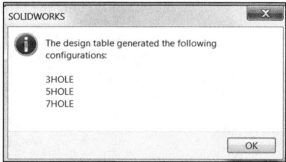

	A	B	C	D	E	F	G	H
1	Design Table for: FLATBAR							
2		D1@Sketch1	D2@Sketch1	D1@Boss-Extrude1	D1@Sketch2	D3@LPattern1	D1@LPattern1	
3	Default	4	0.25	0.09	0.19	0.5	9	
4	3HOLE	1					3	
5	5HOLE	2					5	
6	7HOLE	3					7	
7								
8								
9								
10								

Sheet1

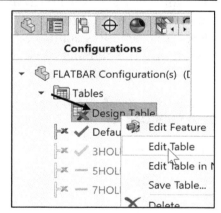

SOLIDWORKS

The design table generated the following configurations:

3HOLE
5HOLE
7HOLE

OK

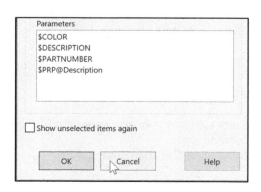

Configurations

FLATBAR Configuration(s) (D
  Tables
    Design Table
    Defau   Edit Feature
    3HOL   Edit Table
    5HOL   Edit Table in N
    7HOL   Save Table...
                  Delete

Parameters

$COLOR
$DESCRIPTION
$PARTNUMBER
$PRP@Description

☐ Show unselected items again

OK   Cancel   Help

Columns C through G are filled with the default FLATBAR values.

Enter parameters for DESCRIPTION and PARTNUMBER. Custom Properties begin with the prefix "$PRP@". SOLIDWORKS Properties begin with the prefix "$".

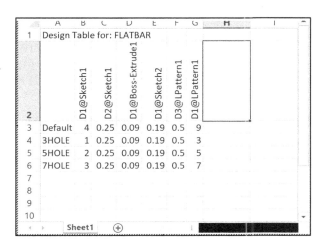

Enter DESCRIPTION custom Property.
407) Double-click **Cell H2**.

408) Enter **$PRP@DESCRIPTION**.

409) Click **Cell H3**. Enter **9HOLES**.

410) Click **Cell H4**. Enter **3HOLES**.

411) Click **Cell H5**. Enter **5HOLES**.

412) Click **Cell H6**. Enter **7HOLES**.

Enter the PARTNUMBER Property.
413) Double-click **Cell I2**.

414) Enter **$PARTNUMBER**.

415) Click **Cell I3**.

416) Enter **GIDS-SC-10009-9**.

417) Click **Cell I4**.

418) Enter **GIDS-SC-10009-3**.

419) Click **Cell I5**.

420) Enter **GIDS-SC-10009-5**.

421) Click **Cell I6**.

422) Enter **GIDS-SC-10009-7**.

423) Click a **position** in the Graphics window to update the Design Table.

424) View the updated ConfigurationManager.

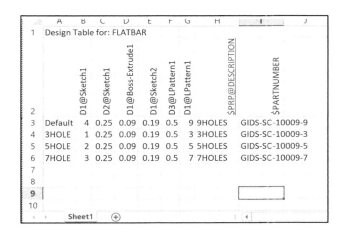

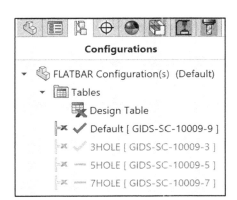

Activity: FLATBAR Drawing - Sheet2

Select configurations in the drawing. The Properties option in the Drawing view displays a list of configuration names.

Open the FLATBAR drawing.

425) Click **Open** 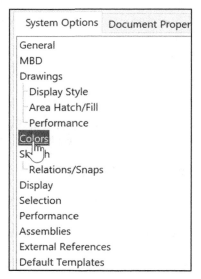 from the Menu bar.

426) Double-click the **FLATBAR drawing** from the SW-TUTORIAL-2020 folder. The FLATBAR drawing is displayed. The dimensions tied to the Design table are displayed in a different color. The dimension color is controlled from the System Options, Colors section.

Copy the Front view.
427) Click inside the FLATBAR **Front view** boundary.

428) Press **Ctrl C**.

429) **Right-click** in the Graphics window.

430) Click **Add Sheet**. Sheet2 is displayed.

431) Right-click **Properties**. Note: Click the down arrow if needed.

432) Click **Browse** from the Sheet Properties dialog box.

433) Double-click **CUSTOM-A** Sheet Format from the SW-TUTORIAL-2020/MY-TEMPLATES folder.

434) Click **OK** from the Sheet Properties box. CUSTOM-A was created at the beginning of the chapter. If you did not create CUSTOM A, accept the default sheet format and fill in all of the needed document properties and custom properties.

Paste the Front view from Sheet1.
435) Click a **position** inside the Sheet2 boundary.

436) Press **Ctrl V**. The Front view is displayed.

Display the 3HOLE FLATBAR configuration on Sheet2.
437) Right-click inside the **Front view** boundary.

438) Click **Properties**. Click the down arrow if needed.

439) Select **3HOLE** from the Use named configuration list.

440) Click **OK** from the Drawing View Properties dialog box. The 3HOLE FLATBAR configuration is displayed.

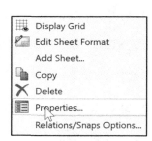

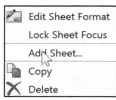

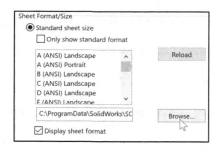

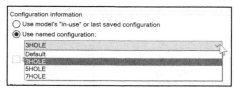

441) Right-click inside the **Drawing View** boundary.

442) Click **Open Part**.

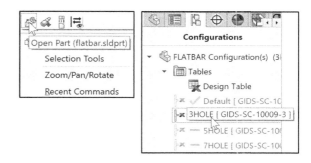

443) View the **model configuration**. 3 HOLE is selected by the drawing.

Return to the FLATBAR Drawing Sheet2.
444) Click **Window**, **FLATBAR - Sheet2** from the Menu bar.

445) Click on the **9X** dimension text in the Graphics window. The Dimension PropertyManager is displayed. Note: You can also use the Pop-up menu in this procedure.

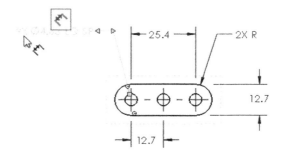

446) Replace the 9X dimension text with **3X** in the Dimension Text box as illustrated.

447) Click **OK** ✔ from the Dimension PropertyManager.

448) **Align** all dimension text as illustrated below.

449) Address **all extension line gaps and any other needed information**.

Save the FLATBAR Sheet2 drawing.

450) Click **Save** 💾.

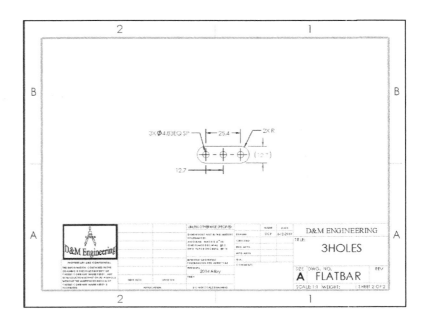

The 5HOLE and 7HOLE configurations are explored as an exercise. Combine the FLATBAR configurations with the SHAFT-COLLAR part to create three different assemblies. Select configuration in the assembly. The Properties in the FeatureManager option displays a list of configuration names.

The FLATBAR-SHAFTCOLLAR assembly contains a FLATBAR fixed to the assembly Origin and a SHAFTCOLLAR mated to the FLATBAR left hole. The default configuration utilizes the FLATBAR-9HOLE part.

Design Tables exist in the assembly. Utilize a Design Table to control part configurations 3HOLE, 5HOLE and 7HOLE. Utilize the Design Table to Control Suppress/Resolve state of a component in an assembly. Insert the parameter $STATE into the Design Table.

Activity: FLATBAR - SHAFTCOLLAR Assembly

Return to the Default FLATBAR configuration.
451) Right-click in the **3HOLE FLATBAR** view boundary.

452) Click **Open Part**.

453) Click the **ConfigurationManager** tab.

454) Double-click the **Default** configuration.

455) Click the **FeatureManager** tab. The FLATBAR (Default) FeatureManager is displayed.

Save the model.
456) Click **Save** .

Create the FLATBAR-SHAFTCOLLAR assembly.
457) Click **New** from the Menu bar.

458) Double-click **Assembly** from the Templates tab. The Begin Assembly PropertyManager is displayed. FLATBAR is an active document. FLATBAR is displayed in the Open documents box.

459) Click **FLATBAR** from the Open documents box.

460) Click **OK** from the Begin Assembly PropertyManager. The FLATBAR is fixed to the Origin.

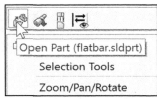

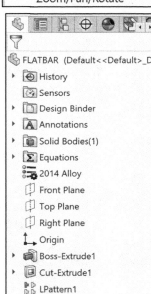

Save the FLATBAR-SHAFTCOLLAR assembly.

461) Click **Save** 🖫 .

462) Enter **FLATBAR-SHAFTCOLLAR** for Assembly name. Click **Save**.

Insert the SHAFTCOLLAR part.

463) Click the **Insert Components** 🖇 tool from the Assembly toolbar.

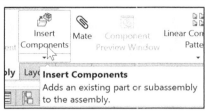

464) Double-click the **SHAFT-COLLAR** part from the SW-TUTORIAL-2020 folder.

465) Click a **position** to the front left of the FLATBAR as illustrated.

Fit the model to the Graphics window. Save the model.
466) Press the **f** key.

467) Click **Save** 🖫 .

Mate the SHAFTCOLLAR. Insert a Concentric mate.
468) Click the **left inside hole face** of the FLATBAR.

469) Hold the **Ctrl** key down.

470) Click the inside **cylindrical face** of the SHAFT-COLLAR.

471) Release the **Ctrl** key. The Mate pop-up menu is displayed.

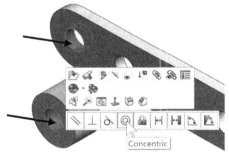

472) Click **Concentric** from the Mate pop-up menu.

Insert a Coincident mate.
473) Click the **back flat face** of the SHAFT-COLLAR.

474) **Rotate** to view the front face of the FLATBAR.

475) Hold the **Ctrl** key down.

476) Click the **front face** of the FLATBAR.

477) Release the **Ctrl** key.

478) Click **Coincident** from the Mate pop-up menu.

Insert and mate the second SHAFT-COLLAR to the right hole.

479) Click the **Insert Components**  tool from the Assembly toolbar.

480) Double-click the **SHAFT-COLLAR** part from the SW-TUTORIAL-2020 folder.

481) Click a **position** to the front right of the FLATBAR as illustrated.

Mate the second SHAFTCOLLAR. Insert a Concentric mate.
482) Click the **right inside hole face** of the FLATBAR.

483) Hold the **Ctrl** key down.

484) Click the **inside cylindrical face** of the SHAFT-COLLAR.

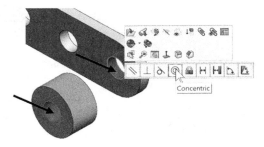

485) Release the **Ctrl** key.

486) Click **Concentric** from the Mate pop-up menu.

Insert a Coincident mate.
487) Click the **back flat face** of the SHAFT-COLLAR.

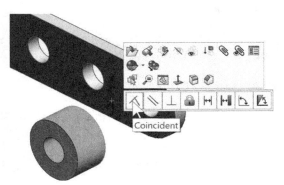

488) Rotate to view the front face of the FLATBAR.

489) Hold the **Ctrl** key down. Click the **front face** of the FLATBAR.

490) Release the **Ctrl** key.

491) Click **Coincident** from the Mate pop-up menu.

Save the FLATBAR-SHAFTCOLLAR assembly.
492) Click **Save**.

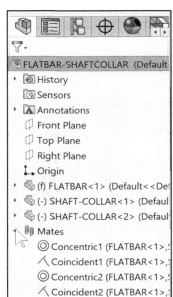

The FLATBAR-SHAFTCOLLAR FeatureManager displays the Default configuration of the FLATBAR in parentheses, FLATBAR<1> (Default).

The instance number <1> indicates the first instance of the FLATBAR. Note: Your instance number will be different if you delete the FLATBAR and then reinsert into the assembly. The exact instance number is required for the Design Table.

Create a Design Table that contains three new configurations. Each configuration utilizes a different FLATBAR configuration. Control the Suppress/Resolve State of the second SHAFT-COLLAR.

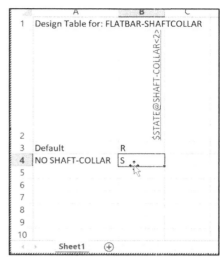

Insert a Design Table.
493) Click **Insert**, **Tables**, **Design Table** from the Menu bar. The Auto-create option is selected by default.

494) Click **OK** ✅ from the Design Table PropertyManager.

Enter the Design Table values.
495) Default is displayed in Cell A3. Click **Cell A4**.

496) Enter **NO SHAFT-COLLAR**.

497) Double-click **CELL B2**.

498) Enter **$STATE@SHAFT-COLLAR<2>**.

499) Click **Cell B3**. Enter **R** for Resolved.

500) Click **Cell B4**.

501) Enter **S** for Suppressed.

502) Click a **position** outside the Design Table in the Graphics window.

503) Click **OK** to display the NO SHAFT-COLLAR configuration.

Display the configurations.
504) If needed, click the **ConfigurationManager** 🔧 tab.

505) Double-click the **NO SHAFT-COLLAR** configuration. The second SHAFT-COLLAR is suppressed in the Graphics window.

506) Double-click the **Default** configuration. The second SHAFT-COLLAR is resolved. Both SHAFT-COLLARs are displayed in the Graphics window.

Insert FLATBAR configurations.
507) Right-click **Design Table**.

508) Click **Edit Table**.

509) Click **Cancel** from the Add Rows and Columns dialog box.

Enter Configuration names.
510) Click **Cell A5**. Enter **3HOLE FLATBAR**.

511) Click **Cell A6**. Enter **5HOLE FLATBAR**.

512) Click **Cell A7**. Enter **7HOLE FLATBAR**.

Enter STATE values.
513) Click **Cell B5**. Enter **S** for Suppress.

514) Click Cell **B6**. Enter **S** for Suppress.

515) Click **Cell B7**. Enter **S** for Suppress.

Enter Design Table values.
516) Double-click **Cell C2**.

517) Enter **$CONFIGURATION@FLATBAR<1>**.

518) Click **Cell C5**.

519) Enter **3HOLE**.

520) Click **Cell C6**.

521) Enter **5HOLE**.

522) Click **Cell C7**.

523) Enter **7HOLE**.

524) Click a **position** in the Graphics window to exit.

525) Click **OK** to create the three configurations.

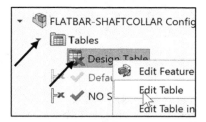

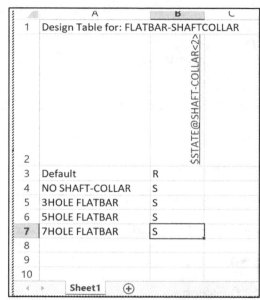

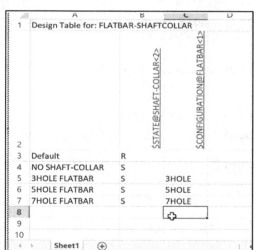

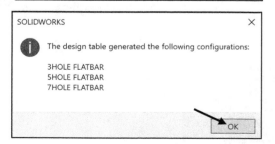

Display the configurations.

526) Double-click the **3HOLE FLATBAR** configuration.

527) Double-click the **5HOLE FLATBAR** configuration.

528) Double-click the **7HOLE FLATBAR** configuration.

529) Double-click the **Default** configuration.

530) Click the **Assembly FeatureManager** 🔲 tab.

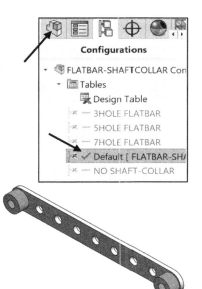

Display an Isometric view. Save the assembly.

531) Display an **Isometric view** 🔲.

532) Click **Save** 🔲.

Close all documents.

533) Click **Windows**, **Close All** from the Menu bar.

Always return to the default configuration in the assembly. Control the individual configuration through properties of a view in a drawing and properties of a component in the assembly.

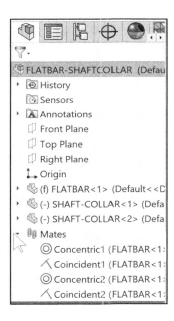

🔍   Additional details on Design Tables and Configurations are available in SOLIDWORKS Help.

Insert a Center of Mass Point

Add a center of mass (COM) point to parts, assemblies or drawings. COM points added in component documents also appear in the assembly document. In a drawing document of parts or assemblies that contain a COM point, you can display and reference the COM point.

Add a COM to a part or assembly by clicking **Center of Mass** (Reference Geometry toolbar) or **Insert, Reference Geometry, Center of Mass** 🔶 or checking the **Create Center of Mass feature** box in the Mass Properties dialog box.

The center of mass of the model is displayed in the Graphics window and in the FeatureManager design tree just below the origin.

The position of the **COM** point
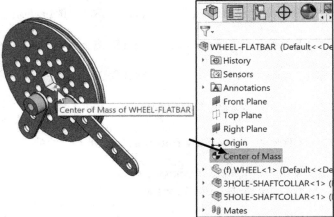 updates when the model's
center of mass changes. The COM
point can be suppressed and
unsuppressed for configurations.

You can measure distances and
add reference dimensions between
the COM point and entities such
as vertices, edges, and faces.

Add a center of mass (COM)
point to a drawing view. The
center of mass is a selectable entity in
drawings, and you can reference it to
create dimensions.

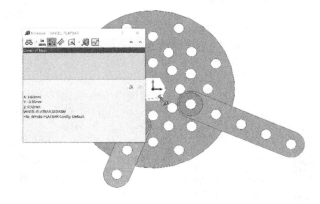

In a drawing document, click **Insert**,
Model Items. The Model Items
PropertyManager is displayed. Under
Reference Geometry, click the **Center of
Mass** icon. Enter any needed additional
information. Click **OK** from the Model
Items PropertyManager. View the results
in the drawing.

The part or assembly **needs to have
a COM before** you can view the COM
in the drawing. To view the center of
mass in a drawing, click **View**,
Hide/Show, **Center of mass**.

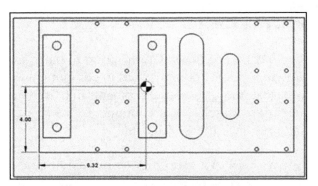

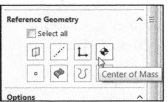

Chapter Summary

You created two drawings: the FLATBAR drawing and the LINKAGE assembly drawing. The drawings contained Standard views, Detail view, Section view and an Isometric view.

The drawings utilized a Custom Sheet Format and a Custom Drawing Template. The Sheet Format contained the Company logo and Title block information.

The FLATBAR drawing consisted of two Sheets: Sheet1 and Sheet2. You obtained an understanding of displaying views with the ability to insert, add, and modify dimensions. You used two major design modes in the drawings: Edit Sheet Format and Edit Sheet.

The LINKAGE assembly drawing contained two sheets. Sheet1 contained an Exploded view and a Bill of Materials. The Properties for the Bill of Materials were developed in each part and assembly. Sheet2 utilized a Detail view and a Section view of the AirCylinder assembly.

You created three configurations of the FLATBAR part with a Design Table. The Design Table controlled parameters and dimensions of the FLATBAR part. You utilized these three configurations in the FLATBAR-SHAFTCOLLAR assembly.

Drawings are an integral part of the design process. Part, assemblies and drawings all work together. From your initial design concepts, you created parts and drawings that fulfilled the design requirements of your customer.

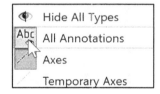 To show a hidden dimension, click **View**, **Hide/Show**, **All Annotations** from the Menu bar.

Questions

1. Describe a Bill of Materials and its contents.

2. Name the two major design modes used to develop a drawing in SOLIDWORKS.

3. True or False. Units, Dimensioning Standards, Arrow size, Font size are modified in the Options, Document Properties section.

4. How do you save a Sheet Format?

5. Identify seven components that are commonly found in a title block.

6. Describe the procedure to insert an Isometric view to the drawing.

7. In SOLIDWORKS, a drawing file name ends with a _____ suffix. A part file name ends with a _____ suffix.

8. True or False. In SOLIDWORKS, if a part is modified, the drawing is updated with a Rebuild command.

9. True or False. In SOLIDWORKS, when a dimension in the drawing is modified, the part is updated with a Rebuild command.

10. Name three guidelines to create General Notes on a drawing.

11. True or False. Most engineering drawings use the following font: Times New Roman - all small letters.

12. What are Leader lines? Provide an example.

13. Describe the key differences between a Detail view and a Section view on a drawing.

14. Identify the procedure to create an Exploded view.

15. Describe the purpose of a Design Table for a part and for an assembly.

16. Review the Design Intent section in the Introduction. Identify how you incorporated design intent into the drawing.

17. Identify how you incorporate design intent into configurations with a Design Table.

Exercises

Exercise 4.1: FLATBAR - 3 HOLE Drawing

Create the A (ANSI) Landscape - IPS - Third Angle 3HOLES drawing as illustrated below. Do not display Tangent Edges. Do not dimension to Hidden Lines.

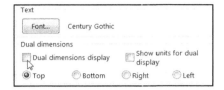

- First create the part from the drawing, then create the drawing. Use the default A (ANSI) Landscape Sheet Format/Size.

- Insert a Front, Top and a Shaded Isometric view as illustrated. Insert dimensions.

	Property Name	Type	Value / Text Expression	Evaluated Va
1	Material	Text	"SW-Material@3HOLE.SLDPRT"	1060 Alloy
2	<Type a new property>			

- Address extension line gap between the feature lines.

- Address proper display mode.

- Add a Smart (Linked) Parametric note for MATERIAL THICKNESS in the drawing as illustrated.

- Hide the dimension in the Top view. Insert Centerlines.

	Property Name	Type	Value / Text Expression	Evaluated Value
2	DrawnBy	Text	DCP	DCP
3	DrawnDate	Text	9-1-2019	9-1-2019
4	CompanyName	Text	D&M ENGINEERING	D&M ENGINEERING
5	Description	Text	3HOLES	3HOLES
6	Revision	Text	A	A
7	SWFormatSize	Text	8.5in*11in	8.5in*11in
8	<Type a new property			

- Modify the Hole dimension text to include 3X THRU EQ. SP. and 2X as illustrated.

- Insert Company and Third Angle projection icons. The icons are available in the homework folder.

- Insert Custom Properties: Material, Number, Description, DrawnBy, DrawnDate, CompanyName, Revision, etc.

- Material is 1060 Alloy.

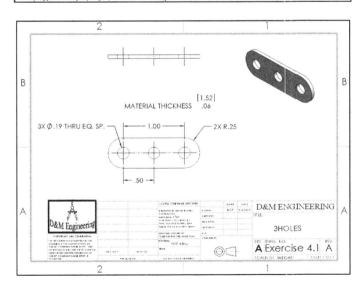

Exercise 4.2: CYLINDER Drawing

Create the A (ANSI) Landscape - IPS - Third Angle
CYLINDER drawing as illustrated below. Do not display
Tangent Edges. Do not dimension to Hidden lines.

- First create the part from the drawing - then create the
  drawing. Use the default A
  (ANSI) Landscape Sheet
  Format/Size.

- Insert the Front, Right and
  Isometric view as illustrated.

- Insert dimensions.

- Address extension line gap
  between the feature lines.

- Insert proper display modes.

- Insert Company and Third
  Angle projection icons. The
  icons are available in the
  homework folder.

- Insert Centerlines, Center
  Marks, and Annotations.

- Insert Custom
  Properties: Material,
  Revision, Description,
  Number, DrawnBy,
  DrawnDate,
  CompanyName.

- Material is AISI 1020.

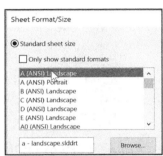

	Property Name	Type	Value / Text Expression	Evaluated Value
1	DrawnDate	Text	9-1-2019	9-1-2019
2	DrawnBy	Text	DCP	DCP
3	CompanyName	Text	D&M ENGINEERING	D&M ENGINEERING
4	Revision	Text	A	A
5	Number	Text	667-888	667-888
6	SWFormatSize	Text	8.5in*11in	8.5in*11in
7				

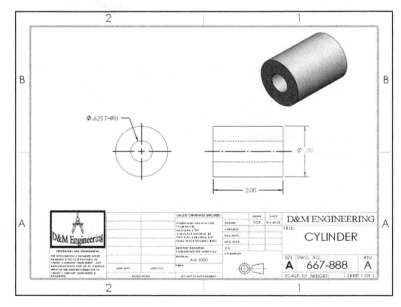

Exercise 4.3: PRESSURE PLATE Drawing

Create the A (ANSI) Landscape - IPS - Third Angle
PRESSURE PLATE drawing. Do not display Tangent edges.
Do not dimension to hidden lines.

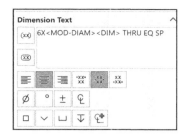

- First create the part from the drawing,
  then create the drawing. Use the default
  A (ANSI) Landscape Sheet Format/Size.

- Insert the Front and Right view as
  illustrated.

- Insert dimensions.

- Address extension line gap
  between the feature lines.

- Address proper display modes.

- Insert Company and Third Angle
  projection icons. The icons are
  available in the homework folder.

- Insert Centerlines, Center Marks,
  and Annotations.

- Insert Custom Properties:
  Material, Revision,
  Description, Number,
  DrawnBy, DrawnDate,
  CompanyName.

- Material is 1060 Alloy.

	Property Name	Type	Value / Text Expression	Evaluated Value
1	DrawnBy	Text	DCP	DCP
2	DrawnDate	Text	9-1-2019	9-1-2019
3	CompanyName	Text	D&M ENGINEERING	D&M ENGINEERING
4	Revision	Text	A	A
5	PartNo	Text	55-568	55-568
6	Description	Text	PRESSURE PLATE	PRESSURE PLATE
7	SWFormatSize	Text	8.5in*11in	8.5in*11in

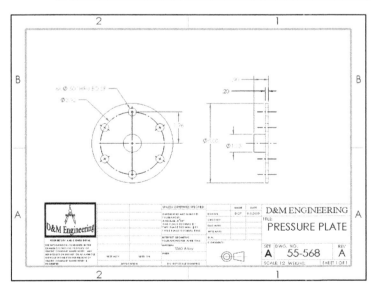

Exercise 4.3B TUBE Drawing

Create the A (ANSI) Landscape - IPS - Third Angle TUBE-2 drawing as illustrated below. Do not display Tangent edges. Phantom lines are fine.

- First create the part from the drawing, then create the drawing. Use the default A (ANSI) Landscape Sheet Format/Size.

- Insert views as illustrated.

- Insert dimensions.

- Address extension line gap between the feature lines.

- Insert Company and Third Angle projection icons. The icons are available in the homework folder.

- Insert Centerlines, Center Marks, and Annotations.

- Address proper display modes.

- Insert Custom Properties: Material, Revision, Description, Number, DrawnBy, DrawnDate, CompanyName.

- Material is 6061 Alloy.

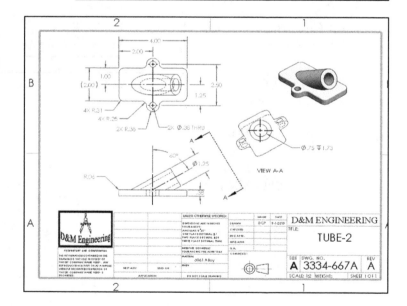

	Property Name	Type	Value / Text Expression	Evaluated Value
1	CompanyName	Text	D&M ENGINEERING	D&M ENGINEERING
2	DrawnDate	Text	9-1-2019	9-1-2019
3	DrawnBy	Text	DCP	DCP
4	Revision	Text	A	A
5	Number	Text	3334-667A	3334-667A
6	SWFormatSize	Text	8.5in*11in	8.5in*11in
7				

Exercise 4.4: LINKS Assembly Drawing

- Create the LINK assembly. Utilize three different FLATBAR configurations and a SHAFT-COLLAR. The parts are located in the chapter folder.

- Create the LINK assembly drawing as illustrated. Use the default A (ANSI) Landscape Sheet Format/Size.

- Insert Company name, logo and Third Angle projection icon. The icon is available in the homework folder.

- Remove all Tangent Edges.

- Insert Custom Properties: Description, Number, Revision, DrawnBy, DrawnDate, and CompanyName.

- Insert a Bill of Materials as illustrated with Balloons.

	Property Name	Type	Value / Text Expression	Evaluated Value
2	DrawnDate	Text	9-1-2019	9-1-2019
3	CompanyName	Text	D&M ENGINEERING	D&M ENGINEERING
4	Revision	Text	A	A
5	PartNo	Text	9998-099	9998-099
6	Description	Text	LINK-2	LINK-2
7	SWFormatSize	Text	8.5in*11in	8.5in*11in
8				

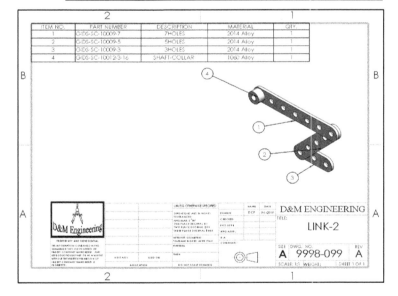

ITEM NO.	PART NUMBER	DESCRIPTION	MATERIAL	QTY.
1	GIDS-SC-10009-7	7HOLES	2014 Alloy	1
2	GIDS-SC-10009-5	5HOLES	2014 Alloy	1
3	GIDS-SC-10009-3	3HOLES	2014 Alloy	1
4	GIDS-SC-10012-3-16	SHAFT-COLLAR	1060 Alloy	1

Exercise 4.5: PLATE-1 Drawing

Create the A (ANSI) Landscape - MMGS - Third Angle PLATE-1 drawing as illustrated below. Do not display Tangent edges. Do not dimension to Hidden lines.

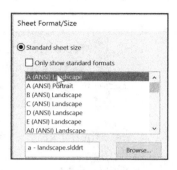

- First create the part from the drawing, then create the drawing. Use the default A (ANSI) Landscape Sheet Format/Size.

- Insert the Front and Right view as illustrated.

- Insert dimensions.

- Address extension line gap between the feature lines.

- Address display modes.

- Insert Company and Third Angle projection icons. The icons are available in the homework folder.

- Insert Centerlines, Center Marks, and Annotations.

- Insert Custom Properties: Material, Description, Number, Revision, DrawnBy, DrawnDate, and CompanyName.

- Material is 1060 Alloy.

	Property Name	Type	Value / Text Expression	Evaluated Value
1	Description	Text	PLATE-1	PLATE-1
2	DrawnBy	Text	DCP	DCP
3	DrawnDate	Text	9-1-2019	9-1-2019
4	CompanyName	Text	D&M ENGINEERING	D&M ENGINEERING
5	Revision	Text	A	A
6	Number	Text	5444-999	5444-999
7	SWFormatSize	Text	215.9mm*279.4mm	215.9mm*279.4mm

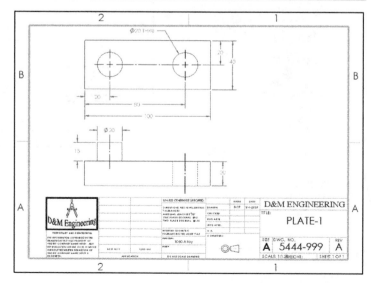

Exercise 4.6: **PLATE Drawing**

Create the A (ANSI) Landscape - IPS - Third Angle FLAT-PLATE drawing. Do not display Tangent edges. Phantom lines are fine.

- First create the part from the drawing, then create the drawing. Use the default A (ANSI) Landscape Sheet Format/Size.

- Insert the Front, Top, Right and Isometric views as illustrated.

- Insert dimensions.

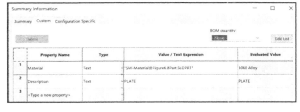

- Address extension line gap between the feature lines.

- Address proper display modes.

- Insert Company and Third Angle projection icons. The icons are available in the homework folder.

- Insert Centerlines, Center Marks, and Annotations.

	Property Name	Type	Value / Text Expression	Evaluated Value
1	DrawnBy	Text	DCP	DCP
2	DrawnDate	Text	9-1-2019	9-1-2019
3	CompanyName	Text	D&M ENGINEERING	D&M ENGINEERING
4	Description	Text	PLATE	PLATE
5	Revision	Text	A	A
6	Number	Text	2445-990	2445-990
7	SWFormatSize	Text	8.5in*11in	8.5in*11in

- Insert Custom Properties: Description, Number, Material, Revision, DrawnBy, DrawnDate, and CompanyName.

- Material is 1060 Alloy.

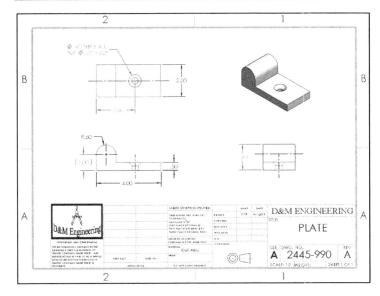

Exercise 4.7: LINKAGE-2 Drawing

- Create the LINKAGE-2 assembly drawing. Utilize the default A (ANSI) Landscape Sheet Format/Size. Note: If needed, open the LINKAGE-2 assembly from the Chapter 4 Homework/AirCylinder 2 folder.

- Insert an Isometric shaded view of the LINKAGE-2 Assembly created in the Chapter 2 exercises.

- Define the PART NO. Property and the DESCRIPTION Property for the AXLE, FLATBAR- 9HOLE, FLATBAR - 3HOLE and SHAFT COLLAR.

	Property Name	Type	Value / Text Expression	Evaluated Value
1	SWFormatSize	Text	215.9mm*279.4mm	215.9mm*279.4mm
2	Revision	Text	A	A
3	DrawnBy	Text	DCP	DCP
4	DrawnDate	Text	9-1-2019	9-1-2019
5	Description	Text	LINKAGE-2	LINKAGE-2
6	CompanyName	Text	D&M ENGINEERING	D&M ENGINEERING
7	PartNo	Text	333-223-22	333-223-22

- Save the LINKAGE-2 assembly to update the properties. Return to the LINKAGE-2 Drawing. Insert a Bill of Materials with Auto Balloons as illustrated.

- Insert the Company and Third Angle Projection icon.

- Insert Custom Properties.

☼ Use the Pack and Go option to save an assembly or drawing with references. The Pack and Go tool saves either to a folder or creates a zip file to e-mail. View SOLIDWORKS help for additional information.

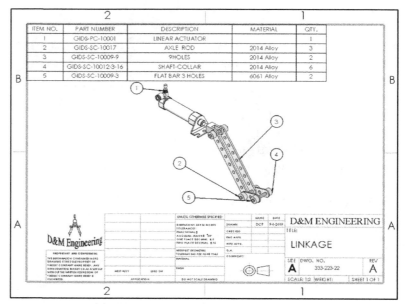

Exercise 4.8: Vertical Section View Drawing

Create the A (ANSI) Landscape - IPS - Third Angle Vertical Section View drawing as illustrated below. Do not display Tangent edges.

Insert Company and Third Angle projection icons. The icons are available in the homework folder.

Insert Custom Properties: Material, Description, DrawnBy, DrawnDate, CompanyName, Revision, etc.

1. Open **Section View** from the Chapter 4 Homework\Section folder. Create the **drawing document**.

2. Insert the **Front view (Drawing View1)**.

3. Click inside the **Drawing View1** view boundary. The Drawing View1 PropertyManager is displayed.

Display the origin on Sheet1.

4. Click **View ➤ Hide/Show ➤ Origins** from the Menu bar.

5. Click the **Section View** ⇄ drawing tool. The Section View PropertyManager is displayed.

6. Click the **Section** tab. Click the **Vertical** Cutting Line button.

7. Click the **origin**. Note: You can select the midpoint vs. the origin.

Place the Section view.

	Property Name	Type	Value / Text Expression	Evaluated Value
1	PartNo	Text	445-777	445-777
2	Revision	Text	A	A
3	DrawnBy	Text	DCP	DCP
4	DrawnDate	Text	9-1-209	9-1-209
5	CompanyName	Text	D&M ENGINEERING	D&M ENGINEERING
6	Description	Text	SECTION VIEW	SECTION VIEW
7	SWFormatSize	Text	8.5in*11in	8.5in*11in

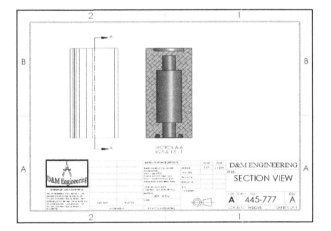

8. Click a **position** to the right of Drawing View1. The Section arrows point to the right. If required, click Flip direction. Check **Auto hatching** from the Section View box.

9. Check the **Shaded With Edges** option from the Display Style box. Click **OK** ✔ from the Section View A-A PropertyManager. Section View A-A is created and is displayed in the Drawing FeatureManager.

10. Click inside the **Drawing View1** view boundary. The Drawing View1 PropertyManager is displayed. Modify the Scale to **1.5:1**. Click **OK** ✔ from the Drawing View1 PropertyManager. Both drawing views are modified (parent and child).

Exercise 4.9: Aligned Section View Drawing

Create an Aligned Section view drawing.

Create the A (ANSI) Landscape - IPS - Third Angle drawing as illustrated below.

Insert Company and Third Angle projection icons. The icons are available in the homework folder.

Insert Custom Properties: Material, Description, DrawnBy, DrawnDate, CompanyName etc.

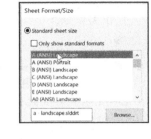

1. Open **Aligned Section** from the Chapter 4 Homework\Aligned Section folder.

2. Create the **drawing document**. Insert the **Front (Drawing View1)** and **Right** view.

4. Click inside the **Drawing View1** view boundary. The Drawing View1 PropertyManager is displayed. Click the **Section View** ⇄ drawing tool. The Section View PropertyManager is displayed.

5. Click the **Half Section** tab.

6. Click the **Leftside Down** Half Section button.

7. Select the **center** of the drawing view as illustrated. The Section View dialog box is displayed.

8. Click the **LENSCAP** in the Graphics window. The LENSCAP component is displayed in the Excluded components box. Check the **Auto hatching** box.

9. Click **OK** from the Section View dialog box.

Place the Section View.

10. Click a **position** above Drawing View1. Click **OK** ✔ from the Section View A-A PropertyManager.

11. Modify the **scale** to **1:4**. **Rebuild** 🔄 the drawing.

12. **Move** the drawing views to fit the Sheet. View the drawing FeatureManager.

13. **Close** the drawing.

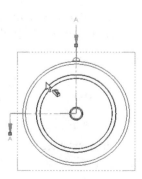

Use Section View Assist, previously called the Section View User Interface, to add offsets to existing section views. To edit existing section views with Section View Assist: Right-click the section view or its cutting line and click **Edit Cutting Line**.

Exercise 4.10: Auxiliary View Drawing

Create an Auxiliary View drawing as illustrated from an existing Front view.

Create the A (ANSI) Landscape - IPS - Third Angle drawing as illustrated.

Insert Company and Third Angle projection icons.

The icons are available in the homework folder.

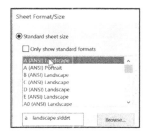

Insert Custom Properties: Material, Description, DrawnBy, DrawnDate, CompanyName etc.

	Property Name	Type	Value / Text Expression	Evaluated Value
1	Material	Text	"SW-Material@Auxiliary View.SLDPRT"	1060 Alloy
2	<Type a new property>			

1. **Copy** the Chapter 4 Homework folder to your hard drive.

2. Open **Auxiliary view** from the Chapter 4 Homework\Auxiliary folder.

3. Create the **drawing** document.

4. Insert the **Front (Drawing View1)**, **Top**, **Right** and **Isometric** view.

5. Click the **angled edge** in Drawing View1.

6. Click the **Auxiliary View** drawing tool.

7. Click a **position** up and to the right.

8. Drag the **A-A arrow** as illustrated.

9. Drag the **text** off the view.

10. Click **OK** from the Auxiliary View PropertyManager.

11. **Close** the drawing.

	Property Name	Type	Value / Text Expression	Evaluated Value
2	Revision	Text	A	A
3	PartNo	Text	3345-88	3345-88
4	DrawnBy	Text	DCP	DCP
5	Description	Text	AUXILIARY	AUXILIARY
6	CompanyName	Text	D&M ENGINEERING	D&M ENGINEERING
7	DrawnDate	Text	9-1-2019	9-1-2019
8				

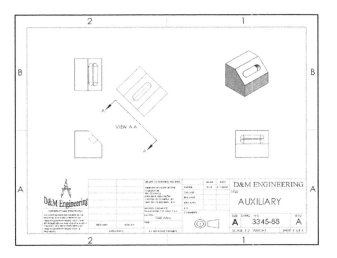

Notes:

Chapter 5

Advanced Features

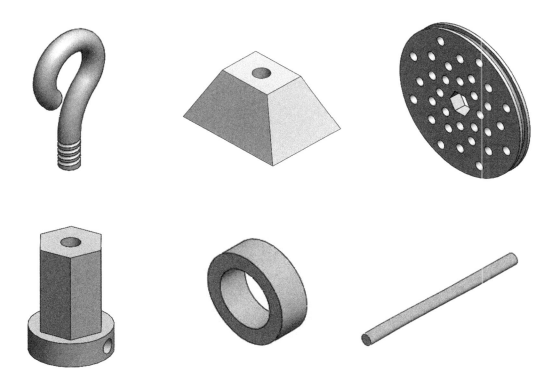

Below are the desired outcomes and usage competencies based on the completion of Chapter 5.

Desired Outcomes:	Usage Competencies:

• Six parts for the PNEUMATIC-TEST-MODULE assembly:
   ○ WEIGHT
   ○ HOOK
   ○ WHEEL
   ○ HEX-ADAPTER
   ○ AXLE-3000
   ○ SHAFTCOLLAR-500
	• Apply the following Advanced modeling features: Plane, Lofted Base, Extruded Cut, Swept Base, Dome, Thread, Extruded Boss/Base, Revolved Cut, Extruded Cut, Circular Pattern, Axis, Instant3D, Hole Wizard, Advanced Hole, Split Line, and Thread Wizard.
	
	• Reuse geometry.
	
	• Modify existing parts to create new parts with the Save as copy command.

Notes:

Chapter 5 - Advanced Features

Chapter Objective

Obtain the ability to reuse geometry by modifying existing parts and to create new parts. Knowledge of the following SOLIDWORKS features: *Plane, Lofted Base, Extruded Cut, Swept Base, Dome, Thread, Extruded Boss/Base, Revolved Cut, Extruded Cut, Circular Pattern, Axis, Instant3D, Hole Wizard, Advanced Hole, Split Line and Thread Wizard.*

Create six individual parts:

- WEIGHT

- HOOK

- WHEEL

- HEX-ADAPTER

- AXLE-3000

- SHAFTCOLLAR-500

Develop a working understanding with multiple documents in an assembly. Build on sound assembly modeling techniques that utilize symmetry, component patterns, and mirrored components.

Create five assemblies:

- 3HOLE-SHAFTCOLLAR assembly

- 5HOLE-SHAFTCOLLAR assembly

- WHEEL-FLATBAR assembly

- WHEEL-AND-AXLE assembly

- PNEUMATIC-TEST-MODULE assembly

On the completion of this chapter, you will be able to:

- Create new parts and copy parts with the Save As command to reuse similar geometry.

- Utilize Construction geometry in a sketch.

- Apply the following SOLIDWORKS features:

    o   Extruded Boss/Base

    o   Extruded Cut

    o   Dome

    o   Plane

    o   Lofted Boss/Base

    o   Swept Boss/Base

    o   Thread

    o   Revolved Cut

    o   Hole Wizard

    o   Axis

    o   Circular Pattern

    o   Instant3D

    o   Advanced Hole

    o   Split Line

Chapter Overview

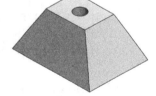

Six additional parts are required for the final PNEUMATIC-TEST-MODULE assembly. Each part explores various modeling techniques.

Create three new parts in this chapter:

- WEIGHT

- HOOK

- WHEEL

The WEIGHT and HOOK parts were applied in the Chapter 3 Homework exercises. View the Chapter 3 Homework folder for the parts if needed.

Utilize the Save As command and modify existing parts that were created in the previous chapters to create three additional parts for the PNEUMATIC-TEST-MODULE assembly.

- HEX-ADAPTER

- AXLE-3000

- SHAFTCOLLAR-500

The HEX-ADAPTER part utilizes modified geometry from the HEX-STANDOFF part.

The AXLE-3000 part utilizes modified geometry from the AXLE part.

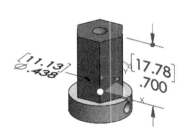

The SHAFTCOLLAR-500 part utilizes modified geometry from the SHAFT-COLLAR part.

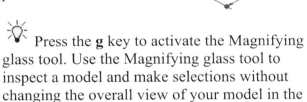

🔆 Press the **g** key to activate the Magnifying glass tool. Use the Magnifying glass tool to inspect a model and make selections without changing the overall view of your model in the Graphics window.

🔆 Download all needed model files (SW-TUTORIAL-2020 folder) from the SDC Publications website (www.sdcpublications.com) to a local hard drive.

WEIGHT Part

The WEIGHT part is a machined part. Utilize the Lofted ⬇ feature. Create a Loft by blending two or more profiles. Each profile is sketched on a separate plane.

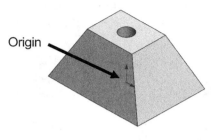

Create Plane1. Offset Plane1 from the Top Plane.

Sketch a rectangle for the first profile on the Top Plane.

Sketch a square for the second profile on Plane1.

Select the corner of each profile to create the Lofted feature.

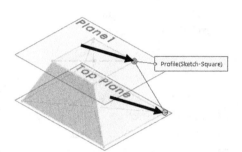

Utilize the Instant3D tool to create an Extruded Cut ▣ feature to create a Through All hole centered on the top face of the Loft feature.

Reference geometry defines the shape or form of a surface or a solid. Reference geometry includes planes, axes, coordinate systems, and points.

🔅 When using the Instant3D tool, you lose the ability to select various End Conditions to maintain design intent.

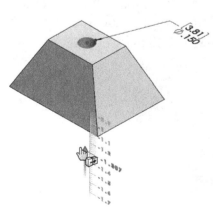

All parts in this chapter utilize a custom part template. Create the custom part template from the default part template. Save the Custom Part template in the SW-TUTORIAL-2020\MY-TEMPLATE folder. If needed, create the SW-TUTORIAL-2020\MY-TEMPLATE folder.

Activity: Create the WEIGHT Part

Create a New part template.

1) Click **New** ⬜ from the Menu bar.

2) Double-click **Part** from the Templates tab. The Part FeatureManager is displayed.

Set Document Properties. Set drafting standard.

3) Click **Options** ⚙ from the Main menu.

4) Click the **Document Properties** tab from the dialog box.

5) Select **ANSI** from the Overall drafting standard drop-down menu.

Set document units and precision.

6) Click **Units**.

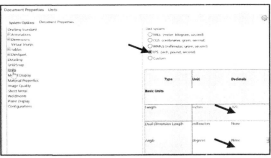

7) Select **IPS**, [**MMGS**] for Unit system.

8) Select **.123**, [**.12**] for linear units Decimal places.

9) Select **None** for Angular units Decimal places.

Set Leader arrow direction.

10) Click **Dimensions**. Check the **Smart** box as illustrated.

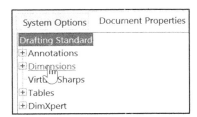

11) Click **OK** from the Document Properties - Detailing - Dimensions dialog box.

Save the Part template. Enter name.

12) Click **Save As** 📓.

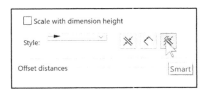

13) Select **Part Templates (*.prtdot)** for Save as type.

14) Select **SW-TUTORIAL-2020\MY-TEMPLATES** for Save in folder.

15) Enter **PART-ANSI-IN**, [PART-ANSI-MM] for File name.

16) Click **Save**.

Close the Part template.

17)   Click **File**, **Close** from the Menu bar.

Create a New part.

18)   Click **New** ⬚ from the Menu bar.

19)   Click the **SW-TUTORIAL-2020\MY-TEMPLATES** tab. Note: Additional templates are displayed.

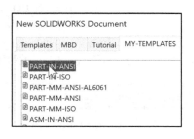

20)   Double-click **PART-ANSI-IN**, [PART-ANSI-MM]. The Part FeatureManager is displayed.

Save the part. Enter name and description.

21)   Click **Save As** 💾.

22)   Select the **SW-TUTORIAL-2020** folder.

23)   Enter **WEIGHT** for File name.

24)   Enter **WEIGHT** for Description.

25)   Click **Save**.

Insert Plane1. Display and Isometric view.

26)   Click the **Isometric view** 📦 icon.

27)   Right-click **Top Plane** from the FeatureManager.

28)   Click **Show**. The Top Plane is displayed in the Graphics window.

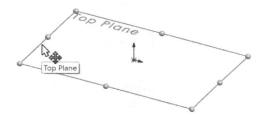

29)   Hold the **Ctrl** key down.

30)   Click the **boundary** of the Top Plane as illustrated.

31)   Drag the **mouse pointer** upward.

32)   Release the **mouse pointer**.

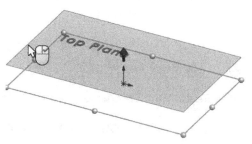

33)   Release the **Ctrl** key. The Plane PropertyManager is displayed. Top Plane is displayed in the First Reference box.

🔆 **Add relations, then dimensions.** This keeps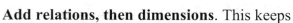
the user from having too many unnecessary dimensions.
This also helps to show the design intent of the model. Dimension what geometry you intend to modify or adjust.

34)   Enter **.500**in, [**12.70**] for Distance.

35)   Click **OK** ✔ from the Plane PropertyManager.

Plane1 is displayed in the Graphics window and is listed in the FeatureManager. Plane1 is offset from the Top Plane.

A Lofted feature requires two sketches. The first sketch, Sketch1, is a rectangle sketched on the Top Plane centered about the

Origin └─. The second sketch, Sketch2, is a square sketched on Plane1 centered about the Origin.

Create Sketch1 in the Top Plane.

36)   Right-click **Top Plane** from the FeatureManager.

37)   Click **Sketch** 🗆 from the Context toolbar. The Sketch toolbar is displayed.

38)   Click **Center Rectangle** 🗆 from the Consolidated Sketch tool. The Center Rectangle icon is displayed.

39)   Click the **Origin** └─.

40)   Click a **position** to the top right as illustrated.

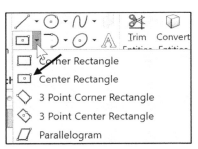

💡 The Center Rectangle tool provides the ability to sketch a rectangle located at a center point, in this case the Origin. This eliminates the need for centerlines to the Origin with a Midpoint geometric relation.

💡 CommandManager and FeatureManager tabs and tree folders will vary depending on system setup and Add-ins.

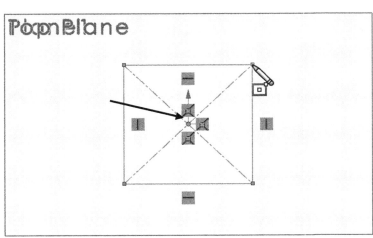

Add dimensions.

41) Click the **Smart Dimension** 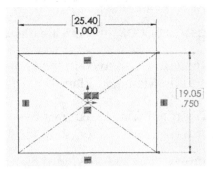 Sketch tool.

42) Click the **top horizontal** line.

43) Click a **position** above the line.

44) Enter **1.000**in, [25.40].

45) Click the **right vertical** line.

46) Click a **position** to the right.

47) Enter **.750**in, [19.05].

48) Click the **Green Check mark** ✔ .

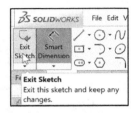

Close Sketch1.
49) Click **Exit Sketch** from the Sketch toolbar. The sketch is fully defined and is displayed in black.

Rename Sketch1.
50) Rename **Sketch1** to **Sketch-Rectangle**.

Save the part.
51) Click **Save** 💾 .

Display an Isometric view.
52) Click **Isometric view** 🔲 from the Heads-up View toolbar.

Create Sketch2 on Plane1. Plane1 is your Sketch plane.
53) Right-click **Plane1** from the FeatureManager. Plane1 is your Sketch plane.

54) Click **Sketch** 🖉 from the Context toolbar. The Sketch toolbar is displayed.

55) Click the **Center Rectangle** ⬛ Consolidated Sketch tool. The Center Rectangle icon is displayed.

56) Click **Top View**.

57) Click the **Origin** ⌐ .

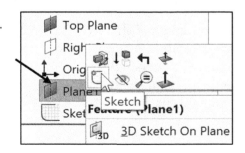

58) Click a **position** as illustrated.

59) Right-click **Select** to de-select the Center Rectangle tool.

If needed, add an Equal relation between the left vertical line and the top horizontal line.
60) Click the **left vertical line** of the rectangle.

61) Hold the **Ctrl** key down.

62) Click the **top horizontal line** of the rectangle.

63) Release the **Ctrl** key.

64) Click **Equal** = from the Add Relations box.

65) Click **OK** ✔ from the Properties PropertyManager.

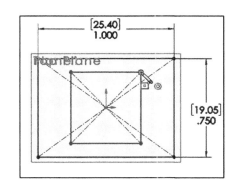

Add a dimension.
66) Click the **Smart Dimension** ✎ Sketch tool.

67) Click the **top horizontal** line. Click a **position** above the line.

68) Enter **.500**in, [**12.70**]. Click the **Green Check mark** ✔ . The sketch is displayed in black.

Close Sketch2. Display an Isometric view.
69) Click **Exit Sketch** from the Sketch toolbar. Sketch2 is fully defined.

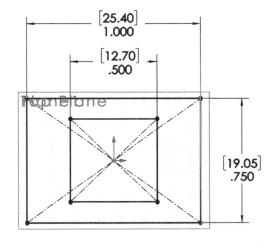

70) Click **Isometric view** 🔲 from the Heads-up View toolbar. View the results in the Graphics window.

If you did not select the Origin, insert a Coincident relation between the rectangle and the Origin to fully define Sketch2.

☀ Think design intent. When do you use the various End Conditions and Geometric sketch relations? What are you trying to do with the design? How does the component fit into an assembly?

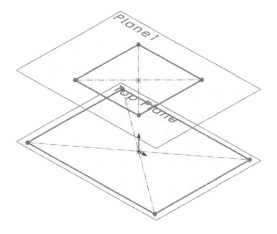

Rename Sketch2.
71)   Rename **Sketch2** to **Sketch-Square**.

Save the WEIGHT part.

72)   Click **Save** 💾.

🔅 Lofted features are comprised of multiple sketches. Name sketches for clarity.

Activity: WEIGHT Part - Lofted Feature

Insert a Lofted feature.

73)   Click the **Features** tab from the CommandManager.

74)   Click the **Lofted Boss/Base** 🪣 Feature tool. The Loft PropertyManager is displayed.

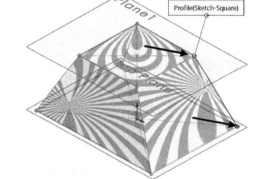

75)   Clear the **Profiles** box.

76)   Click the **back right corner** of Sketch-Rectangle as illustrated.

77)   Click the **back right corner** of Sketch-Square. Sketch-Rectangle and Sketch-Square are displayed in the Profiles box.

78)   Click **OK** ✔ from the Loft PropertyManager. Loft1 is displayed in the FeatureManager.

79)   **Hide** all planes.

🔅 To display the Selection Filter toolbar, right-click in the Graphics window, and click the **Selection Filters** drop-down menu icon. The Selection Filter is displayed.

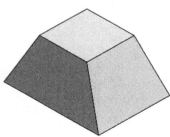

🔅 To clear a Filter icon 🔻, click **Clear All Filters** from the Selection Filter toolbar.

80) **Expand** Loft1 in the FeatureManager. Sketch-Rectangle and Sketch-Square are the two sketches that contain the Loft feature.

81) **Zoom in** on the Loft1 feature.

Activity: WEIGHT Part - Instant3D - Extruded Cut Feature

Insert a New sketch for the Extruded Cut feature.

82) Right-click the **top square face** of the Loft1 feature for the Sketch plane.

83) Click **Sketch** from the Context toolbar. The Sketch toolbar is displayed.

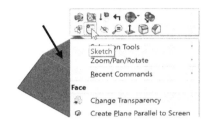

84) Click the **Circle** Sketch tool. The circle icon is displayed.

85) Click the **center** as illustrated.

86) Click a **position** to the right as illustrated.

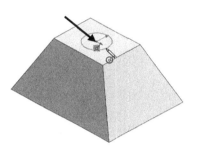

Add a dimension.

87) Click the **Smart Dimension** Sketch tool.

88) Click the **circumference** of the circle.

89) Click a **position** in the Graphics window above the circle to locate the dimension.

90) Enter **.150**in, [3.81] in the Modify box.

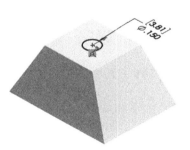

Insert an Extruded feature using the Instant3D tool.

91) **Exit** the Sketch. By default, Instant3D is active.

92) Click the **diameter** of the circle, Sketch3, as illustrated.

93) Click the **Arrowhead** and drag it below the model.

94) Click a **position** on the Instant3D ruler. The Extrude feature is displayed in the FeatureManager.

Display Wireframe style.

95) Click **Wireframe** from the Heads-up View toolbar. View the Extrude feature.

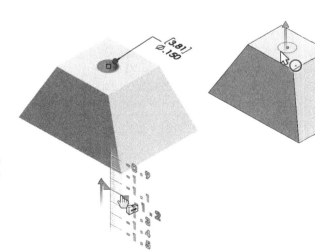

Rename the Extrude feature.

96)   Rename the **Extrude** feature to **Hole-for-Hook**.

Display an Isometric view. Display Shaded With Edges. Save the WEIGHT part.

97)   Click **Isometric view** 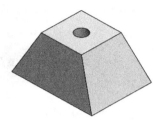 from the Heads-up View toolbar.

98)   Click **Shaded With Edges** from the Heads-up View toolbar.

99)   Click **Save** . The WEIGHT part is complete. Later, apply material to the part.

 Review the WEIGHT Part

The WEIGHT part was created with the Loft feature. The Loft feature required two planes: Top Plane and Plane1. Profiles were sketched on each plane. Profiles were selected to create the Loft feature.

An Extruded Cut feature was created using the Instant3D tool to create a Through All center hole in the WEIGHT.

HOOK Part

The HOOK part fastens to the WEIGHT. The HOOK is created with a Swept Base feature.

The Swept Base feature  adds material by moving a profile along a path.

The Swept Base feature requires two sketches (path and profile) or a sketch path and a circular profile diameter. If the sketch profile is a circle, enter the circular profile diameter in the Swept PropertyManager.

For non-circular sketch profiles, create the sketch on a perpendicular plane to the path and use the pierce relation to locate the profile on the path.

Create the HOOK part with a Swept Base feature.

The Swept Base feature uses:

- A path sketched on the Right Plane.

- A circular profile diameter.

Utilize the Dome feature 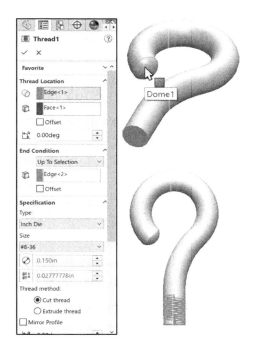 tool to create a spherical feature on a circular face.

Utilize the Thread feature tool to create a right-hand #8-36 thread for the HOOK. The Thread tool can add or removes material.

Reference geometry defines the shape or form of a surface or a solid. Reference geometry includes planes, axes, coordinate systems and points.

Activity: Create the HOOK Part

Create the New part.

100) Click **New** from the Menu bar.

101) Select the **SW-TUTORIAL-2020\MY-TEMPLATES** tab. Additional templates are displayed.

102) Double-click **PART-ANSI-IN**, [PART-ANSI-MM].

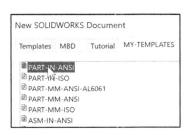

Save the part. Enter name. Enter description.

103) Click **Save As**.

104) Select the **SW-TUTORIAL-2020** folder.

105) Enter **HOOK** for File name.

106) Enter **HOOK** for Description.

107) Click **Save**. The HOOK FeatureManager is displayed.

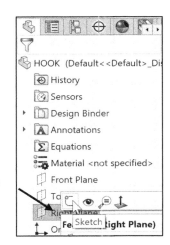

If the sketch profile is a circle, enter the diameter in the Swept PropertyManager. For non-circular sketch profiles, create the sketch on a perpendicular plane to the path and use the pierce relation to locate the profile on the path.

Sketch1 is the Sweep Path sketched on the Right Plane.

Sketch the Sweep Path.
108) Right-click **Right Plane** from the FeatureManager.

109) Click **Sketch** from the Context toolbar.

110) Click the **Line** Sketch tool. The Insert Line PropertyManager is displayed.

111) Sketch a **vertical line** from the Origin as illustrated.

Add a dimension.
112) Click the **Smart Dimension** Sketch tool.

113) Click the **vertical line**.

114) Click a **position** to the right.

115) Enter .**250**in, [**6.35**].

116) Click the **Green Check mark** .

Fit the model to the Graphics window.
117) Press the **f** key.

Create the Centerpoint arc.
118) Click the **Centerpoint Arc** Sketch tool from the Consolidated Sketch toolbar. The Centerpoint Arc icon is displayed.

119) Click the **arc center point** vertically aligned to the Origin as illustrated.

120) Click the **arc start point** as illustrated.

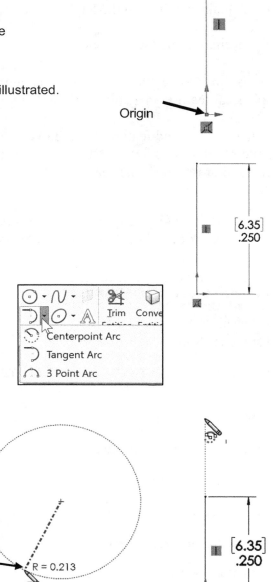

Origin

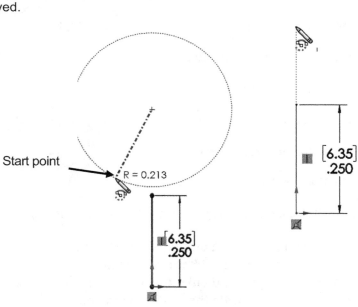

Start point

R = 0.213

121) Move (do not drag) the **mouse pointer** clockwise approximately 270°.

122) Click a point **horizontally aligned** to the arc start point. If needed add a horizontal relationship.

123) Click the **3 Point Arc**  Sketch tool from the Consolidated Sketch toolbar. The Arc PropertyManager is displayed.

124) Click the **vertical line** endpoint.

125) Click the **center point arc** endpoint.

126) Drag and pull the center of the **3 Point Arc downwards**.

127) Click the center of the **center point arc line** as illustrated.

128) Click **OK** ✔ from the Arc PropertyManager.

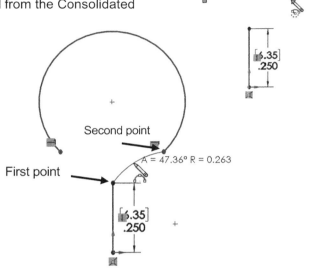

🔅 It is important to draw the correct shape with the 3 Point Arc tool as illustrated.

Add a Vertical relation between the Origin and the center point of the arc.

129) Click the **Origin** ⌞.

130) Hold the **Ctrl** key down.

131) Click the **center point** of the Center point arc.

132) Release the **Ctrl** key.

133) Click **Vertical** | from the Add Relations box.

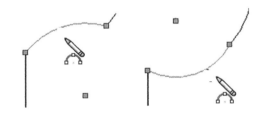

Correct shape   **Incorrect shape**

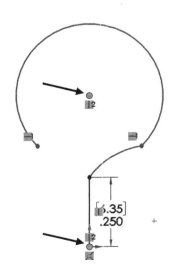

If needed, add a Horizontal relation.
134) Click the **start point** of the Center point arc.

135) Hold the **Ctrl** key down.

136) Click the **end point** of the Center point arc.

137) Release the **Ctrl** key.

138) Click **Horizontal** ▬ from the Add Relations box.

Add a Tangent relation.
139) Click the **vertical line**.

140) Hold the **Ctrl** key down.

141) Click the **3 Point Arc**.

142) Release the **Ctrl** key.

143) Click **Tangent** ⌒ from the Add Relations box.

Add a second Tangent relation.

144) Click the **3 Point Arc**.

145) Hold the **Ctrl** key down.

146) Click the **Center point** arc.

147) Release the **Ctrl** key.

148) Click **Tangent** ⌒ from the Add Relations box.

Add dimensions.
149) Click the **Smart Dimension** ⟋ Sketch tool.

150) Click the **3 Point Arc**.

151) Click a **position** to the left.

152) Enter **.500**in, [**12.70**].

153) Click the **Green Check mark** ✓ .

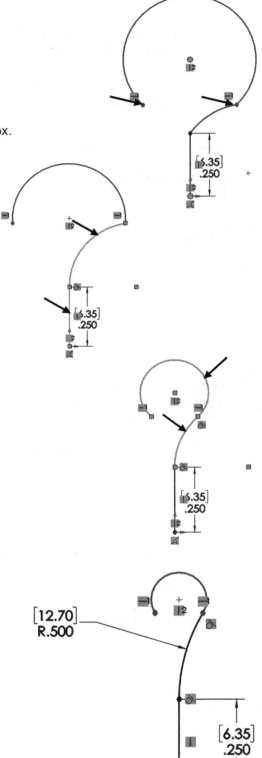

Dimension the overall length of the sketch.
154) Click the **top of the arc**.

155) Click the **Origin** ⌞.

156) Click a **position** to the right of the profile. Accept the default dimension.

157) Click the **Green Check mark** ✔.

Modify the overall length.
158) Double-click the default **dimension**.

159) Enter **1.000**in, [**25.40**].

160) Click the **Green Check mark** ✔.

Fit the model to the Graphics window.
161) Press the **f** key.

162) Move the **dimensions** as illustrated.

By default, the Dimension tool utilizes the center point of an arc or circle. Select the circle profile during dimensioning. Utilize the Leaders tab in the Dimension PropertyManager to modify the arc condition to Minimum or Maximum.

Close the sketch.
163) Click **Exit Sketch** from the Sketch toolbar.

Rename Sketch1.
164) Rename **Sketch1** to **Sketch-Path**.

Save the HOOK.

165) Click **Save** 💾.

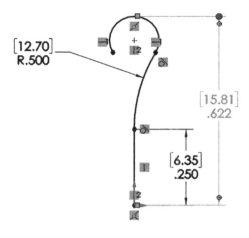

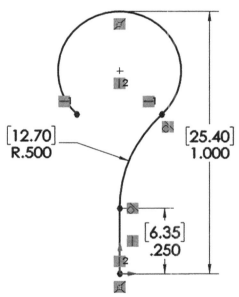

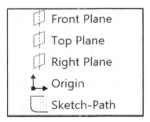

Activity: HOOK Part - Swept Base Feature

Insert the Swept feature.

166) Click the **Swept Boss/Base** 🔩 Features tool. The Sweep PropertyManager is displayed.

167) Select **Circular Profile**.

168) Click **Sketch-Path** from the fly-out FeatureManager. Sketch-Profile is displayed in the Profile box.

169) Enter **0.150**in diameter.

170) Click **OK** ✔ from the Sweep PropertyManager. Sweep1 is displayed in the FeatureManager.

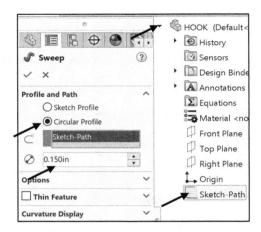

Save the HOOK part.

171) Click **Save** 💾.

Activity: HOOK Part - Dome Feature

Insert a Dome feature.

172) **Rotate** the model with the middle mouse button.

173) Click the **flat face** of the Sweep1 feature in the Graphics window as illustrated.

174) Click the **Dome** 🔵 Features tool (Insert, Features, Dome). The Dome PropertyManager is displayed. Face<1> is displayed in the Parameters box.

175) Enter **.050**in, [**1.27**] for Distance.

176) Click **OK** ✔ from the Dome PropertyManager. Dome1 is displayed in the FeatureManager.

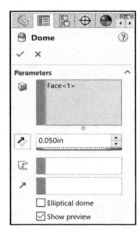

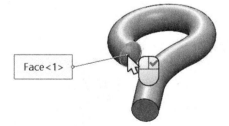

Face<1>

The HOOK requires a simplified right-hand #8-36 thread from the bottom of the Sweep1 feature. Utilize the Thread feature. The Thread feature provides the ability to create helical threads on cylindrical faces using profile sketches. Store custom thread profiles as library features.

The Thread tool Type and Size profiles are nominal thread profiles only. Do not use them for production-quality threads. To create production-quality threads, modify the nominal profiles to meet your design requirements.

Define the start thread location, specify an offset, set end conditions, specify the type, size, diameter, pitch and rotation angle, and choose options such as right-hand or left-hand thread.

Activity: HOOK Part - Thread Feature

Create a right-hand #8-36 thread from the bottom of the Sweep1 feature.

177) Click **Thread** from the Features toolbar.

178) Click **OK**. The Thread PropertyManager is displayed.

179) Click the **bottom circular edge** of Sweep1. Edge <1> is displayed in the Edge of cylinder box.

180) Click inside the **Optional start location** box.

181) Click the **bottom face**. Face<1> is displayed.

Select End Condition.

182) Select **Up To Selection** for End Condition.

183) Click inside the **End Condition** box.

184) Click the **ending edge** as illustrated.
Edge<2> is displayed in the End Condition
box.

Set Thread Specification.

185) Click **Inch Die** from the drop-down menu.

Set Thread Size.

186) Click **#8-36** for size from the drop-down
menu. View the Override diameter and
Override pitch number.

Set the Thread method.

187) Click the **Cut thread** box.

Create a Right-hand thread.

188) Click **Right-hand thread** as illustrated.

189) Click the **Trim with start face** box.

190) Click **OK** ✔ from the Thread
PropertyManager. Thread1 is displayed in
the FeatureManager.

Apply Material.

191) Right-click the **Material** folder in the
FeatureManager.

192) Click **Edit Material**. The Material dialog box
is displayed.

193) Apply **Plain Carbon Steel**.

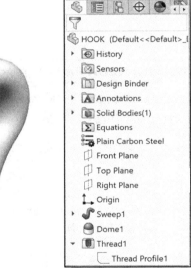

Display an Isometric view. Save the model.

194) Click **Isometric view** 🔲 from the Heads-up
View toolbar.

195) Click **Save** 💾. The HOOK part is finished.

🔆 Utilize the new Search feature in the Material
dialog box to quickly locate the desired material.

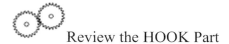 Review the HOOK Part

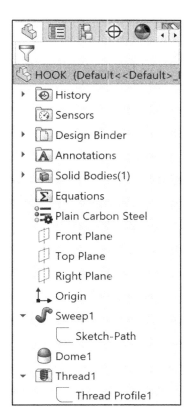

The HOOK part was created with a Swept, Dome, and Thread feature. A Swept Base feature added material by moving a profile along a path. The Swept Base feature requires two sketches (path and profile) or a sketch path and a circular profile diameter.

If the sketch profile is circular, enter the diameter in the Swept PropertyManager. For non-circular sketch profiles, create the sketch on a perpendicular plane to the path and use the pierce relation to locate the profile on the path.

The Dome feature created a spherical face on the end of the Swept Base feature.

The Thread feature provides the ability to create helical threads on cylindrical faces using profile sketches. Store custom thread profiles as library features.

The Thread tool Type and Size profiles are nominal thread profiles only. Do not use them for production quality threads. To create production quality threads, modify the nominal profiles to meet your design requirements.

If you modify a document property from an Overall drafting standard, a modify message is displayed as illustrated.

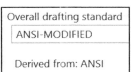

When you create a new part or assembly, the three default Planes (Front, Right and Top) are aligned with specific views. The Plane you select for the Base sketch determines the orientation of the part.

WHEEL Part

The WHEEL part is a machined part.

Create the WHEEL part with the Extruded Boss/Base feature 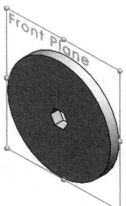 tool. Utilize the Mid Plane option to center the WHEEL on the Front Plane.

Utilize the Revolved Cut feature tool to remove material from the WHEEL and to create a groove for a belt.

The WHEEL contains a complex pattern of holes. Apply the Extruded Cut feature tool.

Simplify the geometry by dividing the four holes into two Extruded Cut features.

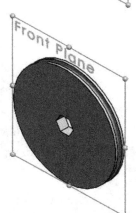

The first Extruded Cut feature contains two small circles sketched on two bolt circles. The bolt circles utilize Construction geometry.

🔆 Utilize the Hole Wizard feature when creating non-Through All complex geometry holes.

The second Extruded Cut feature ▣ utilizes two small circles sketched on two bolt circles. The bolt circles utilize Construction geometry.

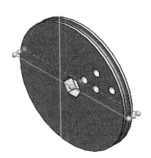

Utilize the Circular Pattern Feature ⬡ tool. The two Extruded Cut features are contained in the Circular Pattern. Revolve the Extruded Cut features about the Temporary Axis located at the center of the Hexagon.

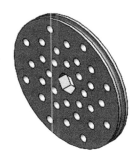

Create a Reference Axis. The Reference Axis is utilized in the WHEEL-AXLE assembly.

Construction geometry is used only to assist in creating the sketch entities and geometry that are ultimately incorporated into the part. Construction geometry is ignored when the sketch is used to create a feature. Construction geometry uses the same line style as centerlines.

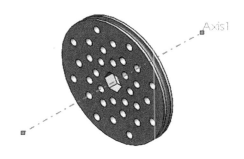

🔆 You can utilize the Hole Wizard feature tool instead of the Cut-Extrude feature tool, or use the Instant3D tool to create a Through All hole for any part. See SOLIDWORKS Help for additional information.

🔆 Slots are available in the Hole Wizard. Create regular slots as well and counterbore and countersink slots. You also have options for position and orientation of the slot. If you have hardware already mated in place, the mates will not be broken if you switch from a hole to a slot.

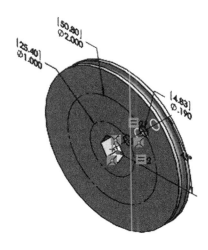

Activity: WHEEL Part

Create the New part.

196) Click **New** ⬚ from the Menu bar.

197) Click the **SW-TUTORIAL-2020\MY-TEMPLATES** tab. Additional templates are displayed.

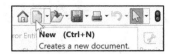

198) Double-click **PART-ANSI-IN**, [PART-ANSI-MM].

Save the part. Enter name. Enter description.

199) Click **Save As** 🖫.

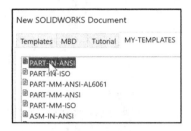

200) Select the **SW-TUTORIAL-2020** folder.

201) Enter **WHEEL** for File name.

202) Enter **WHEEL** for Description.

203) Click **Save**. The WHEEL FeatureManager is displayed.

Insert the sketch for the Extruded Base feature.
204) Right-click **Front Plane** from the FeatureManager.

205) Click **Sketch** ✏ from the Context toolbar. The Sketch toolbar is displayed.

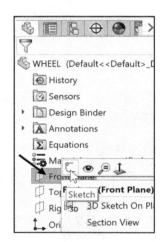

206) Click the **Circle** ⊙ Sketch tool. The Circle PropertyManager is displayed.

207) Click the **Origin** ↳ as illustrated.

208) Click a **position** to the right of the Origin.

Insert a polygon.
209) Click the **Polygon** ⬡ Sketch tool. The Polygon PropertyManager is displayed.

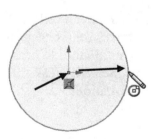

210) Click the **Origin** ↳.

211) Drag and click the **mouse pointer** horizontally to the right of the Origin to create the hexagon as illustrated.

212) Click **OK** ✔ from the Polygon PropertyManager.

De-select the Polygon Sketch tool.
213) Right-click **Select**.

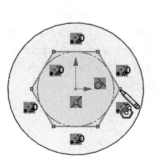

Add a Horizontal relation.

214) Click the **Origin** ⌐.

215) Hold the **Ctrl** key down.

216) Click the **right point** of the hexagon.

217) Release the **Ctrl** key.

218) Click **Horizontal** ▬ from the Add Relations box.

219) Click **OK** ✔ from the Properties PropertyManager.

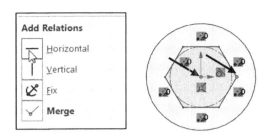

Add dimensions.

220) Click the **Smart Dimension** ✎ Sketch tool. Click the **circumference** of the large circle.

221) Click a **position** above the circle. Enter **3.000**in, [**76.20**].

222) Click the **Green Check mark** ✔.

223) Click the **circumference** of the inscribed circle for the Hexagon.

224) Click a **position** above the Hexagon.

225) Enter **.438**in, [**11.13**].

226) Click the **Green Check mark** ✔.

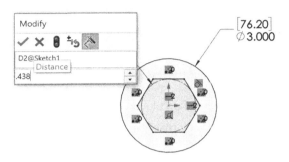

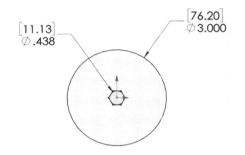

Activity: WHEEL Part - Extruded Boss/Base Feature

Insert an Extruded Boss/Base feature.

227) Click **Extruded Boss/Base** 🗔 from the Features toolbar. The Boss-Extrude PropertyManager is displayed.

228) Select **Mid Plane** for End Condition in Direction 1.

229) Enter **.250**in, [**6.35**] for Depth.

230) Click **OK** ✔ from the Boss-Extrude PropertyManager. Boss-Extrude1 is displayed in the FeatureManager.

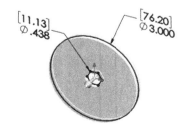

Fit the model to the Graphics window.
231) Press the **f** key.

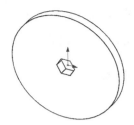

Display Hidden Lines Removed. Save the WHEEL part.
232) Display **Hidden Lines Removed**.

233) Click **Save** .

Activity: WHEEL Part - Revolved Cut Feature

Insert a new sketch for the Revolved Cut feature.
234) Right-click **Right Plane** from the FeatureManager.

235) Click **Sketch** ⌗ from the Context toolbar. The Sketch toolbar is displayed.

236) Click **Right view** 🔲 from the Heads-up View toolbar.

Sketch the axis of revolution.
237) Click the **Centerline** ✐ Sketch tool from the Consolidated Sketch toolbar. The Insert Line PropertyManager is displayed.

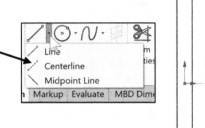

238) Click the **Origin** ⌊.

239) Click a **position** horizontally to the right of the Origin as illustrated.

De-select the sketch tool.
240) Right-click **Select**.

241) **Zoom in** on the top edge.

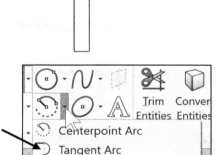

Sketch the profile.
242) Click the **Line** ✐ Sketch tool.

243) Sketch the **first vertical line** as illustrated.

244) Click the **Tangent Arc** ⌒ Sketch tool. The Arc PropertyManager is displayed.

245) Click the **end point** of the vertical line.

246) Sketch a **180° arc** as illustrated.

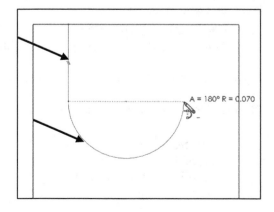

De-select the sketch tool.
247) Right-click **Select** in the Graphics window.

248) Click the **Line** 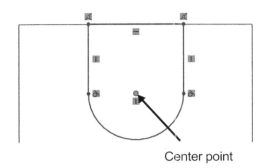 Sketch tool.

249) Sketch the **second vertical line** as illustrated. The end point of the line is Coincident with the top horizontal edge of Extrude1.

250) Sketch a **horizontal line** collinear with the top edge to close the profile.

Add a Vertical relation.
251) Right-click **Select** in the Graphics window.

252) Click the **Origin** from the FeatureManager.

253) Hold the **Ctrl** key down.

254) Click the **center point** of the arc. Release the **Ctrl** key.

255) Click **Vertical** from the Add Relations box.

Add an Equal relation.
256) Click the **left vertical** line.

257) Hold the **Ctrl** key down.

258) Click the **right vertical** line.

259) Release the **Ctrl** key.

260) Click **Equal** from the Add Relations box.

Add dimensions.
261) Click the **Smart Dimension** Sketch tool.

262) Click the **arc**.

263) Click a position to the **left** of the profile.

264) Enter .063in, [**1.6**]. Click the **Green Check mark** .

265) Click the **right vertical** line.

266) Click a position to the **right** of the profile.

267) Enter .078in, [**1.98**]. Click the **Green Check mark** . The sketch should be fully defined.

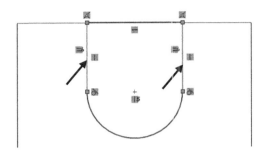

Center point

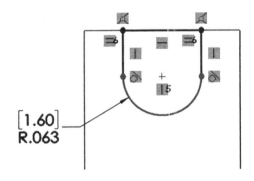

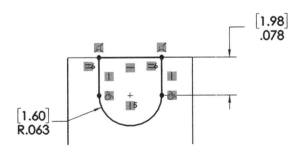

Fit the model to the Graphics window.
268) Press the **f** key.

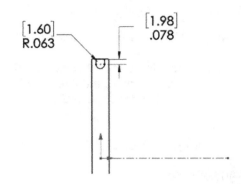

De-select the sketch tool.
269) Right-click **Select** in the Graphics window.

Activity: WHEEL Part - Revolved Cut Feature

Insert a Revolved Cut feature.
270) Select the Axis of Revolution. Click the **centerline** in the Graphics window as illustrated.

271) Click **Revolved Cut** from the Features toolbar. The Cut-Revolve PropertyManager is displayed. The Cut-Revolve PropertyManager displays 360 degrees for Direction 1 Angle.

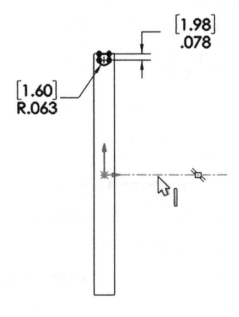

272) Click **OK** from the Cut-Revolve PropertyManager. Cut-Revolve1 is displayed in the FeatureManager.

Save the WHEEL part.
273) Click **Save** .

Four bolt circles, spaced 0.5in, [12.7] apart, locate the 8 - ∅.190, [4.83] holes. Simplify the situation. Utilize two Extruded Cut features on each bolt circle.

Position the first Extruded Cut feature hole on the first bolt circle and third bolt circle.

Position the second Extruded Cut feature hole on the second bolt circle and fourth bolt circle.

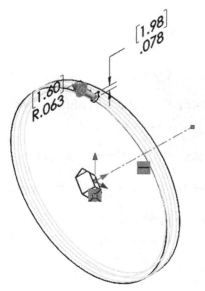

Activity: WHEEL Part - First Extruded Cut Feature

Display the Top Plane.
274) Right-click **Top Plane** from the FeatureManager.

275) Click **Show** 👁 from the Context toolbar.

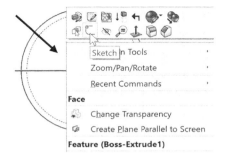

Display a Front view - Hidden Lines Visible.
276) Click **Front view** 🔲 from the Heads-up View toolbar.

277) Click **Hidden Lines Visible** 🔲 from the Heads-up View toolbar.

Insert a new sketch for the first Extruded Cut feature.
278) Right-click the **Boss-Extrude1 front face** as illustrated.

279) Click **Sketch** ⌐ from the Context toolbar. The Sketch toolbar is displayed.

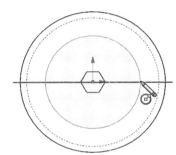

Create the first construction bolt circle.
280) Click the **Circle** ⊙ Sketch tool. The Circle PropertyManager is displayed.

281) Click the **Origin** ↳.

282) Click a **position** to the right of the hexagon as illustrated.

283) Check the **For construction** box.

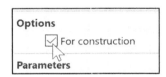

Create the second construction bolt circle.

284) Click the **Origin** ↳.

285) Click a **position** to the right of the first construction bolt circle as illustrated.

286) Check the **For construction** box. The two bolt circles are displayed with Construction style lines.

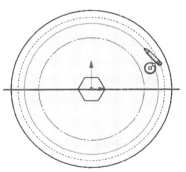

💡 Construction geometry is used only to assist in creating the sketch entities and geometry that are ultimately incorporated into the part. Construction geometry is ignored when the sketch is used to create a feature. Construction geometry uses the same line style as centerlines.

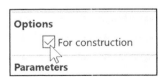

De-select the circle Sketch tool.
287) Right-click **Select**.

Insert a centerline.

288) Click the **Centerline** ⟋ Sketch tool. The Insert Line
PropertyManager is displayed.

289) Sketch a **45° centerline** (approximately) from the
Origin to the second bolt circle as illustrated.

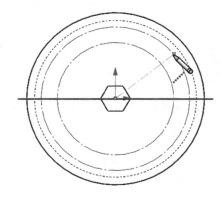

Sketch the two circle profiles.

290) Click the **Circle** ⟳ Sketch tool. The Circle
PropertyManager is displayed.

291) Sketch a **circle** at the intersection of the centerline
and the first bolt circle.

292) Sketch a **circle** at the intersection of the centerline
and the second bolt circle.

De-select the Circle Sketch tool.
293) Right-click **Select** in the Graphics window.

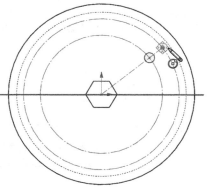

Note: An Intersection relation is created between three
entities: the center point of the small circle, the
centerline, and the bolt circle.

Add an Equal relation.
294) Click the **first circle**.

295) Hold the **Ctrl** key down.

296) Click the **second circle**.

297) Release the **Ctrl** key.

298) Right-click **Make Equal** = from the Context
toolbar.

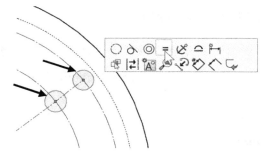

Add dimensions.
299) Click the **Smart Dimension** ⟋ Sketch
tool.

300) Click the **first construction** circle.

301) Click a **position** above the profile.

302) Enter **1.000**in, [25.4].

303) Click the **Green Check mark** ✓ .

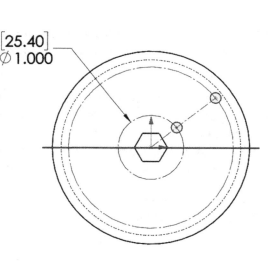

[25.40]
⌀ 1.000

304) Click the **second construction** circle.

305) Click a **position** above the profile.

306) Enter **2.000**in, [**50.80**].

307) Click the **Green Check mark** 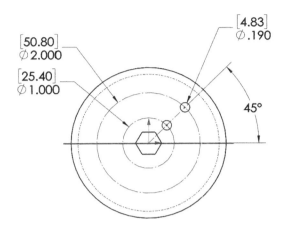.

308) Click the **second small** circle.

309) Click a **position** above the profile.

310) Enter **.190**in, [**4.83**].

311) Click the **Green Check mark** .

312) Click **Top Plane** from the fly-out FeatureManager.

313) Click the **45° centerline**.

314) Click a **position** between the two lines.

315) Enter **45**deg for angle.

316) Click the **Green Check mark** .

Note: If the sketch is not fully defined, you may need to add an Intersection relation between the center point of the small circle, the centerline, and the bolt circle.

Insert an Extruded Cut feature.
317) Click **Extruded Cut** from the Features toolbar. The Cut-Extrude PropertyManager is displayed.

318) Select **Through All** for the End Condition in Direction 1.

319) Click **OK** from the Cut-Extrude PropertyManager. Cut-Extrude1 is displayed in the FeatureManager.

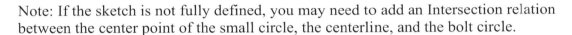

Activity: WHEEL Part - Second Extruded Cut Feature

Insert a new sketch for the second Extruded Cut feature.
320) Right-click the **Boss-Extrude1** front face.

321) Click **Sketch** from the Context toolbar.

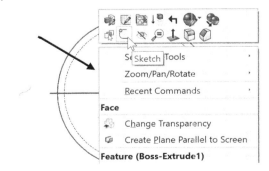

Sketch two additional Construction line bolt circles, 1.500in, [38.1] and 2.500in, [63.5]. Create the first Construction bolt circle.

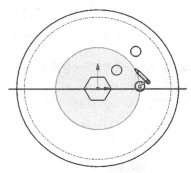

322) Click the **Circle** ⊙ Sketch tool. The Circle PropertyManager is displayed.

323) Click the **Origin** ⌞.

324) Click a **position** between the two small circles.

325) Check the **For construction** box.

Create the second additional construction bolt circle.

326) Click the **Origin** ⌞.

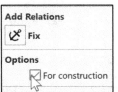

327) Click a **position** to the right of the large construction bolt circle as illustrated.

328) Check the **For construction** box from the Circle PropertyManager. The two bolt circles are displayed with the two construction lines.

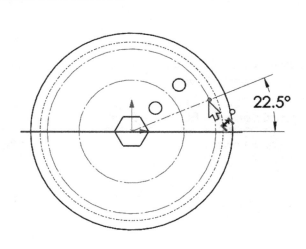

Insert a centerline.

329) Click the **Centerline** ⟋ Sketch tool. The Insert Line PropertyManager is displayed.

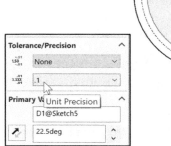

330) Sketch a **22.5° centerline** to the right from the Origin to the second bolt circle as illustrated.

331) Select **.1** from the Unit Precision box.

Display Hidden Lines Removed.

332) Click **Hidden Lines Removed** ⬜ from the Heads-up View toolbar.

Sketch the two circle profiles.

333) Click the **Circle**  Sketch tool. The Circle PropertyManager is displayed.

334) Sketch a **circle** at the intersection of the centerline and the first bolt circle.

335) Sketch a **circle** at the intersection of the centerline and the second bolt circle.

De-select the Circle Sketch tool.
336) Right-click **Select**.

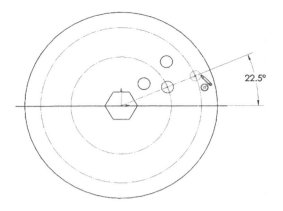

Add an Equal relation.
337) Click the **first circle**.

338) Hold the **Ctrl** key down.

339) Click the **second circle**.

340) Release the **Ctrl** key.

341) Right-click **Make Equal** = from the shortcut toolbar.

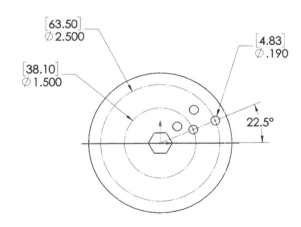

Add dimensions.
342) Click the **Smart Dimension** Sketch tool. The Smart Dimension icon is displayed.

343) Click the **first construction** circle.

344) Click a **position** above the profile.

345) Enter **1.500**in, [**38.1**].

346) Click the **second construction** circle.

347) Click a **position** above the profile.

348) Enter **2.500**in, [**63.5**].

349) Click the **small circle** as illustrated.

350) Click a **position** above the profile.

351) Enter **.190**in, [**4.83**]. The sketch should be fully defined.

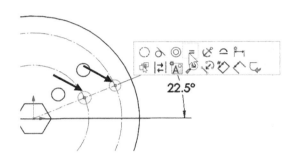

Insert an Extruded Cut feature.

352) Click **Extruded Cut** ⬜ from the Features toolbar. The Cut-Extrude PropertyManager is displayed.

353) Select **Through All** for End Condition in Direction 1.

354) Click **OK** ✔ from the Cut-Extrude PropertyManager. Cut-Extrude2 is displayed in the FeatureManager.

Save the model.

355) Click **Save** 💾 .

View the Temporary Axes.

356) Click **View**, **Hide/Show**, check **Temporary Axes** from the Menu bar.

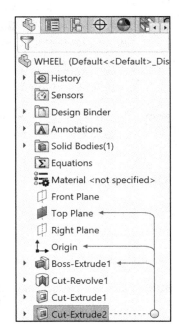

💡 The book is designed to expose the new SOLIDWORKS user to many different tools, techniques and procedures. It may not always use the most direct tool or process.

Activity: WHEEL Part - Circular Pattern Feature

Insert a Circular Pattern.

357) Click **Isometric view** 🔲 from the Heads-up View toolbar.

358) Click **Circular Pattern** 🔹 from the Consolidated Features toolbar. The Circular Pattern PropertyManager is displayed.

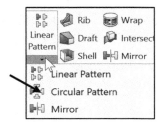

359) Click **inside** the Pattern Axis box.

360) Click the **Temporary Axis** in the Graphics window at the center of the Hexagon. Axis<1> is displayed in the Pattern Axis box.

361) Click the **Equal spacing** box.

362) Enter **360**deg for Angle.

363) Enter **8** for Number of Instances.

364) Click inside the **Features to Pattern** box.

365) Click **Cut-Extrude1** and **Cut-Extrude2** from the fly-out FeatureManager. Cut-Extrude1 and Cut-Extrude2 are displayed in the Features to Pattern box.

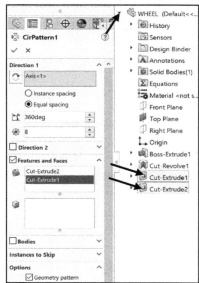

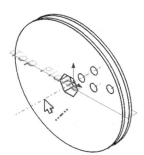

366) Check the **Geometry pattern** box.

367) Click **OK** ✓ from the Circular Pattern PropertyManager. CirPattern1 is displayed in the FeatureManager.

Save the WHEEL part.

368) Click **Save** 🖫.

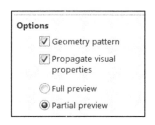

Utilize a Reference Axis to locate the WHEEL in the PNEUMATIC-TEST-MODULE assembly. The Reference Axis is located in the FeatureManager and Graphics window. The Reference Axis is a construction axis defined between two planes.

Insert a two Plane Reference axis.

369) Click the **Axis** ⟍ tool from the Reference Geometry Consolidated Features toolbar. The Axis PropertyManager is displayed.

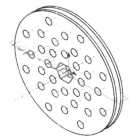

370) Click **Top Plane** from the fly-out FeatureManager.

371) Click **Right Plane** from the fly-out FeatureManager. The selected planes are displayed in the Selections box.

372) Click **Two Planes**.

373) Click **OK** ✓ from the Axis PropertyManager. Axis1 is displayed in the FeatureManager.

Axis1 is positioned through the Hex Cut centered at the Origin ⌞.

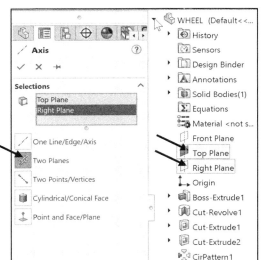

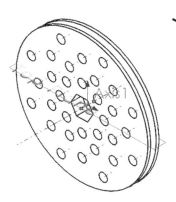

374) Click and drag the **Axis1 handles** outward to extend the length on both sides as illustrated.

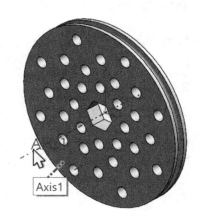

Display an Isometric view - Shaded With Edges. Clear Temporary Axes. Hide all Planes.

375) Click **Isometric view** .

376) Click **View**, **Hide/Show**, un-check **Temporary Axes** from the Menu bar. **Hide** all Planes.

377) Click **Shaded With Edges** from the Heads-up View toolbar.

Save the WHEEL part.

378) Click **Save** .

 Sketched lines, arcs or circles are modified from profile geometry to construction geometry. Select the geometry in the sketch. Check the For construction box option.

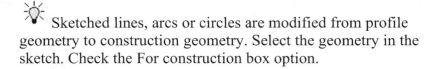 Review the WHEEL Part

The WHEEL part was created with the Extruded Boss/Base feature. You sketched a circular sketch on the Front Plane and extruded the sketch with the Mid Plane option.

A Revolved Cut feature removed material from the WHEEL and created the groove. The Revolved Cut feature utilized an arc sketched on the Right Plane. A sketched centerline was required to create the Revolved Cut feature.

The WHEEL contained a complex pattern of holes. The first Extruded Cut feature contained two small circles sketched on two bolt circles. The bolt circles utilized construction geometry. Geometric relationships and dimensions were used in the sketch. The second Extruded Cut feature utilized two small circles sketched on two bolt circles.

The two Extruded Cut features were contained in one Circular Pattern and revolved about the Temporary Axis. The Reference Axis was created with two perpendicular planes. Utilize the Reference Axis, Axis1 in the WHEEL-AXLE assembly.

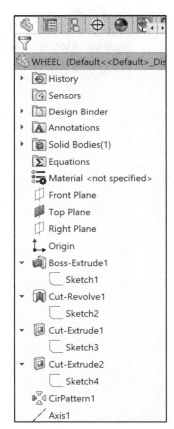

Modify a Part

Conserve design time and cost. Modify existing parts and assemblies to create new parts and assemblies. Utilize the Save as copy tool to avoid updating the existing assemblies with new file names.

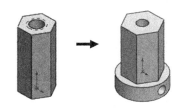

The HEX-STANDOFF part was created in a previous chapter. The HEX-ADAPTER is required to fasten the WHEEL to the AXLE. Start with the HEX-STANDOFF part.

Utilize the Save As command and enter the HEX-ADAPTER for the new file name. Check the Save as copy and continue check box. The HEX-ADAPTER is the new part name. Open the HEX-ADAPTER. Modify the dimensions of the Extruded Base feature.

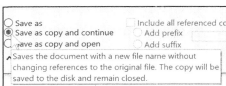

Utilize Edit Definition to modify the Hole Wizard Tap Hole to a Standard Hole. Insert an Extruded Boss/Base feature to create the head of the HEX-ADAPTER.

Insert an Extruded Cut feature. Sketch a circle on the Right Plane. Extrude the circle in Direction1 and Direction2 with the Through All End Condition option. Note: You can use the Hole Wizard feature with a 3D Sketch. Feature order determines the internal geometry of the Hole. If the Hole feature is created before the Extrude2-Head feature, the Through All End Condition will extend through the Boss-Extrude1 feature.

If the Hole feature is created after the Extrude2-Head feature, the Through All End Condition will extend through the Boss-Extrude1 feature and the Extrude2-Head feature.

Modify feature order by dragging feature names in the FeatureManager. Utilize the Save As command to create the AXLE3000 part from the AXLE part.

Utilize the Save As command to create the SHAFTCOLLAR-500 part from the SHAFT-COLLAR part. Save the HEX-STANDOFF as the HEX-ADAPTER part.

Activity: HEX-ADAPTER Part

Create the HEX-ADAPTER.

379) Click **Open** 🗁 from the Menu bar.

380) Double-click **HEX-STANDOFF**. The HEX-STANDOFF FeatureManager is displayed. Note: HEX-STANDOFF was created in a previous chapter. Click **Save As** 🖫.

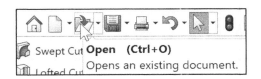

381) Select the **SW-TUTORIAL-2020** folder.

382) Enter **HEX-ADAPTER** for File name.

383) Enter **HEX-ADAPTER 10-24** for Description.

384) Check the **Save as copy and continue** box.

385) Click **Save**.

386) Click **File**, **Close** from the Menu bar.

Open the HEX-ADAPTER.

387) Click **Open** 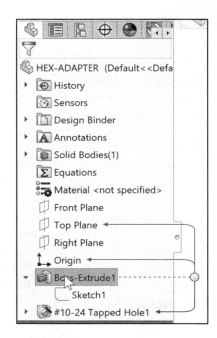 from the Menu bar.

388) Double-click **HEX-ADAPTER**. The HEX-ADAPTER FeatureManager is displayed.

Modify the Boss-Extrude1 dimensions.

389) Double-click **Boss-Extrude1** from the FeatureManager.

390) Move the **dimensions** off the model.

391) Double-click **.735**in, [**18.67**].

392) Enter **.700**in, [**17.78**] for depth.

393) Double-click **.313**in, [**7.95**].

394) Enter **.438**in, [**11.13**] for diameter.

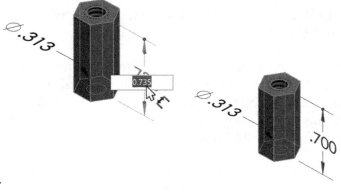

Modify the #10-24 Tapped Hole1 feature.

395) Right-click **#10-24 Tapped Hole1** from the FeatureManager.

396) Click **Edit Feature** 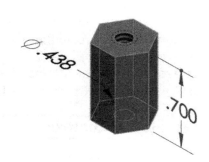 from the Context toolbar. The Hole Specification PropertyManager is displayed.

The Hole Specification PropertyManager is part of the Hole Wizard tool located in the Features toolbar.

The Type tab is selected by default.

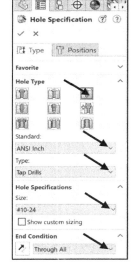

397) Select the **Hole** tab from the Hole Specification box.

398) Select **ANSI Inch** for Standard.

399) Select **Tap Drills** for Type.

400) Select **#10-24** for Size.

401) Select **Through All** for End Condition.

402) Click **OK** ✅ from the Hole Specification PropertyManager. The Tap Hole is modified.

Insert a sketch for the Extruded Boss feature.
403) **Rotate** the model to view the bottom face.

404) Right-click the **bottom hexagonal face** of the Boss-Extrude1 feature as illustrated.

405) Click **Sketch** 🗒 from the Context toolbar.

406) Click **Bottom view** 🔲 from the Heads-up View toolbar.

407) Click the **Circle** ⊙ Sketch tool. The Circle PropertyManager is displayed.

408) Click the **Origin** ⌞ as illustrated.

409) Click a **position** in the Graphics window to the right of the Origin.

Add a dimension.
410) Click the **Smart Dimension** Sketch tool.

411) Click the **circumference** of the circle.

412) Click a **position** above the circle to locate the dimension.

413) Enter **.625**in, [**15.88**] in the Modify dialog box.

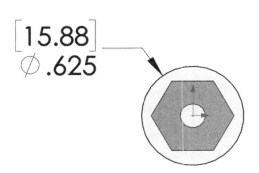

Fit the model to the Graphics window.
414) Press the **f** key.

Activity: HEX-ADAPTER Part - Extruded Boss/Base Feature

Extrude the sketch to create the Extruded Boss/Base feature.

415) Click **Isometric view** from the Heads-up View toolbar.

416) Click **Extruded Boss/Base** from the Features toolbar. The Boss-Extrude PropertyManager is displayed.

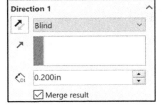

417) Enter **.200**in, **[6.35]** for Depth. The Direction arrow points downward. Flip the **Direction arrow** if required.

418) Click **OK** from the Boss-Extrude PropertyManager. Boss-Extrude2 is displayed in the FeatureManager.

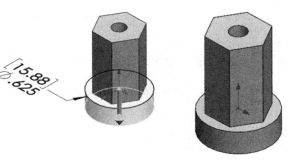

Rename Boss-Extrude2. Save the model.
419) Rename **Boss-Extrude2** to **Extrude2-Head**.

420) Click **Save**.

Rename a feature or sketch for clarity. Slowly click the feature or sketch name twice and enter the new name when the old one is highlighted.

Activity: HEX-ADAPTER Part - Extruded Cut Feature

Insert a new sketch for the Extruded Cut on the Right Plane.
421) Right-click **Right Plane** from the FeatureManager.

422) Click **Sketch** from the Context toolbar.

423) Click **Right view** from the Heads-up View toolbar. Note the location of the Origin.

424) Click **Hidden Lines Visible**.

425) Click the **Circle** Sketch tool. The Circle PropertyManager is displayed.

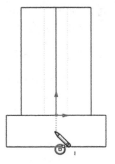

426) Sketch a **circle** below the Origin . The center point is vertically aligned to the Origin (vertical relation).

427) If required, add a **Vertical relation** between the center point of the circle and the Origin.

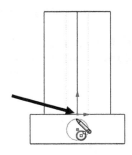

Add dimensions.

428) Click the **Smart Dimension** 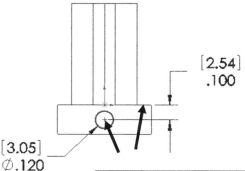 Sketch tool.

429) Click the **middle horizontal** edge.

430) Click the **center point** of the circle.

431) Click a **position** to the right of the profile.

432) Enter **.100**in, [**2.54**].

433) Click the **Green Check mark** .

434) Click the **circumference** of the circle.

435) Click a **position** below the profile.

436) Enter **.120**in, [**3.95**].

437) Click the **Green Check mark** .

Insert an Extruded Cut feature.

438) Click **Extruded Cut** from the Features toolbar. The Cut-Extrude PropertyManager is displayed.

439) Select **Through All - Both** for End Condition in Direction 1.

440) Click **OK** from the Cut-Extrude PropertyManager.

Display an Isometric view. Rename the feature.

441) Click **Isometric view** .

442) Rename the Cut-Extrude# feature to **Extrude3-SetScrew**.

Save the HEX-ADAPTER part.

443) Click **Save** .

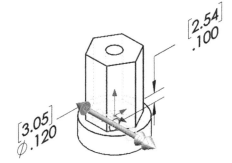

The Through All End Condition is required to penetrate both the Boss-Extrude1 and Extrude2 features. Reorder features in the FeatureManager. Position the Extrude2 feature before the Tap Drill for # 10-24 Tap 1 feature in the FeatureManager.

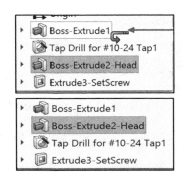

Reorder the Features.
444) Click and drag **Extrude2-Head** from the FeatureManager upward as illustrated.

445) Click a **position** below Boss-Extrude1. The Through All End Condition option for the Tap Drill for # 10-24 Tap 1 feature creates a hole through both Boss-Extrude1 and Boss-Extrude2.

Display a Section view.
446) Click **Front Plane** from the FeatureManager.

447) Click **Section view** 📖 from the Heads-up View toolbar in the Graphics window. The Section View PropertyManager is displayed. View the results.

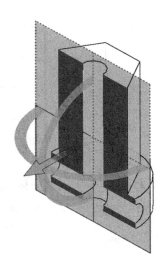

448) Click **OK** ✔ from the Section View PropertyManager.

Display the full view.
449) Click **Section view** 📖 from the Heads-up View toolbar in the Graphics window.

450) Click **Shaded With Edges** 🔲 from the Heads-up View toolbar.

Save the HEX-ADAPTER.
451) Click **Save** 💾. Note the location of the Origin in the model.

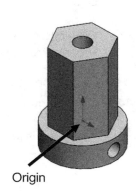

Origin

Close all documents.
452) Click **Windows**, **Close All** from the Menu bar.

🔆 Utilize the Save As command and work on the copied version of the document before making any changes to the original. Keep the original document intact.

🔆 Tangent edges and Origins are displayed for educational purposes in this book.

Review the HEX-ADAPTER Part

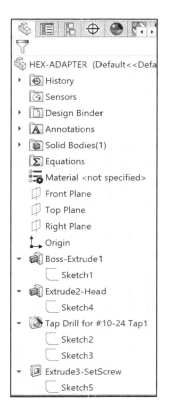

The HEX-ADAPTER part was created by utilizing the Save As and the Save as copy command with the HEX-STANDOFF part. The Boss-Extrude1 feature dimensions were modified. Edit Definition was utilized to modify the Hole type from the Hole Wizard feature.

An Extruded Boss feature added material. An Extruded Cut feature, sketched on the Right Plane with the Through All End Condition for both Direction1 and Direction2, created a hole through the Extruded Boss feature. Reordering features in the FeatureManager modified the Hole. Utilizing existing geometry saved time with the Save as copy command. The original part and its references to other assemblies are not affected with the Save as copy command.

You require additional work before completing the PNEUMATIC-TEST-MODULE assembly. The AXLE and SHAFT-COLLAR were created in Chapter 2. Utilize the Save as copy command to save the parts.

Additional details on Save (Save As copy), Reorder (features), Section View PropertyManager are available in SOLIDWORKS Help.

Utilize Design Table configurations for the AXLE part and SHAFT-COLLAR part developed in the previous chapter.

The AXLE-3000 part and SHAFT-COLLAR-500 part utilize the Save As option in the next section.

Utilize the Save As components or the configurations developed with Design Tables to create the WHEEL-AXLE assembly.

Activity: AXLE-3000 Part

Create the AXLE-3000 part from the AXLE part.

453) Click **Open** from the Menu bar.

454) Double-click **AXLE** from the SW-TUTORIAL-2020 folder. The AXLE FeatureManager is displayed.

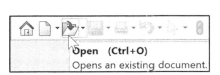

455) Click **Save As** .

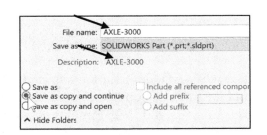

456) Select the **SW-TUTORIAL-2020** folder.

457) Enter **AXLE-3000** for File name.

458) Enter **AXLE-3000** for Description.

459) Check the **Save as copy and continue** check box. Note: You can use the Save as copy and open command. This will open the new copy.

460) Click **Save**.

Close the AXLE part.
461) Click **File**, **Close** from the Menu bar.

Open AXLE-3000 part.
462) Click **Open** from the Menu bar.

463) Double-click **AXLE-3000** from the SW-TUTORIAL-2020 folder. The AXLE-3000 FeatureManager is displayed.

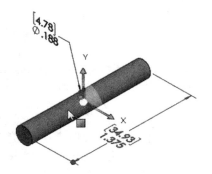

Modify the depth dimension.
464) Double-click the **cylindrical face** in the Graphics window.

465) **Move** the dimensions off the model.

466) Click **1.375**in, [**34.93**]. Enter **3.000**in, [**76.20**].

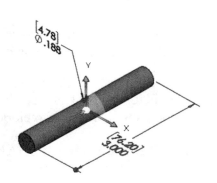

Fit the model to the Graphics window.
467) Press the **f** key.

Save the AXLE-3000 part.

468) Click **Save** .

469) Click **inside** the Graphics window.

Activity: SHAFTCOLLAR-500 Part

Create the SHAFTCOLLAR-500 part.
470) Click **Open** from the Menu bar.

471) Double-click **SHAFT-COLLAR** from the SW-TUTORIAL-2020 folder.

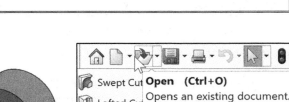

472) Click **Save As** 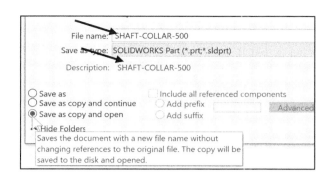.

473) Enter **SHAFT-COLLAR-500** for File name.

474) Enter **SHAFT-COLLAR-500** for Description.

475) Click the **Save as copy and open** check box.

476) Click **Save**. Both models are open at this time.

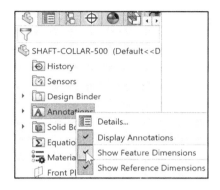

Modify the diameter dimensions of the SHAFT-COLLAR-500 Part.

477) Right-click **Annotations** in the FeatureManager.

478) Check the **Show Feature Dimensions** box.

479) Press the **f** key to fit the model to the graphics area.

480) Click **.438**in, [**11.11**].

481) Enter **.750**in, [**19.05**] for outside diameter.

482) Click **.190**in, [**4.83**].

483) Enter **.500**in, [**12.70**] for inside diameter. View the results.

484) Right-click **Annotations** in the FeatureManager.

485) Un-check the **Show Feature Dimensions** box.

486) Rebuild the model.

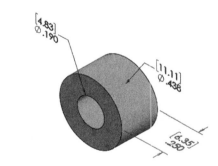

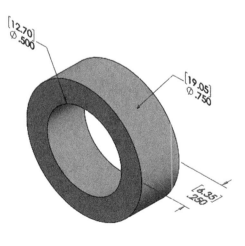

Fit the model to the Graphics window.
487) Press the **f** key.

Save the SHAFT-COLLAR-500 part.

488) Click **Save** .

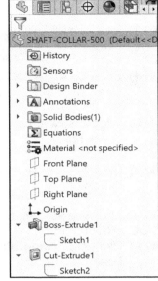

Close all documents.
489) Click **Windows**, **Close All** from the Menu
bar.

Press the **s** key in the Graphics window.
A Context Pop-up features toolbar is
displayed. The features toolbar displays the
last few feature tools applied.

Select the types of Annotations that you
want to display and set text scale and other
Annotations options. In the FeatureManager
design tree, right-click the **Annotations** folder,
and click **details**. View the options from the
Annotation Properties dialog box.

Rename a feature or sketch for clarity.
Slowly click the feature or sketch name twice and
enter the new name when the old one is
highlighted.

Add relations, then dimensions. This will
keep the user from having too many unnecessary
dimensions. This helps to show the design intent
of the model. Dimension what geometry you
intend to modify or adjust.

Chapter Summary

In this chapter, you created six parts. The WEIGHT part utilized the Plane feature, Lofted Base feature and the Extruded Cut (Instant3D tool) feature. The HOOK part utilized the Swept Base feature, Dome feature, and Thread feature. The WHEEL part utilized the Extruded Base feature, Revolved Cut feature, Extruded Cut feature, Circular Pattern feature and Axis feature.

The second three parts utilized existing parts created in an early chapter. The HEX-ADAPTER part, AXLE-3000 part and the SHAFTCOLLAR-500 part utilized existing part geometry along with the Hole Wizard feature.

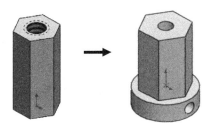

Conserve design time and cost. Modify existing parts and assemblies to create new parts and assemblies. Utilize the Save As/Save as command to save the file in another file format.

Utilize the Save as copy and continue command to save the document to a new file name without replacing the active document. Utilize the Save as copy and open command to save the document to a new file name that becomes the active document. The original document remains open. References to the original document are not automatically assigned to the copy.

You applied the following Sketch tools in this chapter: Circle, Line, Centerline, Tangent Arc, Polygon, Smart Dimension, Center Rectangle, Centerpoint Arc, 3 Point Arc and Convert Entities.

The book is designed to expose the new SOLIDWORKS user to many different tools, techniques and procedures. It may not always use the most direct tool or process.

Questions

1. What is the minimum number of profiles and planes required for a Loft feature?

2. A Swept Boss/Base feature requires a _____ and a _____.

3. Describe the differences between a Loft feature and a Swept feature.

4. Identify the three default reference planes in an assembly.

5. True or False. A Revolved-Cut feature requires an axis of revolution.

6. True or False. A Circular Pattern Feature does not require a seed feature to pattern.

7. Describe the difference between a Swept Boss/Base feature and a Swept Cut feature.

8. Identify the type of Geometric relations that can be added to a sketch.

9. What function does the Save as option perform when saving a part under a new name?

10. True or False. Never reuse geometry from one part to create another part.

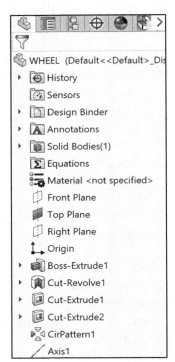

Exercises

Exercise 5.1: Advanced Part

Build the illustrated model. Calculate the volume of the part and locate the Center of mass with the provided information.

Three holes are displayed with an Ø1.00in.

Set the document properties for the model.

☀ Tangent edges and Origin are displayed for educational purposes.

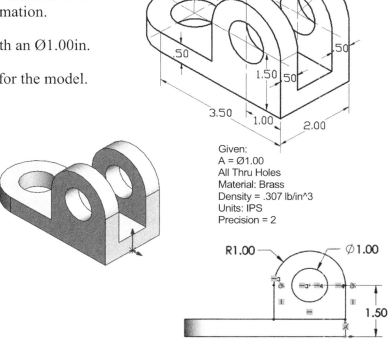

Given:
A = Ø1.00
All Thru Holes
Material: Brass
Density = .307 lb/in^3
Units: IPS
Precision = 2

Exercise 5.2: Advanced Part

Build the illustrated model. Calculate the overall mass of the part and locate the Center of mass with the provided information. Insert a Revolved Base feature and Extruded Cut feature to build this part.

Note: Select the Front Plane as the Sketch plane. Apply the Centerline Sketch tool for the Revolve1 feature. Insert the required geometric relations and dimensions. Sketch1 is the profile for the Revolve1 feature.

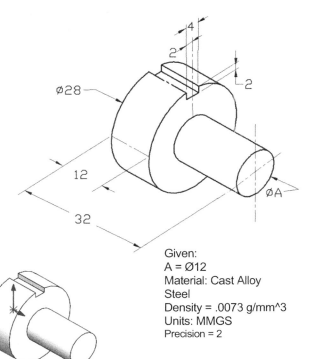

Given:
A = Ø12
Material: Cast Alloy Steel
Density = .0073 g/mm^3
Units: MMGS
Precision = 2

Exercise 5.3: Advanced Part

Build the illustrated model. Calculate the overall mass of the part and locate the Center of mass with the provided information. Insert the required geometric relations and dimensions. Precision = 2.

Insert two features: Extruded Base and Revolved Boss.

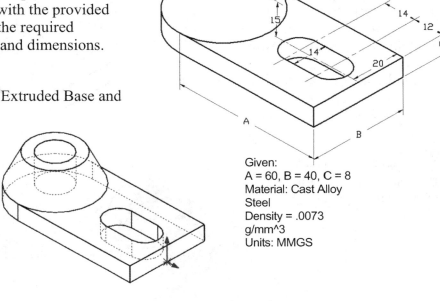

Given:
A = 60, B = 40, C = 8
Material: Cast Alloy Steel
Density = .0073 g/mm^3
Units: MMGS

Exercise 5.4: Advanced Part

Build the illustrated model. Calculate the overall mass of the part and locate the Center of mass with the provided information.

Think about the various features that create the model. Insert the required geometric relations and dimensions.

Apply symmetry. Create the left half of the model first, and then apply the Mirror feature. Precision = 2. Thru ALL holes.

Tangent Edges and Origin are displayed for educational purposes.

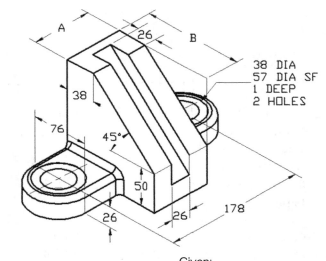

Given:
A = 76, B = 127
Material: 2014 Alloy
Density: .0028 g/mm^3
Units: MMGS
ALL ROUNDS
EQUAL 6MM

Exercise 5.5: Advanced Part

Build the illustrated model. Calculate the overall mass of the part and locate the Center of mass with the provided information. Precision = 2. Thru ALL holes.

Think about the various features that create the part.

Apply reference construction planes to build the circular features. Insert the required geometric relations and dimensions.

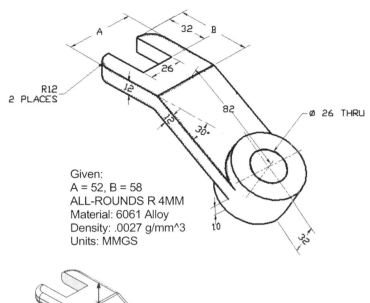

Given:
A = 52, B = 58
ALL-ROUNDS R 4MM
Material: 6061 Alloy
Density: .0027 g/mm^3
Units: MMGS

Exercise 5.6: Advanced Part

Build the illustrated model. Calculate the volume of the part and locate the Center of mass with the provided information. Precision = 2. Thru ALL holes.

Think about the various features that create this model. Insert the required geometric relations and dimensions.

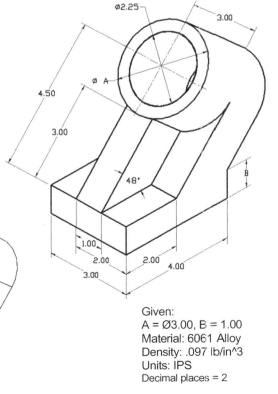

Given:
A = Ø3.00, B = 1.00
Material: 6061 Alloy
Density: .097 lb/in^3
Units: IPS
Decimal places = 2

Exercise 5.7: Advanced Part

Build the illustrated model.

Calculate the overall mass and locate the Center of mass of the illustrated model.

Think about the steps that you would take to build the illustrated part.

Identify the location of the part Origin.

Review the provided dimensions and annotations in the part illustration.

Precision = 2. Thru ALL holes.

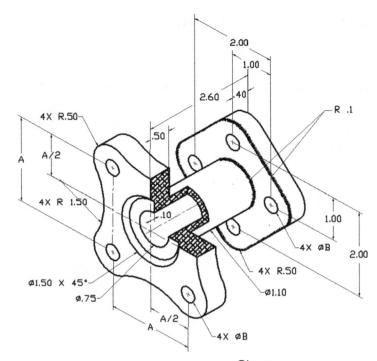

Given:
A = 2.00, B = Ø.35
Material: 1060 Alloy
Density: 0.097 lb/in^3
Units: IPS
Decimal places = 2

Tangent edges and Origin are displayed for educational purposes.

Use Symmetry. When possible and if it makes sense, model objects symmetrically about the origin.

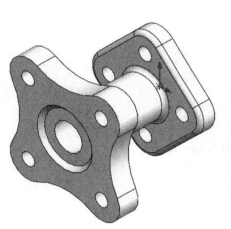

Exercise 5.8: Traditional Ice cream cone

Create a traditional Ice Cream Cone as illustrated.

Create an ANSI - IPS model.

Think about the design features that create this model. Why does the cone use ribs?

Ribs are used for structural integrity.

View the sample FeatureManager. Your FeatureManager can (should) be different. This is just one way to create this part.

Create your own ice cream cone design.

Below are a few sample models from my Freshman Engineering class.

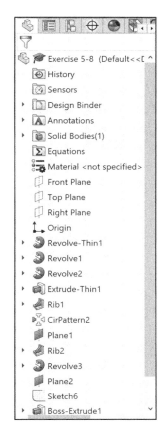

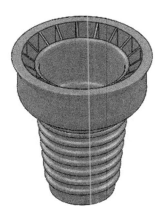

Below are a few sample models from my Freshman Engineering (Cont:).

Exercise 5.9: Gem® Paper clip

Create a simple paper clip. Create an ANSI - IPS model.

Apply material to the model. Precision = 2.

What is your Base Sketch?

The paper clip model uses (lines and arcs) as the path.

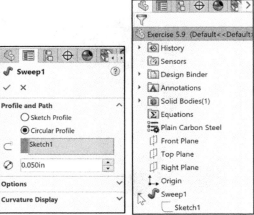

Exercise 5.10: Spring

Create a variable pitch spring (ANSI - IPS) with 6 coils; two active as illustrated.

Sketch a circle, Coincident to the Origin on the Top plane with a .235in dimension.

Create the Helix/Spiral feature.

Create a Region parameters table.

Enter the following information as illustrated. Coils 1, 2, 5, 6 & 7 are the closed ends of the spring. The pitch needs to be slightly larger than the wire.

Enter .021in for the Pitch.

Enter .080in for the free state of the two active coils.

Enter Start angle of 0deg.

Create the Swept Boss feature.

Enter .015in for Depth (Circular Profile).

Add material to the model.

View the results.

Helix/Spiral1 Region parameters table:

	P	Rev	H	Dia
1	0.021in	0	0in	0.235in
2	0.021in	1	0.021i	0.235in
3	0.08in	2	0.0715	0.235in
4	0.08in	3	0.1515	0.235in
5	0.021in	4	0.202i	0.235in
6	0.021in	5	0.223i	0.235in
7	0.021in	6	0.244i	0.235in
8				

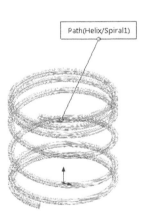

Path(Helix/Spiral1)

Helix/Spiral1

Exercise 5.11: Bottle

Create the container as illustrated.

Create an ANSI - IPS model.

Apply material to the model.

What is your Base Sketch?

What is your Base Feature?

What are the dimensions?

View the sample FeatureManager. Your FeatureManager can (should) be different. This is just one way to create this part. Estimate any needed dimension.

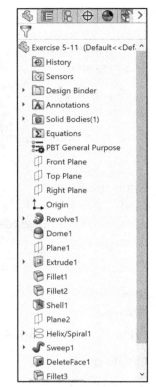

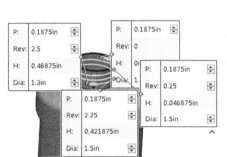

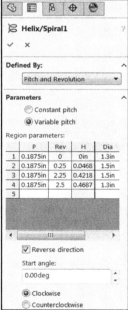

	P	Rev	H	Dia
1	0.1875in	0	0in	1.3in
2	0.1875in	0.25	0.0468	1.5in
3	0.1875in	2.25	0.4218	1.5in
4	0.1875in	2.5	0.4687	1.3in
5				

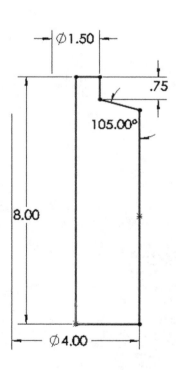

Exercise 5.12: Explicit Equation Driven Curve tool

Create an Explicit Equation Driven Curve on the Front plane. Revolve the curve. Calculate the volume of the solid.

Create a New part. Use the default ANSI, IPS Part template.

Create a 2D Sketch on the Front Plane.

Activate the Equation Driven Curve Sketch f_x tool from the Consolidated drop-down menu.

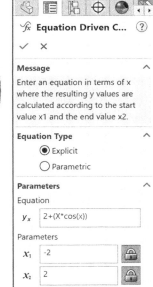

Enter the Equation y_x as illustrated.

Enter the parameters x_1, x_2 that define the lower and upper bounds of the equation as illustrated. View the curve in the Graphics window.

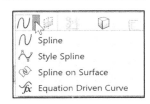

Size the curve in the Graphics window. The Sketch is under defined.

Insert three lines to close the profile as illustrated. Fully define your sketch. Enter dimensions and any needed geometric relation.

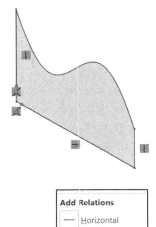

Create the Revolved feature. View the results in the Graphics window. Revolve1 is displayed. Utilize the Section tool parallel with the Right plane to view how each cross section is a circle.

Apply Brass for material.

Precision = 2.

Calculate the volume of the part using the Mass Properties tool. View the results.

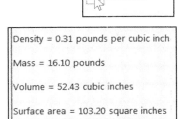

Density = 0.31 pounds per cubic inch

Mass = 16.10 pounds

Volume = 52.43 cubic inches

Surface area = 103.20 square inches

You can create parametric (in addition to explicit) equation-driven curves in both 2D and 3D sketches.

Use regular mathematical notation and order of operations to write an equation. x_1 and x_2 are for the beginning and end of the curve. Use the transform options at the bottom of the PropertyManager to move the entire curve in x-, y- or rotation. To specify $x = f(y)$ instead of $y = f(x)$, use a 90 degree transform.

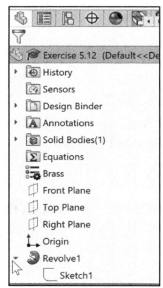

Exercise 5.13: Curve Through XYZ Points tool

The Curve Through XYZ Points  feature provides the ability to either type in (using the Curve File dialog box) or click Browse and import a text file with x-, y-, z-, coordinates for points on a curve.

A text file can be generated by any program which creates columns of numbers. The Curve  feature reacts like a default spline that is fully defined.

Create a curve using the Curve Through XYZ Points tool. Import the x-, y-, z- data.

Verify that the first and last points in the curve file are the same for a closed profile.

1. Create a new part.

2. Click the Curve Through XYZ Points  tool from the Features CommandManager. The Curve File dialog box is displayed.

Import the curve data.

3. Click Browse from the Curve File dialog box.

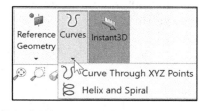

4. Browse to the downloaded folder location: Chapter 5 Homework folder.

5. Set file type to Text Files.

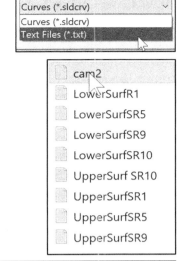

6. Double-click cam2.text. View the data in the Curve File dialog box.

7. Click OK from the Curve File dialog box.

Fix the Curve to the Graphics window. Curve1 is displayed in the FeatureManager. You created a curve using the Curve Through XYZ Points tool with imported x-, y-, z- data from a cam program.

This curve can now be used to create a sketch (closed profile), in this case a cam.

Close the existing model.

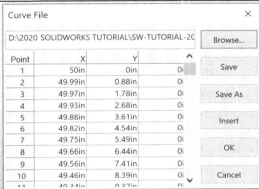

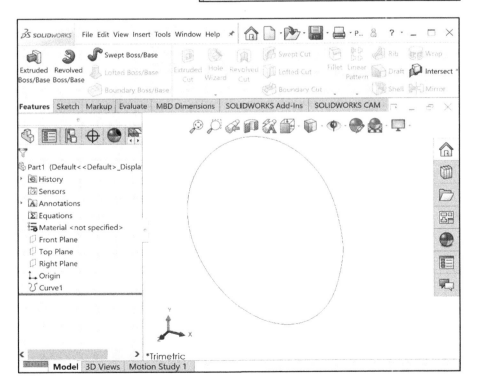

Split Line Curve tool

The Split Line 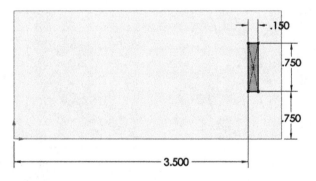 Curve tool projects an entity (sketch, solid, surface, face, plane, or surface spline) to surfaces, or curved or planar faces. It divides a selected face into multiple separate faces. You can split curves on multiple bodies with a single command. You can create the following split lines:

- **Silhouette**. Creates a split line on a cylindrical part.

- **Projection**. Project a sketch on a surface. You can pattern split line features that were created using projected curves. You can create split lines using sketched text. This is useful for creating items such as decals or applying a force or pressure when using SOLIDWORKS Simulation.

- **Intersection**. Splits faces with an intersecting solid, surface, face, plane, or surface spline.

When you create a split line with an open profile sketch, the sketch must span at least two edges of the model.

Exercise 5.15: Split Line

Create a Split Line feature to apply a distributed load to a selected face. The Split Line feature in this case does not add depth to the part.

1. Open **Split Line** from the Chapter 5 Homework folder. The Split Line FeatureManager is displayed.

2. Create a **rectangular sketch** as illustrated. Use the sketch for the Split Line feature.

Create the Split Line feature.

3. Click the **Features** tab.

4. Click the **Split Line** tool from the Curves drop-down menu.

5. Click the **top face** of Boss-Extrude1. Face 1 is displayed in the Faces to Split box.

6. Click **OK** ✅ from the Split Line PropertyManager. Split Line1 is created and is displayed.

7. **Save** the part.

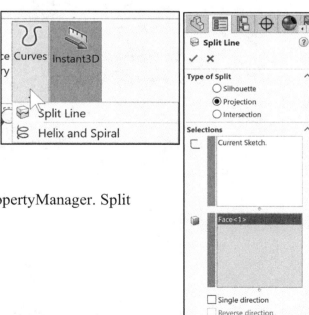

Start a SOLIDWORKS Simulation study and utilize the Split Line feature as your selected face for the applied force on the model.

8. Click **SOLIDWORKS Add-Ins tab ➤ SOLIDWORKS Simulation**. The Simulation tab is displayed.

Create a new Static Study.

9. Click the **Simulation** tab.

10. Click **New Study** from the Study drop-down menu. Static is selected by default. Accept the default study name.

11. Click **OK** ✔ from the Study PropertyManager.

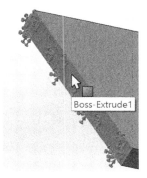

Fix the model at one end and apply a distributed load using the Split Line feature.

12. Right-click the **Fixtures** folder.

13. Click **Fixed Geometry**. The Fixture PropertyManager is displayed.

14. Click the **left face of Boss-Extrude1** as illustrated. The left face is fixed.

15. Click **OK** ✔ from the Fixture PropertyManager. Fixed-1 is displayed.

Apply an External load (force) to the model using the Split Line feature.

16. Right-click the **External Load** folder.

17. Click **Force**. The Force PropertyManager is displayed. In this example, the area of the applied force is normal to the Split Line feature.

18. Click the **Split Line1** feature on the model in the Graphics window.

19. Enter **100lbf** for force.

20. Click **OK** ✔ from the Force PropertyManager. Force-1 is created and is displayed.

21. Apply **Alloy Steel** to the part.

22. **Mesh and Run** the model. View the results.

23. **Close** the part. The Split Line feature divides a selected face into multiple separate faces.

Notes:

Chapter 6

PNEUMATIC-TEST-MODULE and Final ROBOT Assembly

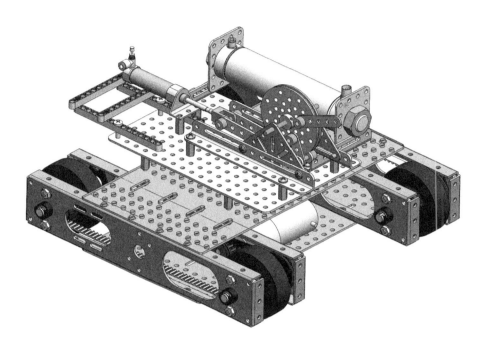

Below are the desired outcomes and usage competencies based on the completion of Chapter 6.

Desired Outcomes:	**Usage Competencies**:
<ul><li>Create five sub-assemblies:<ul><li>WHEEL-AND-AXLE.</li><li>WHEEL-FLATBAR.</li><li>3HOLE-SHAFTCOLLAR.</li><li>5HOLE-SHAFTCOLLAR.</li><li>PNEUMATIC-TEST-MODULE.</li></ul></li><li>Create the final ROBOT assembly.</li></ul>	<ul><li>Reuse geometry.</li><li>Apply Standard Mate types.</li><li>Modify existing assemblies to create new assemblies.</li><li>Utilize the following Assembly tools: Mate, Linear Component Pattern, Feature Driven Component Pattern, Mirror Components, Replace Components, and Performance Evaluation.</li><li>Work with multiple documents in an assembly.</li></ul>

Notes:

Chapter 6 - PNEUMATIC-TEST-MODULE and ROBOT Assembly

Chapter Objective

Develop a working understanding with multiple documents in an assembly. Build on sound assembly modeling techniques that utilize symmetry, component patterns and mirrored components.

Create five sub-assemblies and the final ROBOT assembly:

- 3HOLE-SHAFTCOLLAR sub-assembly.

- 5HOLE-SHAFTCOLLAR sub-assembly.

- WHEEL-FLATBAR sub-assembly.

- WHEEL-AND-AXLE sub-assembly.

- PNEUMATIC-TEST-MODULE sub-assembly.

- ROBOT final assembly.

On the completion of this chapter, you will be able to:

- Utilize various Assembly techniques.

- Suppress and hide components.

- Create new assemblies and copy assemblies to reuse similar parts.

- Use the following Assembly tools:

  o Insert Component.

  o Standard Mates: Concentric, Coincident, and Parallel.

  o Linear Component Pattern.

  o Feature Driven Component Pattern.

  o Circular Component Pattern.

  o Mirror Components.

  o Replace Components.

  o Performance Evaluation.

Chapter Overview

Create the 3HOLE-SHAFTCOLLAR sub-assembly.

Utilize the 3HOLE-SHAFTCOLLAR sub-assembly to create the 5HOLE-SHAFTCOLLAR sub-assembly.

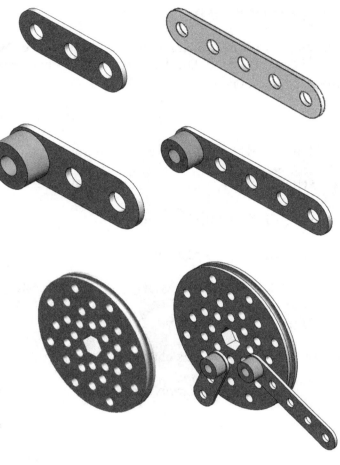

The WHEEL-FLATBAR sub-assembly contains the following items:

- 3HOLE-SHAFTCOLLAR assembly.

- 5HOLE-SHAFTCOLLAR assembly.

- WHEEL part.

Mate to the first component added to the assembly. If you mate to the first component or base component of the assembly and decide to change its orientation later, all the components will move with it.

Remove Tangent edges. Click Display from the Options dialog box, check the Removed box as illustrated.

🔅 Determine the static and dynamic behavior of mates in each sub-assembly before creating the top-level assembly.

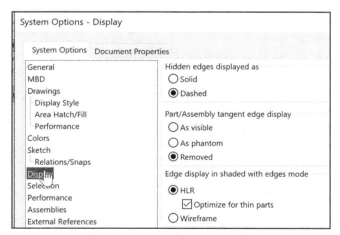

The WHEEL-AND-AXLE assembly contains the following items:

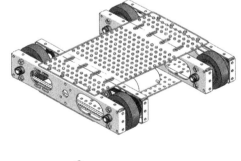

- WHEEL-FLATBAR assembly.

- AXLE-3000 part.

- SHAFTCOLLAR-500 part.

- HEX-ADAPTER part.

Combine the created new assemblies and parts to develop the PNEUMATIC-TEST-MODULE assembly.

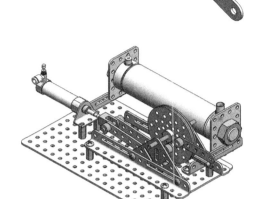

Create the final ROBOT assembly. Insert the Robot-platform assembly, PNEUMATIC-TEST-MODULE assembly, basic_integration assembly and the HEX-STANDOFF components.

All assemblies and components for the final ROBOT assembly are in the Chapter 6 Models folder. Copy the folder to your hard drive.

Tangent edges and Origins are displayed for educational purposes.

Utilize the Pack and Go option to save an assembly or drawing with references and toolbox components. The Pack and Go tool saves either to a folder or creates a zip file to e-mail. View SOLIDWORKS help for additional information.

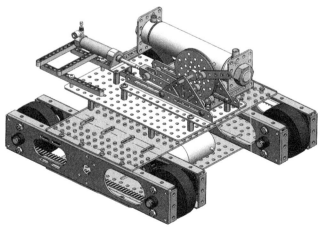

Assembly Techniques

Assembly modeling requires practice and time. Below are a few helpful techniques to address Standard Mates. These techniques are utilized throughout the development of all assemblies.

Mating Techniques:
• Plan your assembly and sub-assemblies in an assembly layout diagram. Group components together to form smaller sub-assemblies.
• Utilize symmetry in an assembly. Utilize Mirror Component and Component Pattern to create multiple instances (copies) of components. Reuse similar components with Save as copy and configurations.
• Use the Zoom and Rotate commands to select the geometry in the Mate process. Zoom to select the correct face.
• Apply various colors to features and components to improve display.
• Activate Temporary Axes and Show Planes when required for Mates, otherwise Hide All Types from the View menu.
• Select Reference planes from the FeatureManager for complex components. Expand the FeatureManager to view the correct plane.
• Remove display complexity. Hide components when visibility is not required.
• Suppress components when Mates are not required. Group fasteners at the bottom of the FeatureManager. Suppress fasteners and their assembly patterns to save rebuild time and file size.
• Utilize Section views to select internal geometry.
• Use the Move Component and Rotate Component commands before Mating. Position the component in the correct orientation.
• Create additional flexibility in a Mate. Distance Mates are modified in configurations and animations. Rename Mates in the FeatureManager.
• Verify the position of the components. Use Top, Front, Right and Section views.

PNEUMATIC TEST MODULE Layout

The PNEUMATIC TEST MODULE assembly is comprised of four major sub-assemblies:

- LINKAGE assembly.

- RESERVOIR assembly.

- FRONT-SUPPORT assembly.

- WHEEL-AND-AXLE assembly.

There are over one hundred components in the PNEUMATIC TEST MODULE assembly. Complex assemblies require planning. The Assembly Layout diagram provides organization for a complex assembly by listing sub-assemblies and parts.

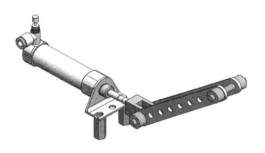

Chapter 2 LINKAGE Assembly

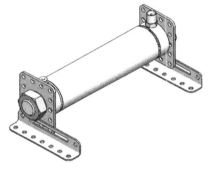

Chapter 3 AIR RESERVOIR Assembly

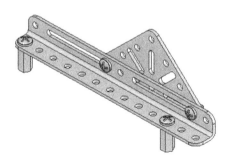

Chapter 3 FRONT SUPPORT Assembly

Chapter 5 WHEEL-AND-AXLE Assembly

Review the Assembly Layout diagram for the PNEUMATIC TEST MODULE assembly.

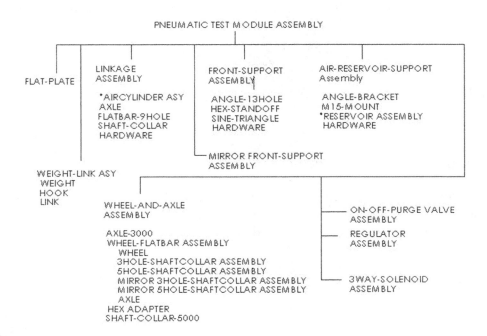

Physical space on the FLAT-PLATE is at a premium. Determine the requirements for hardware and placement after the mechanical components are assembled to the FLAT-PLATE part.

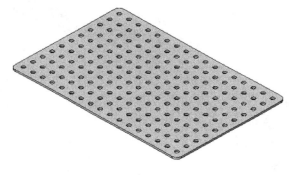

The FLAT-PLATE was created in the Chapter 3 exercises. The FLAT-PLATE is available in the Chapter 6 Models folder.

The LINKAGE assembly, FRONT-SUPPORT assembly and the AIR-RESERVOIR SUPPORT assembly were created in Chapter 2 and 3. The ON-OFF-PURGE VALVE assembly, REGULATOR assembly and the 3WAY SOLENOID VALVE assembly require additional hardware and are addressed in the chapter exercises.

The book is designed to expose the new SOLIDWORKS user to many different tools, techniques and procedures. It may not always use the most direct tool or process.

The WHEEL-FLATBAR assembly consists of the following:

- WHEEL part.

- 3HOLE-SHAFTCOLLAR assembly.

- 5HOLE-SHAFTCOLLAR assembly.

FLATBAR Sub-assembly

There are two similar sub-assemblies contained in the WHEEL-FLATBAR assembly:

- 3HOLE-SHAFTCOLLAR assembly.

- 5HOLE-SHAFTCOLLAR assembly.

Create the 3HOLE-SHAFTCOLLAR assembly. Utilize parts and mating techniques developed in Chapter 2 & 3.

Utilize the Save As command and create the 5HOLE-SHAFTCOLLAR assembly.

Combine the 3HOLE-SHAFTCOLLAR assembly, 5HOLE-SHAFTCOLLAR assembly and the WHEEL part to create the WHEEL-FLATBAR assembly.

The FLATBAR-3HOLE and FLATBAR 5HOLE parts were created in the Chapter 3 exercises. If needed, copy the models from the Chapter 6 Models folder.

Activity: 3HOLE - SHAFTCOLLAR Assembly

Create the 3HOLE-SHAFTCOLLAR assembly.

1)   Click **New** from the Menu bar.

2)   Double-click **Assembly** from the Templates tab.

3)   Double-click **FLATBAR-3HOLE** from the SW-TUTORIAL-2020\Chapter 6 Models folder. The FLATBAR-3HOLE part is in the Chapter 6 Models folder. If you want to create this part, follow the below procedure; otherwise, skip the next few steps to create the part.

4)   Click **Open** from the Menu bar.

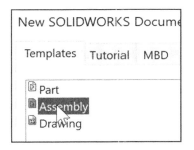

5)   Double-click **FLATBAR** from the SW-
TUTORIAL-2020 folder.

6)   Click **Save As** 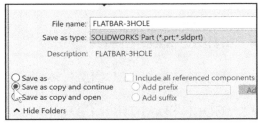.

7)   Select the **SW-TUTORIAL-2020** folder.

8)   Enter **FLATBAR-3HOLE** for File name.

9)   Enter **FLATBAR-3HOLE** for Description.

10) Check the **Save as copy and continue** box.

Save and close the FLATBAR model.
11) Click **Save**.

12) **Close** the FLATBAR model.

Open the FLATBAR-3HOLE model.
13) Click **Open** from the Menu bar.

14) Double-click **FLATBAR-3HOLE** from the SW-
TUTORIAL-2020 folder.

15) Right-click **LPATTERN1** from the
FeatureManager.

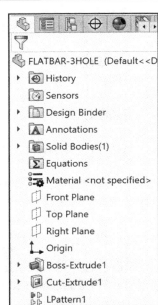

16) Click **Edit Feature** from the Context toolbar.
The LPattern1 PropertyManager is displayed.

17) Enter **3** in the Number of Instances.

18) Click **OK** from the LPattern1
PropertyManager. Note: If needed, delete the
Design Table in the CommandManager.

19) Click **Boss-Extrude1** from the
FeatureManager.

20) **View** the dimensions in the Graphics window.

21) Click the **4.000**in, **[101.60]** dimension.

22) Enter **1.000**in, **[25.4]**.

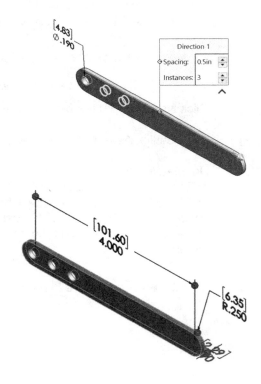

23) Click **Cut-Extrude1** from the FeatureManager.

24) **View** the dimensions in the Graphics window.

25) Right-click **9X** dimension text in the Graphics window. The Dimension PropertyManager is displayed.

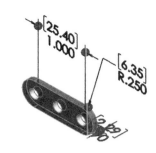

26) Delete the **9X** text in the Dimension Text box.

27) Enter **3X** in the Dimension Text box.

28) Click **OK** ✔ from the Dimension PropertyManager.

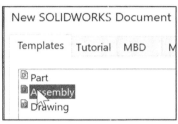

Save the model.

29) Click **Save** 💾.

Create a new assembly.

30) Click **New** ▢ from the Main menu.

31) Double-click **Assembly** from the Templates tab.

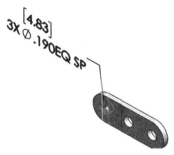

32) Double-click **FLATBAR-3HOLE** from the SW-TUTORIAL-2020\Chapter 6 Models folder. The FLATBAR-3HOLE part is in the Chapter 6 Models folder.

33) Double-click **FLATBAR-3HOLE.**

34) Click **OK** ✔ from the Begin Assembly PropertyManager. The FLATBAR-3HOLE is fixed to the Origin.

Save the assembly. Enter name. Enter description.

35) Click **Save As** from the drop-down Menu bar.

🔆 Tangent edges and Origin are displayed for educational purposes.

🔆 CommandManager and Feature Manager tabs and folder files will vary, depending on system setup and Add-Ins.

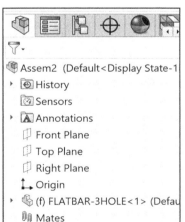

36)   Enter **3HOLE-SHAFTCOLLAR** for File
          name.

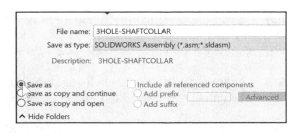

37)   Enter **3HOLE-SHAFTCOLLAR** for
          Description.

38)   Click **Save**.

Save the 3HOLE-SHAFTCOLLAR assembly.

39)   Click **Save** 💾.

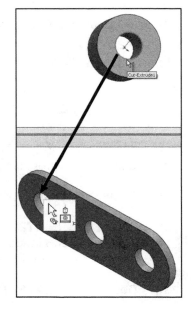

Utilize a Concentric/Coincident 🖱️ SmartMate between the
SHAFT-COLLAR and the FLATBAR-3HOLE.

Open the **SHAFT-COLLAR** part.

40)   Click **Open** 📂 from the Menu bar.

41)   Double-click **SHAFT-COLLAR** from the
          SW-TUTORIAL-2020 folder. SHAFT-COLLAR is the
          current document name.

42)   **Rotate** the SHAFT-COLLAR to view the back circular
          edge.

43)   Click **Window**, **Tile Horizontally** from the Menu bar.

44)   Drag the **back circular edge** of the SHAFT-COLLAR
          to the left circular hole edge of the FLATBAR-3HOLE
          in the Assembly Graphics window as illustrated. The
          mouse pointer displays the Concentric/Coincident
          🖱️ icon.

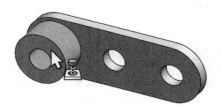

45)   **Release** the mouse button. Note: Select the back
          circular edge of the SHAFT-COLLAR, not the face.

Save the 3HOLE-SHAFTCOLLAR assembly.

46)   **Close** ✖ the SHAFT-COLLAR window.

47)   **Maximize** ▢ the 3HOLE-SHAFTCOLLAR
          assembly.

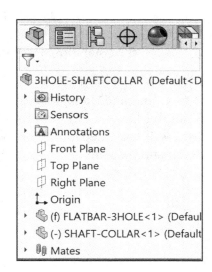

Fit the model to the Graphics window. Save the assembly.

48)   Press the **f** key.

49)   Click **Save** 💾.

Create the 5HOLE-SHAFTCOLLAR assembly.
Utilize the Save As command with the Save as copy
and continue option. Recover from Mate errors.

Save the 3HOLE-SHAFTCOLLAR assembly as the 5HOLE-
SHAFTCOLLAR assembly.

50)   Click **Save As** from the drop-down Menu bar.

51)   Check the **Save as copy and continue** box.

52)   Enter **5HOLE-SHAFTCOLLAR** for File name.

53)   Enter **5HOLE-SHAFTCOLLAR** for Description.

54)   Click **Save**.

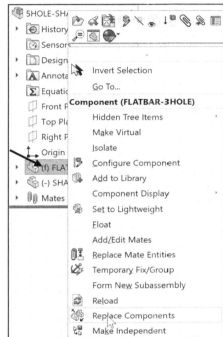

Close the model.

55)   Click **File**, **Close** from the Menu bar.

56)   Click **Save All**.

Open the **5HOLE-SHAFTCOLLAR** assembly.

57)   Click **Open** from the Menu bar.

58)   Double-click **5HOLE-SHAFTCOLLAR**. The
5HOLE-SHAFTCOLLAR FeatureManager is
displayed.

59)   Right-click **FLATBAR-3HOLE** from the
FeatureManager.

60)   **Expand** the Pop-up menu if needed.

61)   Click **Replace Components**. The Replace
PropertyManager is displayed.

62)   Click the **Browse** button.

63)   Double-click **FLATBAR-5HOLE**. Note: The
FLATBAR-5HOLE part is located in the Chapter 6
Models folder.

64)   Check the **Re-attach mates** box.

65)   Click **OK** ✔ from the Replace PropertyManager. The Mate
Entities PropertyManager and the What's Wrong dialog box
is displayed. There are two red Mate error marks displayed in
the Mate Entities box.

66) The What's Wrong dialog box is displayed. Recover from the Mate errors. Click **Close** from the What's Wrong dialog box.

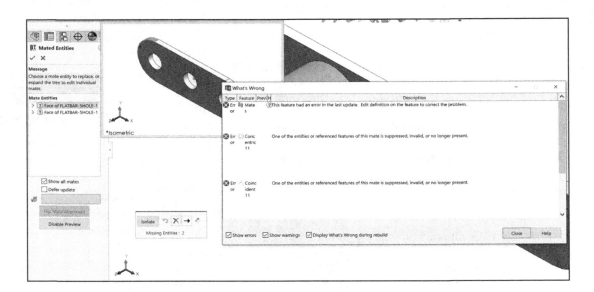

67) Click **OK** ✔ from the Mated Entities PropertyManager.

68) Click **Close** from the What's Wrong dialog box. View the location of the SHAFT-COLLAR in the Graphics window.

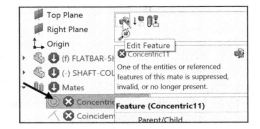

Recover from the Mate errors.

69) **Expand** the Mates folder from the FeatureManager. Right-click the first mate, **Concentric #** from the Mates folder.

70) Click **Edit Feature** from the Context toolbar. The Mate PropertyManager is displayed.

71) Right-click the **Mate Face error** in the Mate Selections box. Click **Delete**.

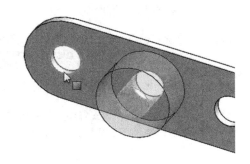

72) Click the **inside face** of the left hole of the FLATBAR as illustrated. Concentric is selected by default.

73) Click **OK** ✔ from the Mate PropertyManager.

74) Right-click the second mate, **Coincident #** from the Mates folder.

75) Click **Edit Feature** from the Context toolbar. The Mate PropertyManager is displayed.

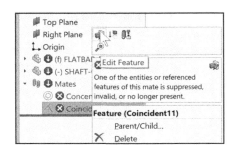

76)   Right-click the **Mate Face error** in the Mate Selections box.

77)   Click **Delete** as illustrated.

78)   Click the **front face** of the FLATBAR as illustrated. The selected faces are displayed in the Mate Selections box. Coincident is selected by default.

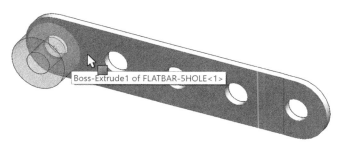

79)   Click **OK** ✔ from the Mate PropertyManager.

80)   **Expand** the Mate folder from the FeatureManager.

81)   **View** the created mates.

🔆 The Mate Entities box will list red X's if the faces, edges or planes are not valid. Expand the Mate Entities and select new references in the Graphics window to redefine the mates.

The FLATBAR-3HOLE is replaced with the FLATBAR-5HOLE part. The Mates are updated.

Fit the model to the Graphics window.
82)   Press the **f** key.

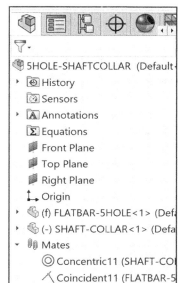

Display an Isometric view. Save the 5HOLE-SHAFTCOLLAR assembly.

83)   Click **Isometric view** 🔲 .

84)   Click **Save** 💾 .

🔆 Incorporate symmetry into the assembly. Divide large assemblies into smaller sub-assemblies.

🔆 CommandManager and Feature Manager tabs and folder files will vary, depending on system setup and Add-Ins.

WHEEL-FLATBAR Assembly

The WHEEL-FLATBAR assembly consists of the following components:

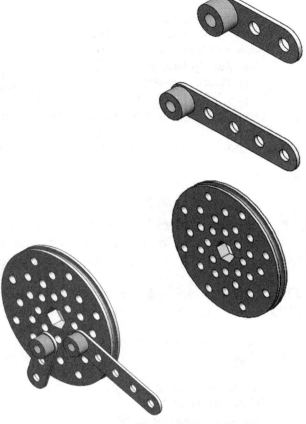

- 3HOLE-SHAFTCOLLAR assembly.

- 5HOLE-SHAFTCOLLAR assembly.

- WHEEL part.

Create the WHEEL-FLATBAR assembly. Mate the 3HOLE-SHAFTCOLLAR assembly 67.5 degrees counterclockwise from the Top Plane.

The 3HOLE-SHAFTCOLLAR assembly is concentric with holes on the second and fourth bolt circle.

Mate the 5HOLE-SHAFTCOLLAR assembly 22.5 degrees clockwise from the Top Plane.

The 5HOLE-SHAFTCOLLAR assembly is concentric with holes on the second and fourth bolt circle.

Use the Quick mate procedure for reference geometry (such as planes, axes, and points) along with model geometry (such as faces, edges and vertices).

Remove the fixed state. Right-click the component name in the FeatureManager. Click **Float**. The component is free to move.

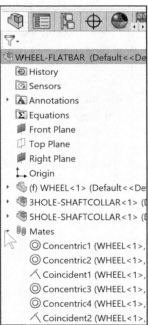

Activity: WHEEL-FLATBAR Assembly

Create the WHEEL-FLATBAR assembly.

85)   Click **New** ☐ from the Menu bar.

86)   Double-click **Assembly** from the Templates tab.

Insert the WHEEL.

87)   Double-click **WHEEL** from the SW-TUTORIAL-2020 folder. The WHEEL was created in Chapter 5. The WHEEL is in the Chapter 6 Models folder.

88)   Click **OK** ✔ from the Begin Assembly PropertyManager. The WHEEL part is fixed to the assembly Origin.

If needed, deactivate the Origins.

89)   Click **View**, **Hide/Show**, uncheck **Origins** from the Menu bar menu.

Save the assembly. Enter name. Enter description.

90)   Click **Save As** from the drop-down Menu bar.

91)   Select the **SW-TUTORIAL-2020** folder.

92)   Enter **WHEEL-FLATBAR** for File name.

93)   Enter **WHEEL-FLATBAR** for Description.

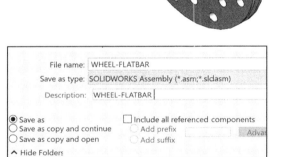

94)   Click **Save**.

Display the Top Plane in the Front view.

95)   Click **Front view** ⬚ from the Heads-up View toolbar. View the WHEEL part.

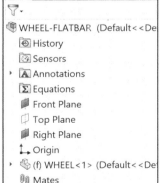

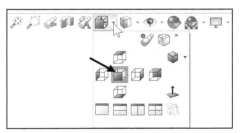

Locate the first set of holes from the Right plane (-Y-axis). Left Hole1 and Left Hole2 are positioned on the second and fourth bolt circle, 22.5° from the Right plane. Select Left Hole1. The x, y, z coordinates, -.287, -.693, .125 are displayed.

$Tan^{-1} (-.287/.693) = 22.5°$

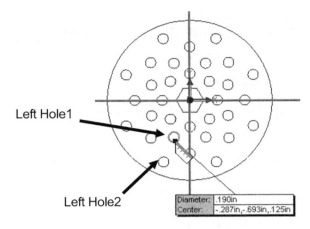

As an exercise, utilize the Measure tool to determine the center-to-center distance between the Left Hole1 and Left Hole2. The center-to-center distance is .500in.

Insert two Concentric mates between Left Hole1 and Left Hole2 and the 3HOLE-SHAFTCOLLAR assembly holes. The FLATBAR-3HOLE center-to-center distance is also .500in.

To determine tolerance issues, utilize two Concentric mates between components with mating cylindrical geometry. If the mating components center-to-center distance is not exact, a Mate error is displayed on the second Concentric mate.

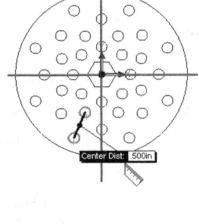

Insert a Coincident mate between the back face of the 3HOLE-SHAFTCOLLAR assembly and the front face of the WHEEL.

Right Hole1 and Right Hole2 are 22.5° from the Top Plane.

Insert two Concentric mates between Right Hole1 and Right Hole2 and the 5HOLE-SHAFTCOLLAR assembly holes.

Insert a Coincident mate between the back face of the 5HOLE-SHAFTCOLLAR assembly and the front face of the WHEEL.

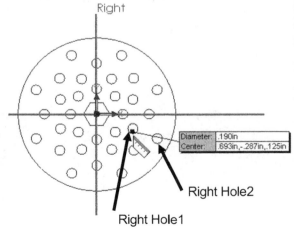

Activity: WHEEL-FLATBAR Assembly - Insert 3HOLE-SHAFTCOLLAR Assembly

Insert the 3HOLE-SHAFTCOLLAR assembly.

96)   Click **Isometric view** from the Heads-up View toolbar.

97)   Click **Insert Components** from the Assembly toolbar.

98)   Double-click the **3HOLE-SHAFTCOLLAR** assembly from the SW-TUTORIAL-2020 folder. The assembly is also located in the Chapter 6 Models folder.

99)   Click a **position** to the left of the WHEEL as illustrated.

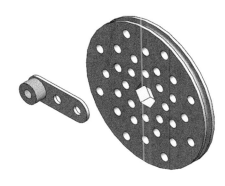

Move and rotate the 3HOLE-SHAFTCOLLAR component.

100)   Click the front face of the **3HOLE-SHAFTCOLLAR**.

101)   Right-click **Move with Triad**.

102)   Hold the **left mouse button** down on the X-axis (red).

103)   Drag the **component** to the left. View the ruler.

104)   Hold the **right mouse button** down on the Z-axis (blue).

105)   Drag the **component** and rotate it about the Z-axis. View the ruler.

106)   **Position** the component until the SHAFT-COLLAR part is approximately in front of the WHEEL Left Hole1.

107)   Release the **right mouse** button.

108)   Click a **position** in the Graphics window to deselect the face.

	Component Display
⌁	Mo_v_e with Triad
🗲	Tempo_r_ary Fix/Group
	_F_orm New Subassembly
🖫	Make Independent
🗗	Copy _w_ith Mates

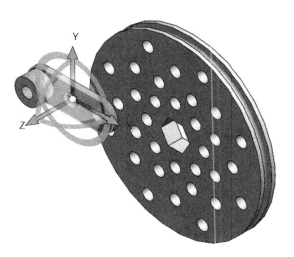

Insert a Concentric mate.

109) Click the **back top inside cylindrical face** of the SHAFT-COLLAR.

110) **Rotate** the model to view the WHEEL.

111) Hold the **Ctrl** key down.

112) Click the **WHEEL Left Hole1** cylindrical face as illustrated.

113) Release the **Ctrl** key. The Mate pop-up menu is displayed.

114) Click **Concentric** from the Mate pop-up menu.

Insert the second Concentric mate.

115) Click the **back middle inside cylindrical face** of the FLATBAR.

116) **Rotate** the model to view the WHEEL.

117) Hold the **Ctrl** key down.

118) Click the **WHEEL Left Hole2** inside cylindrical face as illustrated.

119) Release the **Ctrl** key. The Mate pop-up menu is displayed.

120) Click **Concentric** from the Mate pop-up menu.

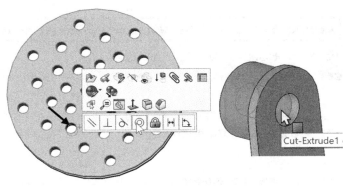

Insert a Coincident mate.

121)   Click the **FLATBAR-3HOLE back** face.

122)   **Rotate** the model to view the WHEEL.

123)   Hold the **Ctrl** key down.

124)   Click the front face of the **WHEEL**.

125)   Release the **Ctrl** key. The Mate pop-up menu is displayed.

126)   Click **Coincident** from the Mate pop-up menu.

Display a Front view.

127)   Click **Front view** .

Save the WHEEL-FLATBAR. assembly.

128)   Click **Save** .

Activity: WHEEL-FLATBAR Assembly - Insert 5HOLE-SHAFTCOLLAR Assembly

Insert the 5HOLE-SHAFTCOLLAR assembly.

129)   Click **Insert Components** from the Assembly toolbar.

130)   Double-click the **5HOLE-SHAFTCOLLAR** assembly from the SW-TUTORIAL-2020 folder. The assembly is in the Chapter 6 Models folder.

131)   Click a **position** to the right of the WHEEL.

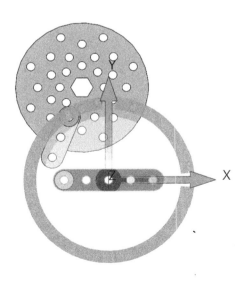

Move the 5HOLE-SHAFTCOLLAR component.

132)   Click the **5HOLE-SHAFTCOLLAR** front face.

133)   Right-click **Move with Triad**.

134)   Hold the **left mouse button** down on the X-axis (red).

135)   Drag the **component** to the right. View the results.

136)   Click **Isometric view** .

137)   Click **inside** the Graphics window.

Mate the 5HOLE-
SHAFTCOLLAR assembly.
Insert a Concentric mate.

138) Click the **back inside cylindrical face** as illustrated on the FLATBAR-5HOLE assembly.

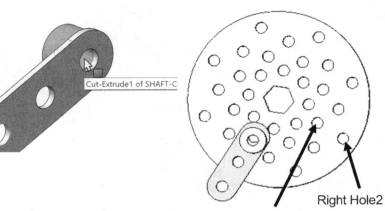

139) **Rotate** the model to view the WHEEL.

140) Hold the **Ctrl** key down.

141) Click the **WHEEL Right Hole1** cylindrical face.

Right Hole2

Right Hole1

142) Release the **Ctrl** key. The Mate pop-up menu is displayed.

143) Click **Concentric** from the Mate pop-up menu.

Insert a Concentric mate.

144) **Move** the 5HOLE-SHAFTCOLLAR to view the back side.

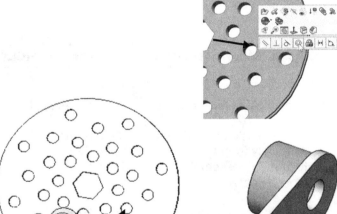

145) Click the **back inside cylindrical** face of the second hole on the FLATBAR-5HOLE assembly.

146) **Rotate** the model to view the WHEEL.

147) Hold the **Ctrl** key down.

148) Click the **WHEEL Right Hole2** cylindrical face.

Right Hole2

Right Hole1

149) Release the **Ctrl** key. The Mate pop-up menu is displayed.

150) Click **Concentric** from the Mate pop-up menu.

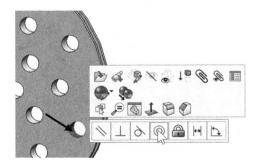

Insert a Coincident mate.

151) Click the **back face** of the FLATBAR-5HOLE.

152) **Rotate** the model to view the WHEEL.

153) Hold the **Ctrl** key down.

154) Click the **front face** of the WHEEL.

155) Release the **Ctrl** key. The Mate pop-up menu is displayed.

156) Click **Coincident** from the Mate pop-up menu.

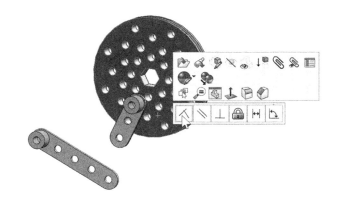

Measure the angle between the 3HOLE-SHAFTCOLLAR assembly and the 5HOLE-SHAFTCOLLAR assembly.

157) Click **Front view** .

158) Click the **Measure** tool from the Evaluate tab in the CommandManager. The Measure dialog box is displayed.

159) Select the **two inside edges** of the FLATBAR assemblies. View the results.

160) If required, click the **Show XYZ Measurements button**. The items are perpendicular.

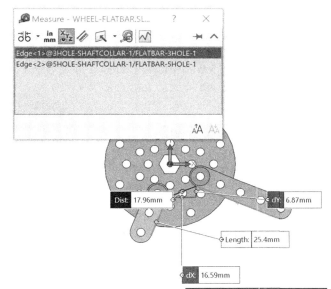

The Measure tool provides the ability to display dual units. Click the **Units/Precision** icon from the Measure dialog box. Set the desired units.

Utilize the Measure tool to measure distances and add reference dimensions between the COM point and entities such as vertices, edges, and faces.

161) **Close** ✖ the Measure dialog box.

Apply the Performance Evaluation tool. The Performance Evaluation tool displays statistics and checks the health of the current assembly.

162) Click the **Performance Evaluation** tool in the Evaluate toolbar. The Performance Evaluation dialog box is displayed. Review the Status and description for the assembly.

163) Click **OK** from the Performance Evaluation dialog box.

Fit the model to the Graphics window. Display an Isometric view.

164) Press the **f** key.

165) Click **Isometric view** 🔲 from the Heads-up View toolbar.

Save the WHEEL-FLATBAR assembly.

166) Click **Save** 💾.

💡 To remove the fixed state, Right-click the component name in the FeatureManager. Click **Float**. The component is free to move.

💡 Mate to the first component added to the assembly. If you mate to the first component or base component of the assembly and decide to change its orientation later, all the components will move with it.

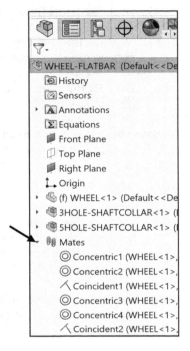

WHEEL-AND-AXLE Assembly

The WHEEL-AND-AXLE assembly contains the following items:

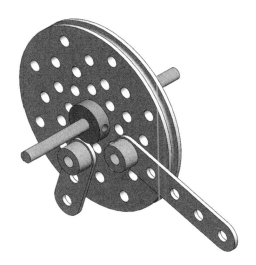

- WHEEL-FLATBAR assembly.

- AXLE-3000 part.

- SHAFTCOLLAR-500 part.

- HEX-ADAPTER part.

Create the WHEEL-AND-AXLE assembly. The AXLE-3000 part is the first component in the assembly. A part or assembly inserted into a new assembly is called a component. The WHEEL-FLATBAR assembly rotates about the AXLE part.

Combine the created new assemblies and parts to develop the PNEUMATIC-TEST-MODULE assembly.

Activity: WHEEL-AND-AXLE Assembly

Create the WHEEL-AND-AXLE assembly.

167) Click **New** from the Menu bar.

168) Double-click **Assembly** from the Templates tab.

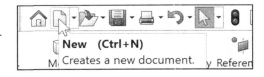

Insert the AXLE-3000 part.

169) Double-click **AXLE-3000** from the SW-TUTORIAL-2020 folder. AXLE-3000 is in the Chapter 6 Models folder if you did not create it in Chapter 5.

170) Click **OK** from the Begin Assembly PropertyManager. The AXLE-3000 part is fixed to the assembly Origin.

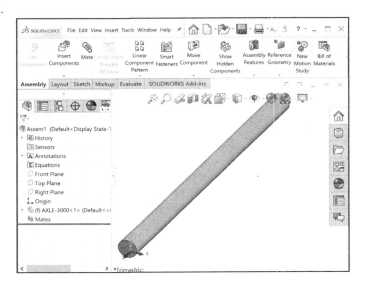

Save the assembly. Enter name. Enter description.

171) Click **Save As** from the drop-down Menu bar.

172) Enter **WHEEL-AND-AXLE** for File name in the SW-TUTORIAL-2020 folder.

173) Enter **WHEEL-AND-AXLE** for Description.

174) Click **Save**.

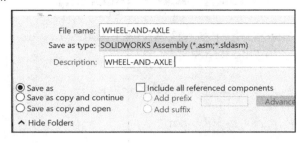

Insert a Coincident mate between the Axis of the AXLE-3000 and the Axis of the WHEEL. Insert a Coincident mate between the Front Plane of the AXLE-3000 and the Front Plane of the WHEEL. The WHEEL-FLATBAR assembly rotates about the AXLE-3000 axis.

Display the Temporary Axes.
175) Click **View**, **Hide/Show**, check **Temporary Axes** from the Menu bar.

Insert the WHEEL-FLATBAR assembly.

176) Click **Insert Components** from the Assembly toolbar.

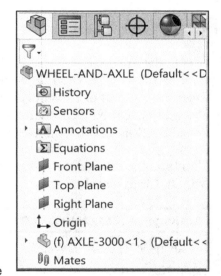

177) Double-click the **WHEEL-FLATBAR** assembly from the SW-TUTORIAL-2020 folder. The assembly is in the Chapter 6 Models folder.

178) Click a **position** to the right of AXLE-3000.

View the Reference WHEEL Axis.
179) Click **View**, **Hide/Show**, check **Axes** from the Menu bar.

Reference geometry defines the shape or form of a surface or a solid. Reference geometry includes planes, axes, coordinate systems, and points.

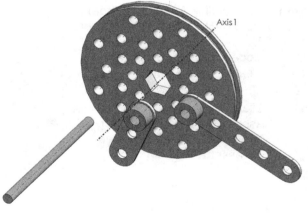

The book is designed to expose the new SOLIDWORKS user to many different tools, techniques and procedures. It may not always use the most direct tool or process.

Insert a Coincident mate.

180) Click **Axis1** in the Graphics window.

181) Hold the **Ctrl** key down.

182) Click the **AXLE-3000 Temporary Axis**.

183) Release the **Ctrl** key. The Pop-up Mate menu is displayed.

184) Click **Coincident** from the Mate Pop-up menu.

185) Click **OK** ✔ from the Coincident PropertyManager.

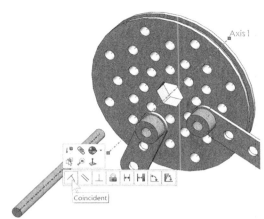

Insert a Coincident mate.

186) **Expand** WHEEL-AND-AXLE from the fly-out FeatureManager.

187) **Expand** AXLE-3000 from the fly-out FeatureManager.

188) Click **Front Plane** of AXLE-3000<1>.

189) **Expand** WHEEL from the fly-out FeatureManager.

190) Hold the **Ctrl** key down.

191) Click **Front Plane** of the WHEEL.

192) Release the **Ctrl** key. The Mate Pop-up menu is displayed.

193) Click **Coincident** from the Mate Pop-up menu.

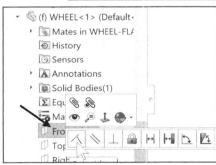

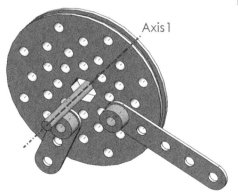

Rotate the WHEEL-FLATBAR assembly about AXLE-3000.

194)   Click and drag the **WHEEL** around AXLE-3000.

Save the WHEEL-AND-AXLE assembly.

195)   Click **Save** 💾.

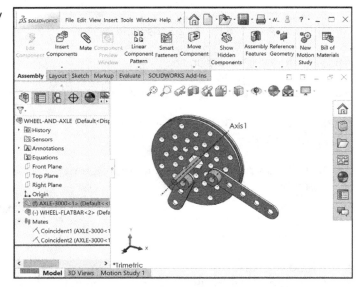

Activity: WHEEL-AND-AXLE Assembly - Insert the HEX-ADAPTER Part

Insert the HEX-ADAPTER part.

196)   Click the **Insert Components** 🗗 tool from the Assembly toolbar.

197)   Double-click **HEX-ADAPTER** from the SW-TUTORIAL-2020 folder.

198)   Click a **position** near the assembly as illustrated. The HEX-ADAPTER is displayed in the FeatureManager.

199)   **Expand** the Mates folder.

200)   **View** the created mates for the assembly. View the inserted components: AXLE-3000, WHEEL-FLATBAR and HEX-ADAPTER.

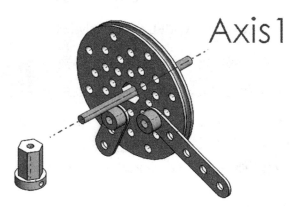

Insert a Concentric mate. Mate the HEX-ADAPTER.

201)   Click the **Mate** 🖉 tool from the Assembly toolbar. The Mate PropertyManager is displayed.

202)   Click the **HEX-ADAPTER** cylindrical face as illustrated.

203)   Click the **AXLE-3000** cylindrical face as illustrated. Concentric is selected by default. The selected faces are displayed in the Mate Selections box.

204)   Click **Aligned** 🔃 from the Concentric1 PropertyManager to flip the HEX-ADAPTER, if required.

205)   Click **OK** ✔ from the Concentric PropertyManager.

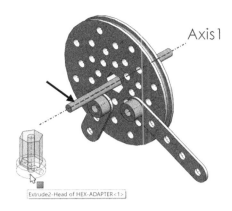

Insert a Coincident mate.

206)   Click the **front face** of the WHEEL.

207)   **Rotate** the assembly to view the flat back circular face of the HEX-ADAPTER as illustrated.

208)   Hold the **Ctrl** key down.

209)   Click the **flat back circular face** of the HEX-ADAPTER.

210)   Release the **Ctrl** key. The Mate pop-up menu is displayed.

211)   Click **Coincident** from the Mate pop-up menu.

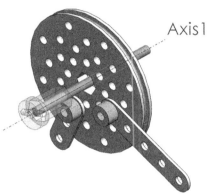

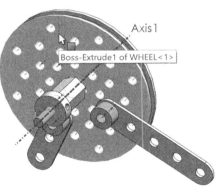

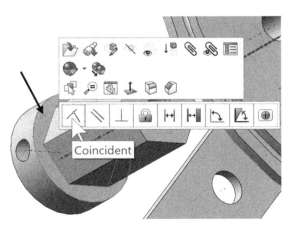

Insert a Parallel mate.

212) **Rotate** the WHEEL-AND-AXLE assembly to view the back-bottom edge of the HEX-ADAPTER.

213) **Zoom in** on the back-bottom edge of the HEX-ADAPTER.

214) Click the **back-bottom edge** of the WHEEL. Do not select the midpoint.

215) Hold the **Ctrl** key down.

216) Click the **top edge** of the HEX-ADAPTER. Do not select the midpoint.

217) Release the **Ctrl** key. The Mate pop-up menu is displayed.

218) Click **Parallel** ⟍ from the Mate pop-up menu.

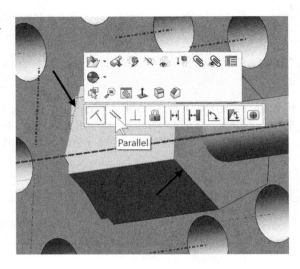

Fit the model to the Graphics window.
219) Press the **f** key.

Display an Isometric view.

220) Click **Isometric view** 🧊.

View the created mates.
221) **Expand** the Mates folder.

Save the WHEEL-AND-AXLE assembly.

222) Click **Save** 💾.

Activity: WHEEL-AND-AXLE Assembly - Insert SHAFTCOLLAR-500 Part

Insert the SHAFTCOLLAR-500 part.

223) Click the **Insert Components** tool from the Assembly toolbar.

224) Double-click **SHAFTCOLLAR-500** from the SW-TUTORIAL-2020 folder.

225) Click a **position** behind the WHEEL-AND-AXLE assembly.

Insert a Concentric mate.

226)   Click **View**, **Hide/Show**, un-check **Temporary Axes** from the Menu bar.

227)   Click **View**, **Hide/Show**, un-check **Axes** from the Menu bar.

228)   Hold the **Ctrl** key down.

229)   Click the **inside cylindrical face** of the SHAFTCOLLAR-500 part.

230)   Click the **cylindrical face** of the AXLE-3000 part.

231)   Release the **Ctrl** key. The Mate pop-up menu is displayed.

232)   Click **Concentric** from the Mate pop-up menu.

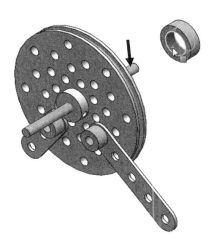

Insert a Coincident mate.

233)   Click the **front face** of the SHAFTCOLLAR-500 part.

234)   **Rotate** the WHEEL to view the back face.

235)   Hold the **Ctrl** key down.

236)   Click the **back face** of the WHEEL.

237)   Release the **Ctrl** key. The Mate pop-up menu is displayed.

238)   Click **Coincident** from the Mate pop-up menu.

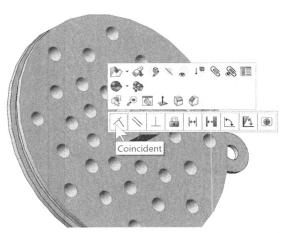

Display an Isometric view.

239)   Click **Isometric view** .

View the created Mates.

240)   **Expand** the Mates folder. View the created mates.

Save the WHEEL-AND-AXLE assembly.

241)   Click **Save** .

Close all files.

242)   Click **Windows**, **Close All** from the Menu bar.

Determine the static and dynamic behavior of mates in each sub-assembly before creating the top-level assembly.

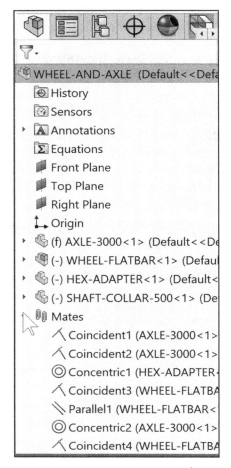

Review the WHEEL-AND-AXLE Assembly

You combined the WHEEL-FLATBAR sub assembly, the AXLE-3000 part, the HEX-ADAPTER part and the SHAFTCOLLAR-500 part to create the WHEEL-AND-AXLE assembly.

The WHEEL-FLATBAR sub-assembly rotated about the AXLE-3000 part. The WHEEL-FLATBAR assembly combined the 3HOLE-SHAFTCOLLAR assembly and the 5HOLE-SHAFTCOLLAR assembly. The 5HOLE-SHAFTCOLLAR assembly was created from the 3HOLE-SHAFTCOLLAR assembly by replacing the FLATBAR component and recovering from two Mate errors. Additional components are added to the WHEEL-AND-AXLE assembly in the chapter exercises.

PNEUMATIC-TEST-MODULE Assembly

Create the PNEUMATIC-TEST-MODULE assembly.

The first component is the FLAT-PLATE. The FLAT-PLATE is fixed to the Origin ↳. The FLAT-PLATE part was created in the Chapter 3 exercises.

Modify the LINKAGE assembly. Insert the HEX-STANDOFF part. Insert the LINKAGE assembly into the PNEUMATIC-TEST-MODULE assembly.

Insert the AIR-RESERVOIR-SUPPORT assembly. Utilize the Linear Component Pattern and Feature Driven Component Pattern tools. Insert the FRONT-SUPPORT assembly. Utilize the Mirror Components tool to create a mirrored version of the FRONT-SUPPORT assembly.

Insert the WHEEL-AND-AXLE assembly. Utilize Component Properties and create a Flexible State. Utilize the ConfigurationManager to select the Flexible configuration for the AirCylinder assembly.

 Work between multiple part and sub-assembly documents to create the final assembly.

Activity: PNEUMATIC-TEST - MODULE Assembly

Create the PNEUMATIC-TEST-MODULE assembly.

243) Click **New** ⬜ from the Menu bar.

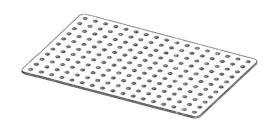

244) Double-click **Assembly** from the Templates tab.

Insert the FLAT-PLATE part.
245) Double-click **FLAT-PLATE** from the SW-TUTORIAL-2020\Chapter 6 Models folder.

246) Click **OK** ✔ from the Begin Assembly PropertyManager. FLAT-PLATE is fixed to the Origin.

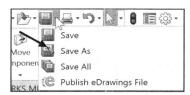

Save the assembly. Enter name. Enter description.
247) Click **Save As** from the drop-down Menu bar.

248) Browse to the **SW-TUTORIAL-2020** folder.

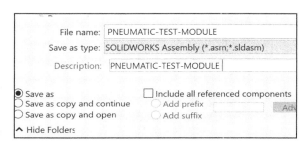

249) Enter **PNEUMATIC-TEST-MODULE** for File name.

250) Enter **PNEUMATIC-TEST-MODULE** for Description.

251) Click **Save**.

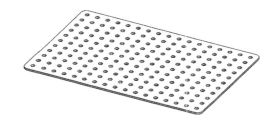

Activity: Modify the LINKAGE Assembly

Modify the LINKAGE assembly created in Chapter 2. Note: The LINKAGE assembly is in the Chapter 6 Models folder\LINKAGE Assembly folder.

252) Click **Open** 🖰 from the Menu bar.

253) Double-click **LINKAGE** from the Chapter 6 Models folder\LINKAGE Assembly folder. The LINKAGE assembly is displayed.

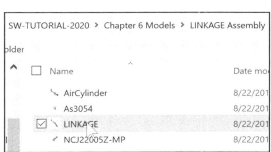

Insert the HEX-STANDOFF part into the LINKAGE assembly.

254) Click the **Insert Components** tool from the Assembly toolbar.

255) Double-click **HEX-STANDOFF** from the SW-TUTORIAL-2020 folder.

256) Click a **position** to the front of the half Slot Cut as illustrated.

View the Temporary Axes.
257) Click **View**, **Hide/Show**, check **Temporary Axes** from the Menu bar.

258) Click **View**, **Hide/Show**, check **Axes** from the Menu bar.

Insert a Coincident mate.
259) **Zoom-in** to view the Temporary Axis of the HEX-STANDOFF part.

260) Click the **HEX-STANDOFF** tapped hole Temporary Axis as illustrated.

261) **Zoom-in** to view the half Slot Cut Axis1 as illustrated.

262) Hold the **Ctrl** key down.

263) Click the **half Slot Cut Axis1**.

264) Release the **Ctrl** key. The Mate pop-up menu is displayed.

265) Click **Coincident** from the Mate pop-up menu.

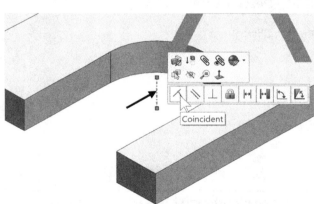

Reference geometry defines the shape or form of a surface or a solid. Reference geometry includes planes, axes, coordinate systems, and points.

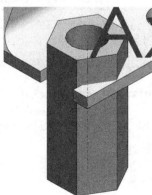

Insert a Coincident mate.

266) Click the **HEX-STANDOFF** top face.

267) **Rotate** to view the BRACKET bottom face.

268) Hold the **Ctrl** key down.

269) Click the **BRACKET bottom face**.

270) Release the **Ctrl** key. The Mate pop-up menu is displayed.

271) Click **Coincident** from the Mate pop-up menu.

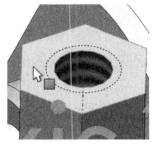

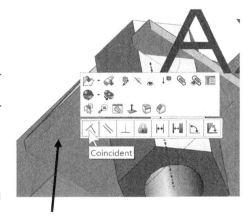

Insert a Parallel mate.

272) **Rotate** to view the model as illustrated.

273) Click the **HEX-STANDOFF** front face.

274) Hold the **Ctrl** key down.

275) Click the **BRACKET front face**.

276) Release the **Ctrl** key. The Mate pop-up menu is displayed.

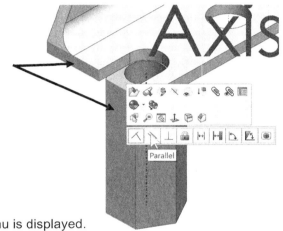

277) Click **Parallel** ⟍ from the Mate pop-up menu.

Fit the model to the Graphics window.

278) Press the **f** key.

Display an Isometric view. Save the model.

279) Click **Isometric view** ⬚.

280) Click **Save** 💾.

View the created Mates.

281) **Expand** the Mates folder. View the created mates.

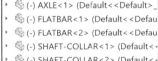

Insert the second HEX-STANDOFF.

282) Click the **Insert Components** 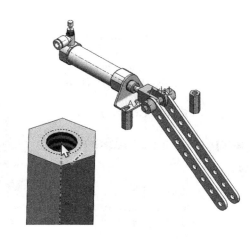 tool from the Assembly toolbar.

283) Double-click **HEX-STANDOFF** from the SW-TUTORIAL-2020 folder.

284) Click a **position** to the back right of the back half Slot Cut as illustrated.

Insert a Concentric mate.
285) **Rotate** the model to view the second HEX-STANDOFF.

286) **Zoom-in** on the Temporary Axis.

287) Click the **second HEX-STANDOFF** Tapped Hole Temporary Axis.

288) **Rotate** to view the back half Slot Cut Axis1.

289) Hold the **Ctrl** key down.

290) Click the **back half Slot Cut Axis1**.

291) Release the **Ctrl** key. The Mate pop-up menu is displayed.

292) Click **Coincident** from the Mate pop-up menu.

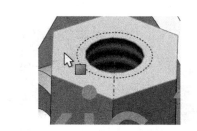

Insert a Coincident mate.
293) Click the **second HEX-STANDOFF top** face.

294) **Rotate** to view the BRACKET bottom face.

295) Hold the **Ctrl** key down.

296) Click the **BRACKET bottom face**.

297) Release the **Ctrl** key. The Mate pop-up menu is displayed.

298) Click **Coincident** from the Mate pop-up menu.

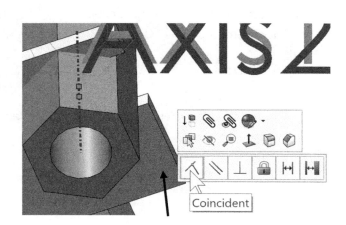

Insert a Parallel mate.

299) Click the **second HEX-STANDOFF** face as illustrated.

300) Hold the **Ctrl** key down.

301) Click the **BRACKET back right** face.

302) Release the **Ctrl** key. The Mate pop-up menu is displayed.

303) Click **Parallel** ⟍ from the Mate pop-up menu.

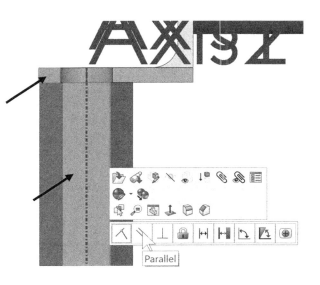

🔆 Use Selection Filter toolbar to select the correct face, edge, Axis, etc.

Fit the model to the Graphics window. Display an Isometric view. Hide Axis and Temporary Axis.
304) Press the **f** key.

305) Click **View, Hide/Show**, check **Temporary Axes** from the Menu bar.

306) Click **View, Hide/Show**, check **Axes** from the Menu bar.

307) Click **Isometric view** 📦.

Save the assembly.
308) Click **Save** 💾.

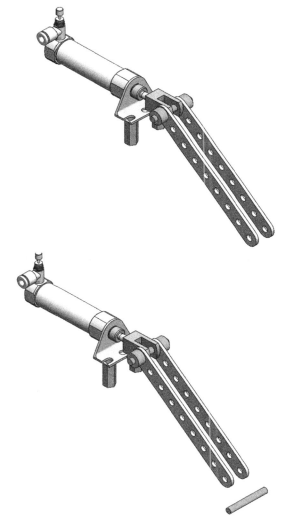

Insert the second AXLE.

309) Click the **Insert Components** 🔧 tool from the Assembly toolbar.

310) Double-click **AXLE** from the SW-TUTORIAL-2020 folder.

311) Click a **position** in the assembly as illustrated.

Insert a Concentric mate.

312) Click the **second AXLE cylindrical face**.

313) Hold the **Ctrl** key down.

314) Click the **FLATBAR bottom hole** face as illustrated.

315) Release the **Ctrl** key. The Mate pop-up menu is displayed.

316) Click **Concentric** from the Mate pop-up menu.

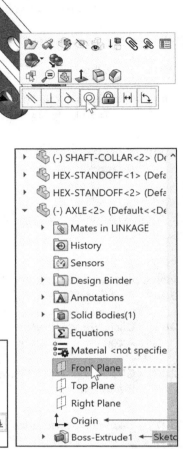

Insert a Coincident mate.

317) Click the second **AXLE Front Plane** from the FeatureManager.

318) Hold the **Ctrl** key down.

319) Click the **LINKAGE assembly Front Plane** from the FeatureManager.

320) Release the **Ctrl** key. The Mate pop-up menu is displayed.

321) Click **Coincident** from the Mate pop-up menu.

Fit the model to the Graphics window.

322) Press the **f** key.

Display an Isometric view. Save the assembly.

323) Click **Isometric view** from the Heads-up View toolbar.

Save the LINKAGE assembly.

324) Click **Save** .

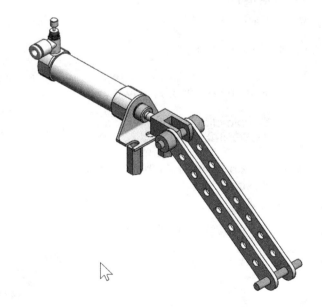

Insert the first SHAFT-COLLAR on the second AXIS.

325) Click the **Insert Components** 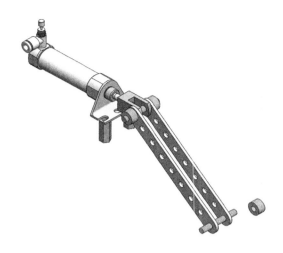 Assembly tool.

326) Double-click **SHAFT-COLLAR** from the SW-TUTORIAL-2020 folder.

327) Click a **position** to the back of the second AXLE as illustrated.

Enlarge the view.
328) Zoom-in on the **SHAFT-COLLAR** and the **AXLE** to enlarge the view.

Insert a Concentric mate.
329) Click the inside **hole face** of the SHAFT-COLLAR.

330) Hold the **Ctrl** key down.

331) Click the **long cylindrical face** of the AXLE.

332) Release the **Ctrl** key. The Mate pop-up menu is displayed.

333) Click **Concentric** from the Mate pop-up menu.

☼ The book is designed to expose the new SOLIDWORKS user to many different tools, techniques and procedures. It may not always use the most direct tool or process.

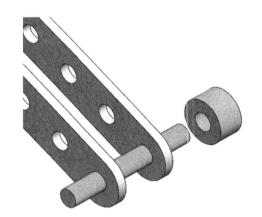

Insert a Coincident mate.

334) Click the **front face** of the SHAFT-COLLAR as illustrated.

335) **Rotate** the assembly to view the back face of the FLATBAR.

336) Hold the **Ctrl** key down.

337) Click the **back face** of the FLATBAR.

338) Release the **Ctrl** key. The Mate pop-up menu is displayed.

339) Click **Coincident** from the Mate pop-up menu.

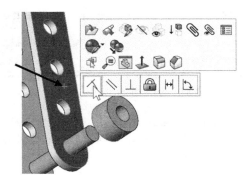

Display the Isometric view.

340) Click **Isometric view** .

Insert the second SHAFT-COLLAR.

341) Click the **Insert Components** Assembly tool.

342) Double-click **SHAFT-COLLAR** from the SW-TUTORIAL-2020 folder.

343) Click a **position** to the front of the AXLE.

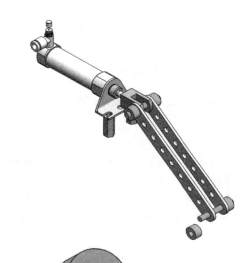

Enlarge the view.

344) **Zoom in** on the second SHAFT-COLLAR and the AXLE.

Insert a Concentric mate.

345) Click the **inside hole face** of the second SHAFT-COLLAR. Note the icon face feedback symbol.

346) Hold the **Ctrl** key down.

347) Click the **long cylindrical face** of the AXLE.

348) Release the **Ctrl** key. The Mate pop-up menu is displayed.

349) Click **Concentric** from the Mate pop-up menu.

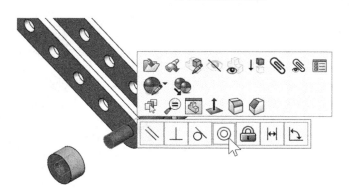

Insert a Coincident mate.

350)   Press the **f** key to fit the model to the
            Graphics window.

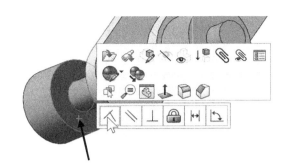

351)   **Zoom in** on the front face of the FLATBAR
            as illustrated.

352)   Click the **front face** of the FLATBAR.

353)   **Rotate** to view the back face of the second
            SHAFT-COLLAR.

354)   Hold the **Ctrl** key down.

355)   Click the **back face** of the second SHAFT-
            COLLAR.

356)   Release the **Ctrl** key. The Mate pop-up
            menu is displayed.

357)   Click **Coincident** from the Mate pop-up
            menu.

Display an Isometric view.

358)   Click **Isometric view** .

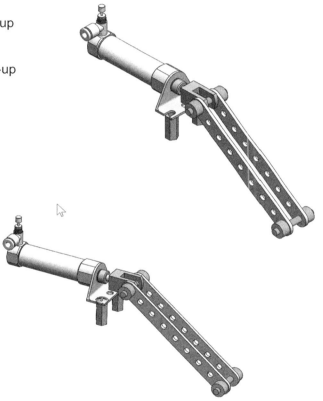

Save the LINKAGE assembly.

359)   **Rebuild** the model.

360)   Click **Save** . As an exercise,
            insert SCREWs between the
            AirCylinder assembly and the
            two HEX-STANDOFFs as
            illustrated.

Activity: PNEUMATIC-TEST-MODULE - Insert LINKAGE Assembly

Insert the LINKAGE assembly into the PNEUMATIC-TEST-MODULE assembly.

361) Click **Window**, **Tile Horizontally** from the Menu bar. The PNEUMATIC-TEST-MODULE assembly should be open. If not, open the PNEUMATIC-TEST MODULE assembly document.

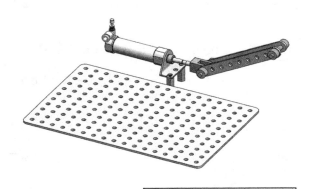

362) Rotate the two **FLATBARs** approximately 45°.

363) Click and drag the **LINKAGE**
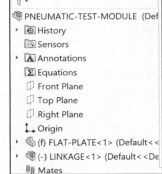 LINKAGE<1> assembly icon into the PNEUMATIC-TEST-MODULE assembly.

364) Click a **position** above the FLAT-PLATE. The LINKAGE assembly should be in a rigid state.

365) **Maximize** the PNEUMATIC-TEST-MODULE Graphics window.

366) Click **View**, **Hide/Show**, un-check **Origins** from the Menu bar. If required, click View, Hide/Show, un-check Planes from the Menu bar.

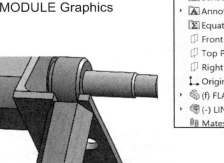

Rotate model.

367) **Rotate** the model to view the Front HEX-STANDOFF part as illustrated.

Insert a Concentric mate.

368) Click the **Front HEX-STANDOFF Tapped** inside hole face as illustrated.

369) **Rotate** the model to view the FLAT-PLATE hole as illustrated. Hold the **Ctrl** key down.

370) Click the **FLAT-PLATE Hole** face in the 5th row, 4th column as illustrated.

371) Release the **Ctrl** key. The Mate pop-up menu is displayed.

372) Click **Concentric** from the Mate pop-up menu.

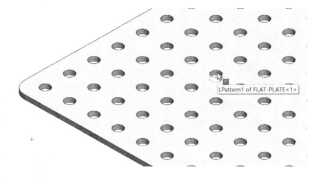

Insert a Parallel mate between two Planes.

373) Click the **PNEUMATIC-TEST-MODULE Front Plane** from the FeatureManager.

374) Hold the **Ctrl** key down.

375) Click the **LINKAGE assembly Front Plane** from the FeatureManager.

376) Release the **Ctrl** key. The Mate pop-up menu is displayed.

377) Click **Parallel** from the Mate pop-up menu.

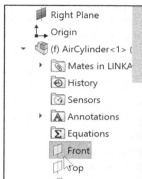
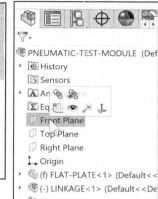

Insert a Coincident mate.

378) Click the **Front HEX-STANDOFF** bottom face.

379) Hold the **Ctrl** key down.

380) Click the **FLAT-PLATE top face**.

381) Release the **Ctrl** key. The Mate pop-up menu is displayed.

382) Click **Coincident** from the Mate pop-up menu.

Display an Isometric view.

383) Click **Isometric view** .

Save the PNEUMATIC-TEST-MODULE assembly.

384) Click **Save** .

The LINKAGE assembly is fully defined and located on the FLAT-PLATE part. Insert the AIR-RESERVOIR-SUPPORT assembly. The AIR-RESERVOIR-SUPPORT assembly was created in the Chapter 3 exercises. The AIR-RESERVOIR SUPPORT is located in Chapter 6 Models folder.

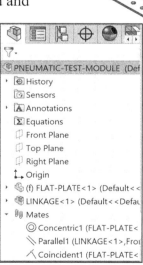

Activity: PNEUMATIC-TEST-MODULE - Insert AIR-RESERVOIR-SUPPORT

Insert the AIR-RESERVOIR-SUPPORT assembly.

385)   Click the **Insert Components** Assembly tool. The Insert Component PropertyManager is displayed.

386)   Double-click the **AIR-RESERVOIR-SUPPORT** assembly from the SW-TUTORIAL-2020\Chapter 6 Models folder. The AIR-RESERVOIR SUPPORT assembly is in the Chapter 6 Models folder. Work from your local hard drive.

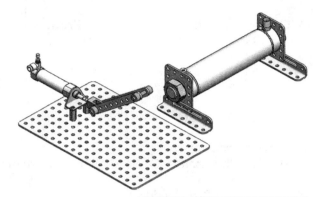

387)   Click a **position** above the FLAT-PLATE as illustrated.

388)   Click the **Rotate Component** Assembly tool. The Rotate Component PropertyManager is displayed. The Rotate icon is displayed in the Graphics window.

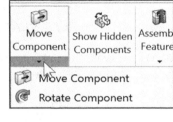

389)   Click and drag the **AIR-RESERVOIR-SUPPORT** until the tank is parallel with the AirCylinder assembly as illustrated.

390)   Click **OK** from the Rotate Component PropertyManager.

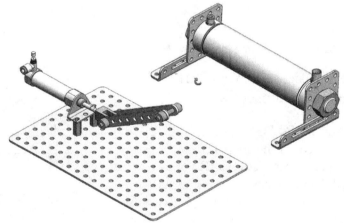

Insert a Concentric mate.

391)   Click the **FLAT-PLATE back left inside hole face** as illustrated.

392)   Hold the **Ctrl** key down.

393)   Click the **fourth ANGLE-BRACKET inside hole** face.

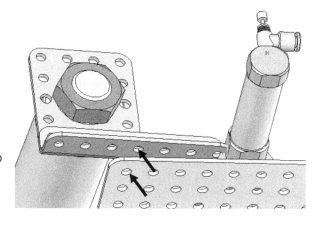

394)   Release the **Ctrl** key. The Mate pop-up menu is displayed.

395)   Click **Concentric** from the Mate pop-up menu.

Insert a Coincident mate.

396)   Click the **ANGLE-BRACKET bottom** face.

397)   **Rotate** the assembly to view the FLAT-PLATE top face. Hold the **Ctrl** key down.

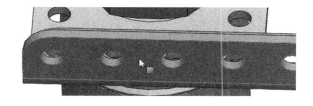

398)   Click the **FLAT-PLATE top** face.

399)   Release the **Ctrl** key. The Mate pop-up menu is displayed.

400)   Click **Coincident** from the Mate pop-up menu.

Insert a Parallel mate.

401)   Click **Left view** ⬛ from the Heads-up View toolbar.

402)   Click the **ANGLE-BRACKET** narrow face.

403)   Hold the **Ctrl** key down.

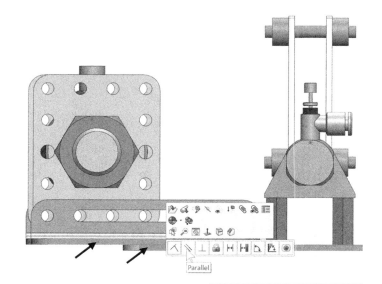

404)   Click the **FLAT-PLATE narrow** face.

405)   Release the **Ctrl** key. The Mate pop-up menu is displayed.

406)   Click **Parallel** ╲ from the Mate pop-up menu. The second Parallel mate is created.

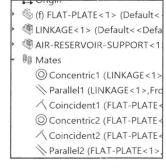

Display an Isometric view.

407) Click **Isometric view** 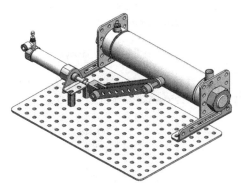 from the Heads-up View toolbar.

Save the PNEUMATIC-TEST-MODULE assembly.

408) Click **Save** .

Component Patterns in the Assembly

There are various methods to define a pattern in an assembly:

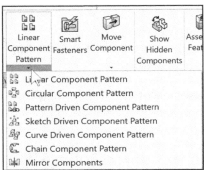

- Linear Component Pattern.

- Circular Component Pattern.

- Pattern Driven Component (Feature).

- Sketch Driven Component Pattern.

- Curve Driven Component Pattern.

- Chain Component Pattern.

- Mirror Components.

A Linear/Circular Component Pattern tool utilizes geometry in the assembly to arrange instances in a Linear or Circular pattern.

A Pattern Driven Component Pattern utilizes an existing feature pattern.

The SCREW part fastens the ANGLE-BRACKET part to the FLAT-PLATE part. Mate one SCREW to the first instance on the ANGLE-BRACKET Linear Pattern.

Utilize the Pattern Driven Component Pattern tool to create instances of the SCREW. Suppress the instances.

Utilize the Linear Component Pattern tool to copy the Pattern Driven Component Pattern of SCREWS to the second ANGLE-BRACKET part.

Drag the part by specific geometry to create a mate.

Activity: PNEUMATIC-TEST-MODULE - Component Pattern

Open the SCREW part.

409) Click **Open** ⃞ from the Menu bar.

410) Double-click **SCREW** from the Chapter 6 Models folder.

411) Un-suppress the **Fillet1** and **Chamfer1** feature.

412) Click **Window**, **Tile Horizontally** to display the SCREW and the PNEUMATIC-TEST-MODULE assembly.

Insert and mate the SCREW.

413) Click the **bottom circular edge** of the SCREW.

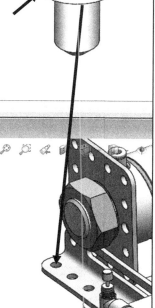

414) Drag the **SCREW** into the PNEUMATIC-TEST-MODULE assembly window. Note: Zoom in on the top circular edge of the ANGLE-BRACKET left hole.

415) Release the mouse pointer on the **top circular edge** of the ANGLE-BRACKET left hole. The mouse pointer displays the Coincident/Concentric feedback symbol.

416) **Return** to the PNEUMATIC-TEST-MODULE assembly window.

417) Click **Left view** ⃞ from the Heads-up View toolbar.

The SCREW part is positioned in the left hole with a Coincident/Concentric mate.

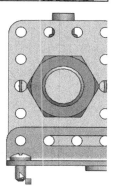

Create a Pattern Driven Component Pattern.

418) Click the **Pattern Driven Component Pattern** ⃞ tool from the Consolidate Assembly toolbar. The PropertyManager is displayed.

419) Click the **SCREW** component in the Graphics window. SCREW<1> is displayed in the Components to Pattern box.

420) Click inside the **Driving Feature or Component** box. **Expand** the AIR-RESERVOIR-SUPPORT assembly in the PNEUMATIC-TEST-MODULE fly-out FeatureManager.

421) **Expand** ANGLE-BRACKET <1>.

422) Click **LPattern1** in the fly-out FeatureManager. The SCREW is the seed feature.

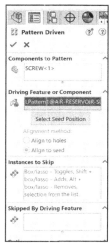

423) Click **OK** ✔ from the PropertyManager. Six instances are displayed in the Graphics window. DerivedLPattern1 is displayed in the FeatureManager.

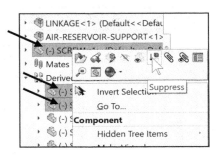

424) Click **SCREW<1>** from the FeatureManager.

425) Hold the **Ctrl** key down.

426) **Expand** DerivedLPattern1 from the FeatureManager.

427) Click the first two entries: **SCREW<2>** and **SCREW<3>**. Release the **Ctrl** key.

428) Right-click **Suppress**. The first three SCREWs are not displayed.

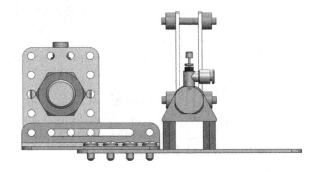

💡 The **Pattern Driven Component Pattern** PropertyManager contains an option to skip instances of a feature component in a pattern.

Activity: PNEUMATIC-TEST-MODULE - Linear Component Pattern

Create a Linear Component Pattern.

429) Click **Front view** 🔲 from the Heads-up View toolbar.

430) Click the **Linear Component Pattern** 🔳 tool from the Assembly toolbar. The Linear Pattern PropertyManager is displayed.

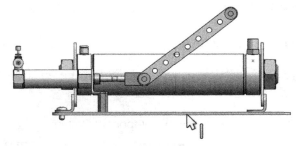

431) Click inside the **Pattern Direction** box for Direction 1.

432) Click the long front edge of the **FRONT-PLATE** as illustrated. Edge<1> is displayed in the Pattern Direction box. The direction arrow points to the right.

433) Enter **177.80**mm, [7in] for Spacing.

434) Enter **2** for Instances.

435) Click inside the **Components to Pattern** box.

436) Click **DerivedLPattern1** from the fly-out FeatureManager.

437) Click **OK** ✔ from the Linear Pattern PropertyManager.

438) Click **Top view** from the Heads-up View toolbar. View the results.

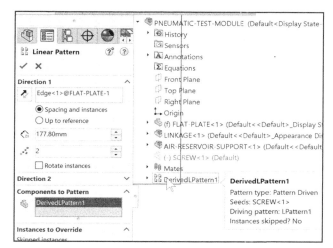

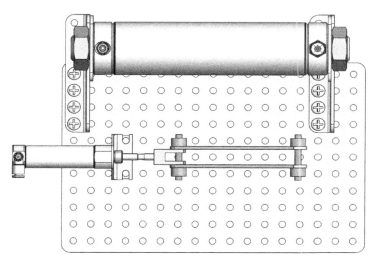

The Linear Component Pattern feature is displayed in the second ANGLE-BRACKET part. LocalLPattern1 is displayed in the FeatureManager.

Display an Isometric view. Save the PNEUMATIC TEST MODULE assembly.

439) Click **Isometric view** .

440) **View** the FeatureManager and the created features in their states.

441) Click **Save** .

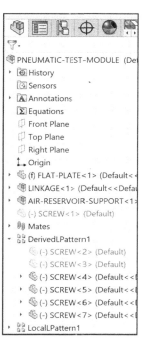

Hide the AIR-RESERVOIR-SUPPORT assembly.

442) Right-click **AIR-RESERVOIR-SUPPORT<1>** in the FeatureManager.

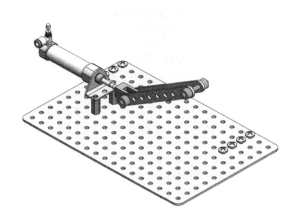

443) Click **Hide Components** . Utilize the Show components tool to display a component that has been hidden.

Hide the LINKAGE assembly.

444) Right-click **LINKAGE<1>** in the FeatureManager.

445) Click **Hide Components** .

Insert the FRONT-SUPPORT assembly into the PNEUMATIC-TEST-MODULE assembly. Mate the FRONT-SUPPORT assembly to the FLAT-PLATE part.

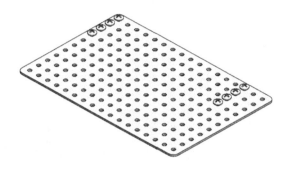

Utilize the Mirror Components tool to create a mirrored copy of the FRONT-SUPPORT assembly.

You can create new components by mirroring existing part or sub-assembly components. The new components can either be a copy or a mirror of the original components. A mirrored component is sometimes called a "right-hand" version of the original "left-hand" version.

Access the Mirror Components tool from the Consolidated Linear Component Pattern drop-down menu.

Activity: PNEUMATIC-TEST-MODULE - Insert FRONT-SUPPORT Assembly

---

Insert and mate the FRONT-SUPPORT assembly.

446) Click the **Insert Components** tool from the Assembly toolbar.

447) Double-click **FRONT-SUPPORT** from the SW-TUTORIAL-2020 folder.

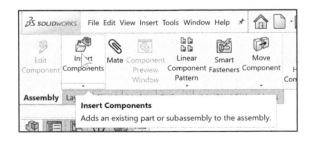

Hide the SCREWs in the FRONT-SUPPORT.

448) Click a **position** above the FLAT-PLATE as illustrated.

449) Right-click **FRONT-SUPPORT** in the FeatureManager.

450) Click **Open Assembly**.

451) **Hide** the SCREW parts and HEX-NUTS if required in the assembly.

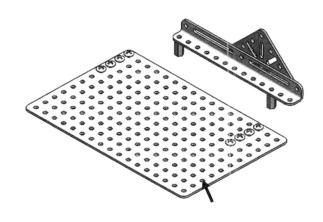

Save the model.

452) Click **Save** 💾.

Return to the PNEUMATIC-TEST-MODULE assembly.

453) Press **Ctrl Tab**.

454) Select the **PNEUMATIC-TEST-MODULE** document.

Save the model.

455) Click **Save** 💾.

456) Click **Yes** to the Message, "Save the document and referenced models now?"

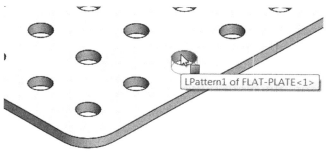

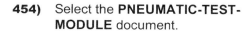

The PNEUMATIC-TEST-MODULE is displayed. The SCREW and HEX-NUT components are hidden in the FRONT-SUPPORT assembly.

Insert a Concentric mate.

457) Click the **FLAT-PLATE hole face** in the 3rd row, right most column (17th column).

458) Hold the **Ctrl** key down.

459) Click the **HEX-STANDOFF Tapped Hole** as illustrated.

460) Release the **Ctrl** key. The Mate pop-up menu is displayed.

461) Click **Concentric** from the Mate pop-up menu. The mate is created.

Insert a Coincident mate.

462) Click the bottom face of the **HEX-STANDOFF** part.

463) **Rotate** the model to view the top face of the FLAT-PLATE.

464) Hold the **Ctrl** key down.

465) Click the top face of the **FLAT-PLATE**.

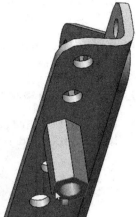

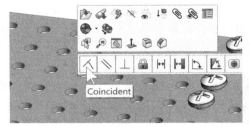

466) Release the **Ctrl** key. The Mate pop-up menu is displayed.

467) Click **Coincident** from the Mate pop-up menu.

Insert a Parallel mate.

468) Click the **FRONT-SUPPORT Front Plane** from the FeatureManager.

469) **Show** the LINKAGE assembly.

470) Hold the **Ctrl** key down.

471) Click the **LINKAGE Front Plane** from the FeatureManager.

472) Release the **Ctrl** key. The Mate pop-up menu is displayed.

473) Click **Parallel** ⟍ from the Mate pop-up menu.

474) **Hide** the LINKAGE assembly.

Save the PNEUMATIC-TEST-MODULE assembly.

475) Click **Save** 💾.

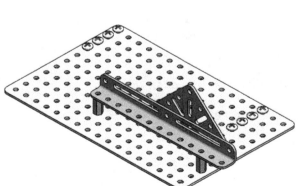
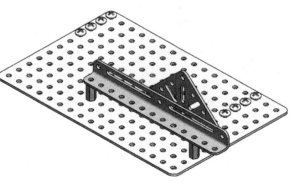

Mirrored Components

Create new components by mirroring existing parts or sub-assemblies. New components are created as copied geometry or mirrored geometry.

A mirrored component is sometimes called a "right-hand" version of the original "left hand" version.

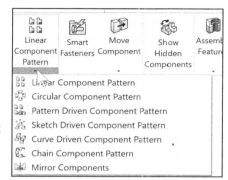

The copied or mirrored component changes when the original component is modified.

A mirrored component creates a new document. The default document prefix is mirror. A copied component does not create a new document.

Suppressed components in the original sub-assembly are not mirrored or copied. The SCREW parts in the FRONT-SUPPORT assembly are not copied.

Left Hand Version (Original)   Right Hand Version (Mirrored)

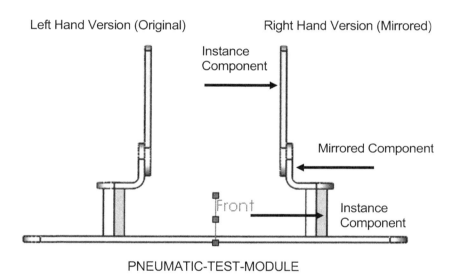

PNEUMATIC-TEST-MODULE

Activity: PNEUMATIC-TEST-MODULE Assembly - Mirrored Component

Insert a Mirrored Component.

476) Click the **Mirror Components** ⬚ tool from the Consolidated Assembly toolbar. The Mirror Components PropertyManager is displayed.

Step 1: The Selections box is displayed. Select face/plane to mirror about and the components to be mirrored.

477) **Expand** the PNEUMATIC-TEST-MODULE fly-out FeatureManager. Click the **FLAT-PLATE Front Plane**. The Front Plane is displayed in the Mirror plane box.

478) Click the **FRONT-SUPPORT** assembly from the fly-out FeatureManager.

479) Click **Next** . The FRONT-SUPPORT assembly is displayed in the Components to Mirror box.

Step 2: Verify the orientation of the components to be mirrored and adjust accordingly using the buttons below.

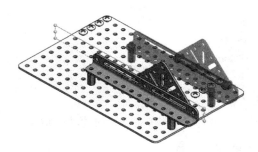

480) Click the **Create opposite hand version** button. View the results in the Graphics window. View your options.

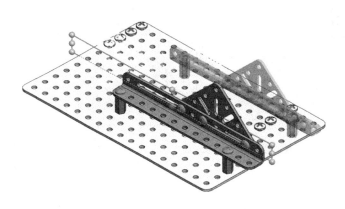

481) Click **Next** . Accept the default settings.

Step 3: The mirrored components selected in step 2 need new geometries. This may be new files or new configurations in existing files. Specify the new configuration or file name using the options below.

482) Click **Next** . This is for any Import Features. Accept the default settings.

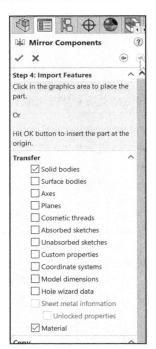

483) Click **OK** from the Mirror Components PropertyManager. Click Yes if needed. View the results in the Graphics window.

Select suppressed components and their mates to be mirrored in the Components to Mirror box.

The MirrorFRONT-SUPPORT, TRIANGLE, ANGLE-13HOLE, and SCREW components are mated.

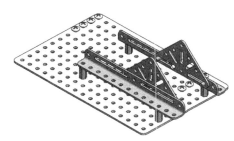

Utilize the Fix option to mate the MirrorFRONT-SUPPORT to its current location in the PNEUMATIC-TEST-MODULE. No other mates are required.

Activity: PNEUMATIC-TEST-MODULE - Fix MIRRORFRONT-SUPPORT

Fix the MirrorFRONT-SUPPORT.

484) Right-click **MirrorFRONT-SUPPORT** from the FeatureManager.

485) Click **Fix**. The MirrorFRONT assembly is fixed in the PNEUMATIC-TEST-MODULE assembly. The MirrorFRONT-SUPPORT does not move or rotate.

Display the LINKAGE assembly.

486) Right-click **LINKAGE** in the FeatureManager.

487) Click **Show Components**.

Save the PNEUMATIC-TEST-MODULE assembly.

488) Click **Save** .

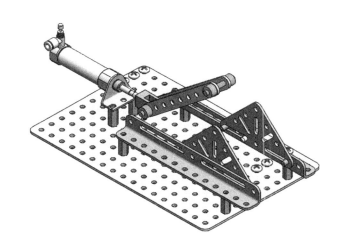

Reuse geometry in the assembly. Utilize the Mirror Component tool to create a left and right version of parts and assemblies.

Component Properties

Component Properties control the flexibility of the sub-assembly when inserted into an assembly. Components do not translate or rotate after insertion into the assembly.

The FLATBAR parts in the LINKAGE assembly do not rotate after insertion into the PNEUMATIC TEST MODULE assembly. The LINKAGE assembly is in the Rigid State.

Insert the WHEEL-AND-AXLE assembly into the PNEUMATIC-TEST-MODULE assembly. Modify the Mate state of the LINKAGE assembly from the Rigid State to the Flexible State. The LINKAGE assembly is free to rotate.

Modify the Mate State of the AirCylinder assembly from the Rigid State to the Flexible State. The AirCylinder assembly is free to translate.

⚬ By default, when you create a sub-assembly, it is rigid. Within the parent assembly, the sub-assembly acts as a single unit and its components do not move relative to each other.

Activity: PNEUMATIC-TEST-MODULE Assembly - Insert WHEEL-AND-AXLE Assembly

Insert the WHEEL-AND-AXLE assembly.

489) Click the **Insert Components** tool from the Assembly toolbar.

490) Double-click the **WHEEL-AND-AXLE** assembly from the SW-TUTORIAL-2020 folder.

491) Click a **position** above the FLAT-PLATE part as illustrated.

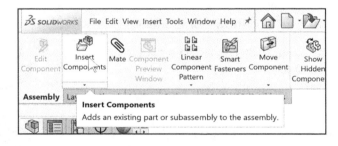

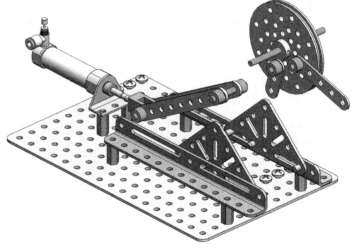

Insert a Concentric mate.
492) Click the **AXLE-3000 cylindrical** face.

493) **Rotate** the model to view the inside TRIANGLE top hole face as illustrated.

494) Hold the **Ctrl** key down.

495) Click the inside **TRIANGLE top hole** face.

496) Release the **Ctrl** key. The Mate pop-up menu is displayed.

497) Click **Concentric** from the Mate pop-up menu.

Insert a Coincident mate.
498) Click the **WHEEL-AND-AXLE Front Plane** from the FeatureManager.

499) Hold the **Ctrl** key down.

500) Click the **LINKAGE Front Plane** from the FeatureManager.

501) Release the **Ctrl** key. The Mate pop-up menu is displayed.

502) Click **Coincident** from the Mate pop-up menu.

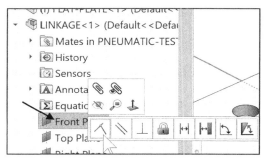

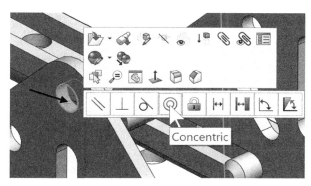

A Concentric mate is required between the left hole of the FLATBAR-3HOLE and right AXLE of the LINKAGE assembly.

DO NOT INSERT A CONCENTRIC MATE AT THIS TIME.

A Concentric mate will result in Mate errors.

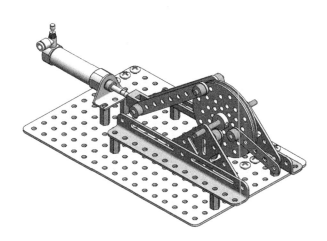

The WHEEL-AND-AXLE is free to rotate in the PNEUMATIC-TEST-MODULE assembly. The LINKAGE assembly interferes with the WHEEL-AND-AXLE. The LINKAGE assembly is not free to rotate or translate. Mate errors will occur.

Sub-assemblies within the LINKAGE assembly are in a Rigid Mate state when inserted into the PNEUMATIC-TEST-MODULE assembly. Remove the Rigid Mate state and insert a Concentric mate between the LINKAGE assembly and the WHEEL-AND-AXLE.

Activity: PNEUMATIC-TEST-MODULE Assembly - Remove Rigid state

Remove the Rigid State.

503)   Right-click the **LINKAGE** assembly from the FeatureManager.

504)   Click **Component Properties** from the Context toolbar. The Component Properties dialog box is displayed.

505)   Check **Flexible** in the Solve as box.

506)   Click **OK** from the Component Properties dialog box. Note the icon next to the LINKAGE assembly in the FeatureManager.

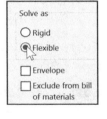

507)   Right-click the **AirCylinder** assembly from the FeatureManager.

508)   Click **Component Properties**. **Flexible** is selected in the Solve as box.

509)   **Flexible** is selected in the Referenced configuration box.

510)   Click **OK** from the Component Properties dialog box.

Hide the FRONT-SUPPORT assembly.

511)   Right-click **FRONT-SUPPORT** from the FeatureManager.

512)   Click **Hide Components**.

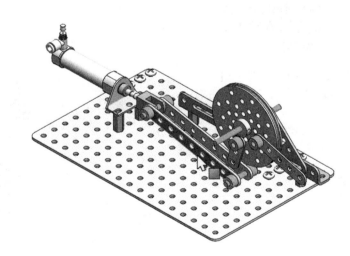

Move the LINKAGE assembly.

513)   Click and drag the **front FLATBAR-9HOLE** downward. The FLATBAR-9HOLE rotates about the left AXLE.

514)   Click a **position** below the WHEEL-AND-AXLE.

515)   Click **inside** the Graphics window to deselect.

Insert a Concentric mate.

516) Click the second bottom **AXLE** face of the LINKAGE assembly.

517) Hold the **Ctrl** key down.

518) Click the inside face of the **bottom hole** of the FLATBAR-3HOLE as illustrated.

519) Release the **Ctrl** key. The Mate pop-up menu is displayed.

520) Click **Concentric** from the Mate pop-up menu.

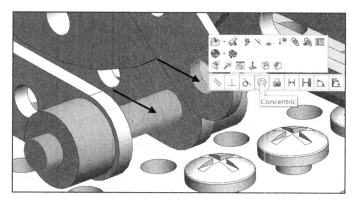

Display an Isometric view. Save the PNEUMATIC-TEST-MODULE assembly.

521) Click **Isometric view** .

522) Click **Save** .

The AirCylinder assembly is inserted in a Rigid state by default. The AirCylinder contains three configurations:

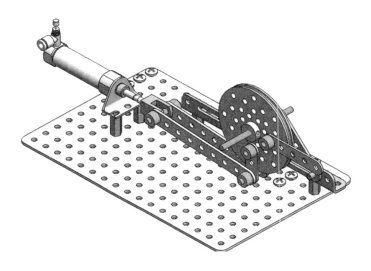

- Default

- Extended

- Flexible

Open the AirCylinder assembly. Modify the configuration to Flexible.

Open the LINKAGE assembly. Modify the AirCylinder Component Properties from Default to Flexible.

Activity: PNEUMATIC-TEST-MODULE Assembly - Review AirCylinder Configurations

Review the AirCylinder configurations.

523) Right-click **LINKAGE** from the PNEUMATIC-TEST-MODULE FeatureManager.

524) Click **Open Subassembly** from the Context toolbar.

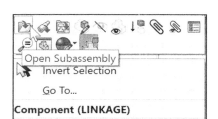

525)   Right-click **AirCylinder** in the LINKAGE
FeatureManager.

526)   Click **Open Sub-assembly**.

527)   Click the **Configuration**
Manager 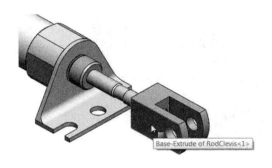 tab at the top of
the AirCylinder
FeatureManager. Three
configurations are displayed:
Default, *Extended* and *Flexible*.
The current Default
configuration sets the Piston
Rod at 0mm.

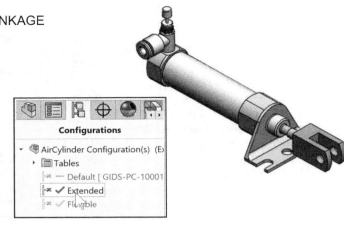

Display the Extended Configuration.
528)   Double-click **Extended**. The
Piston Rod of the AirCylinder
extends 25mm.

Display the Flexible Configuration.
529)   Double-click **Flexible**.

530)   Click the **Rod-Clevis** in the
Graphics window.

531)   Click and drag the **Piston Rod** from left to right.

532)   Click a **position** near its original location. The
AirCylinder remains in the Flexible configuration
for the rest of this project.

Update the LINKAGE assembly.
533)   Click **Window**, **LINKAGE** from the Menu bar.
The current configuration of the AirCylinder part
in the LINKAGE assembly is Default. Modify the
configuration.

534)   Right-click **AirCylinder** from the LINKAGE assembly
FeatureManager.

535)   Click **Component Properties**.

536)   **Flexible** is selected in the Solve as section.

537)   **Flexible** is selected in the Referenced configuration
section.

538)   Click **OK**. The AirCylinder displays the Flexible
configuration next to its name in the FeatureManager.

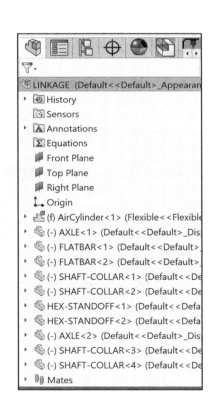

Update the PNEUMATIC-TEST-MODULE.
539)   Click **Window**, **PNEUMATIC-TEST-MODULE** assembly from the Menu bar.

Move the ROD-CLEVIS.
540)   Click and drag the **ROD-CLEVIS** to the right. The WHEEL rotates in a counterclockwise direction.

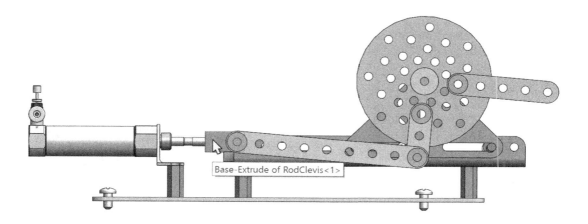

541)   Click and drag the **ROD-CLEVIS** to the left. The WHEEL rotates in a clockwise direction.

Display the AIR-RESERVOIR-SUPPORT assembly.
542)   Right-click **AIR-RESERVOIR-SUPPORT** in the FeatureManager.

543)   Click **Show Components**.

Display the second FRONT-SUPPORT assembly.
544)   Right-click **FRONT-SUPPORT** in the FeatureManager.

545)   Click **Show Components**.

546)   **Display** the SCREWs in the FRONT-SUPPORT.

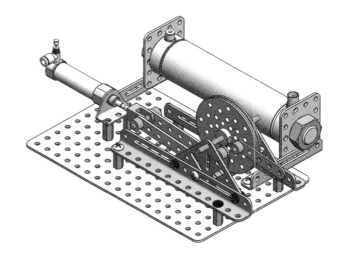

Display an Isometric view. Save the PNEUMATIC-TEST-MODULE.

547)   Click **Isometric view** .

548)   Click **Save** .

549)   Click **Yes** to update referenced
              documents.

Close all documents.
550)   Click **Windows**, **Close All** from
              the Menu bar.

Explore additional parts and
assemblies at the end of this
project. Add the WEIGHT, HOOK
and FLATBAR parts.

Additional details on Show
components, Hide components,
Linear Component Pattern, Circular
Component Pattern, Pattern Driven
Component Pattern, Sketch Driven
Component Patter, Curve Driven
Component Pattern, Mirror
Components, Configurations, and
Configuration Manager are
available in SOLIDWORKS Help.
Select Help, SOLIDWORKS Help
topics.

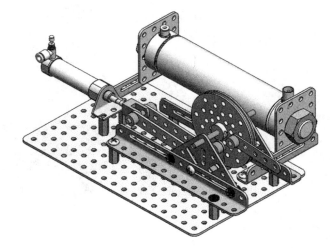

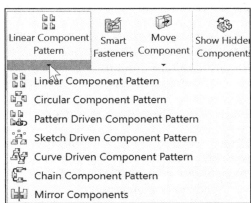

Review the PNEUMATIC-TEST-MODULE Assembly

The PNEUMATIC TEST MODULE assembly was created by combining five major mechanical sub-assemblies.

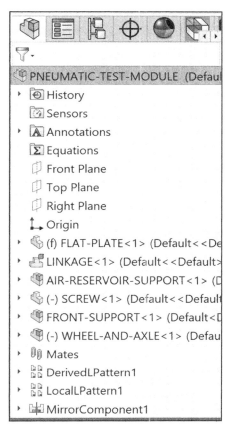

The PNEUMATIC TEST MODULE assembly utilized the FLAT-PLATE as the first component. The LINKAGE assembly, AIR-RESERVOIR-SUPPORT assembly, FRONT-SUPPORT assembly, MirrorFRONT-SUPPORT assembly and the WHEEL-AND-AXLE assembly were mated to the FLAT-PLATE part.

Work at the lowest assembly level. The LINKAGE assembly required two HEX-STANDOFF parts. The HEX-STANDOFF components were inserted into the LINKAGE assembly. The LINKAGE assembly was inserted into the PNEUMATIC-TEST-MODULE assembly.

The AIR-RESERVOIR-SUPPORT assembly was inserted into the PNEUMATIC-TEST-MODULE assembly. The SCREW component utilized a Feature Driven Component Pattern tool and the Linear Component Pattern tool to create multiple copies.

The FRONT-SUPPORT assembly was inserted into the PNEUMATIC-TEST-MODULE assembly. The Mirror Components tool created a mirrored version of the FRONT-SUPPORT assembly.

The WHEEL-AND-AXLE assembly was inserted into the PNEUMATIC-TEST-MODULE assembly. Component Properties created a Flexible state mate.

The LINKAGE assembly "AirCylinder" is in a Flexible state.

In the next section, create the final ROBOT assembly. Either use your created assemblies and components, or use the created assemblies and components provided in the Chapter 6 Models folder.

Final ROBOT Assembly

The final ROBOT assembly is comprised of three major sub-assemblies and the HEX-STANDOFF component:

- Robot Platform assembly.

- PNEUMATIC-TEST-MODULE assembly.

- Basic_integration assembly.

- 4 - HEX-STANDOFF components.

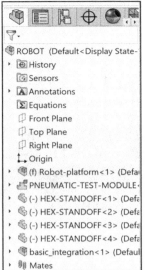

☼ The LINKAGE assembly is in a Flexible state.

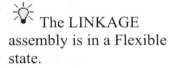

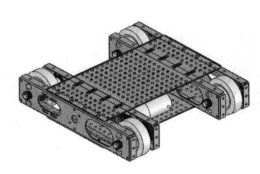

☼ A Rebuild icon may be displayed in the final ROBOT Assembly FeatureManager and a flexible rebuild 🔁 icon at the sub-assembly level when the AirCylinder assembly is in the Flexible state.

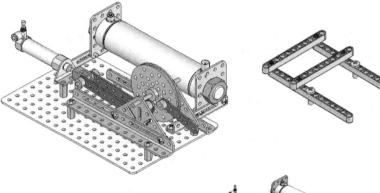

All assemblies and components for the final ROBOT assembly are provided in the Chapter 6 Models folder.

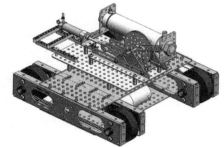

☼ As an exercise apply PhotoView 360 to render the final ROBOT assembly.

Combine the created new assemblies and components to develop the final ROBOT assembly. In this next section, work from the Chapter 6 Models folder on your hard drive. Create the SW-TUTORIAL-2020\Chapter 6 Models folder.

Note: Step-by-step instructions are not provided in this section for the mates.

Activity: Create the ROBOT Assembly

Create the final ROBOT assembly.

551) Click **New** from the Menu bar.

552) Double-click **Assembly** from the Templates tab. The Begin Assembly PropertyManager is displayed.

Insert the Robot-platform assembly.
553) Double-click **Robot-platform** from the SW-TUTORIAL-2020\Chapter 6 Models folder. All files and folders should have been copied from the book.

554) Click **OK** ✔ from the Begin Assembly PropertyManager. The Robot-platform is fixed to the assembly Origin.

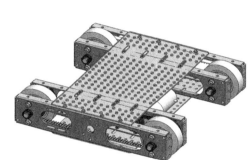

Save the Assembly.
555) Click **Save As** from the drop-down Menu bar.

556) Enter **ROBOT** for File name in the SW-TUTORIAL-2020\Chapter 6 Models folder.

557) Enter **Final ROBOT Assembly** for Description.

558) Click **Save**.

Insert the PNEUMATIC-TEST-MODULE assembly.

559) Click **Insert Components** from the Assembly toolbar. The Insert Components PropertyManager is displayed.

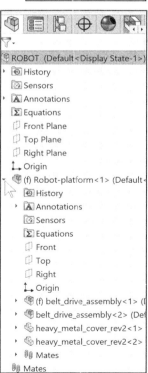

560) Double-click the **PNEUMATIC-TEST-MODULE** assembly from the SW-TUTORIAL-2020\Chapter 6 Models folder.

561) Click a **position** above the Robot-platform.

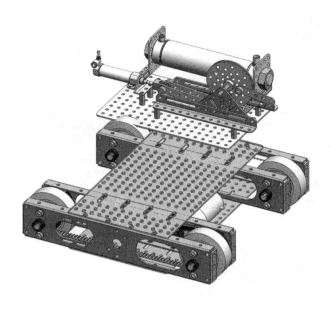

If an assembly or component is loaded in a Lightweight state, right-click the **assembly name** or **component name** from the FeatureManager. Click **Set Lightweight to Resolved**.

Insert 4 HEX-STANDOFF components.

562) Click **Insert Components** from the Assembly toolbar.

563) Double-click the **HEX-STANDOFF** part from SW-TUTORIAL-2020\Chapter 6 models folder.

564) **Pin** the Insert Components PropertyManager.

565) Click **four positions** as illustrated for the four HEX-STANDOFF components.

566) **Un-Pin** the Insert Components PropertyManager.

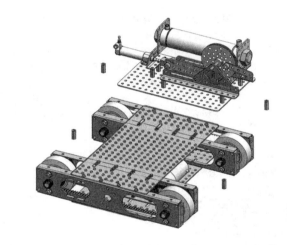

567) Click **OK** from the Insert Components PropertyManager.

Insert Coincident and Concentric mates between the four HEX-STANDOFF components and the PNEUMATIC-TEST-MODULE and the Robot-platform as illustrated. View the location of the HEX-STANDOFF components on the PNEUMATIC-TEST-MODULE assembly.

View the location of the HEX-STANDOFF
components on the Robot-platform as
illustrated.

568) **Expand** the Mates folder. View the
created mates.

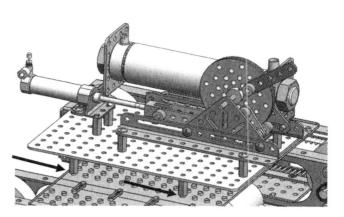

Insert the basic_integration assembly.

569) Browse to the **SW-TUTORIAL-
2020\Chapter 6 Models** folder.

570) Double-click the
basic_integration assembly.

571) Click a position to the **left** of the
assembly as illustrated.

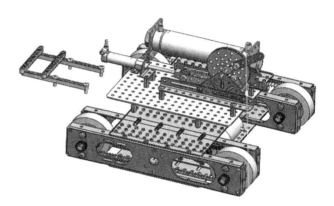

Insert Coincident and Concentric
mates between the two screws in the
basic_integration assembly and the
PNEUMATIC-TEST-MODULE
assembly.

572) **Expand** the Mates folder. View the
created mates.

Save the ROBOT assembly.

573) Click **Save** 📷. You are finished
with the final ROBOT assembly.

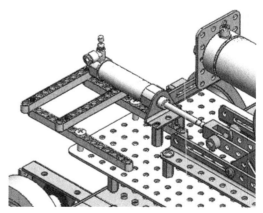

Chapter Summary

In this chapter, you created and worked with multiple documents in sub-assemblies and assemblies. You developed sound assembly modeling techniques that utilized Standard mates, fixed components, symmetry, component patterns and mirrored components.

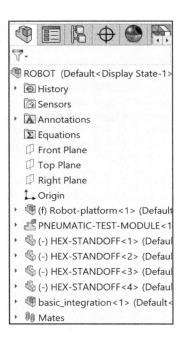

The WHEEL-AND-AXLE assembly combined the WHEEL-FLATBAR assembly with the AXLE-3000 part, SHAFTCOLLAR-500 part and HEX-ADAPTER part.

The WHEEL-FLATBAR assembly combined the 3HOLE-SHAFTCOLLAR assembly and 5HOLE-SHAFTCOLLAR assembly.

The PNEUMATIC-TEST-MODULE assembly combined four major mechanical sub-assemblies. The final ROBOT assembly was comprised of three major sub-assemblies and the HEX-STANDOFF component.

Organize your assemblies. For an exercise, create an assembly Layout Diagram of the final ROBOT assembly to determine the grouping of components and assemblies.

Utilize the Quick mate procedure on Standard mates, Cam mate, Profile Center mate, Slot mate, Symmetric mate and Width mate. To activate the Quick Mates functionality, click Tools, Customize. On the toolbars tab, under Context toolbar settings, select Show quick mates. Quick Mates is selected by default.

As an exercise apply PhotoView 360 to render the final ROBOT assembly.

Questions

1. What function does the Save as copy and continue check box perform when saving a part under a new name?

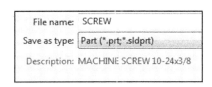

2. True or False. Never reuse geometry from one part to create another part.

3. Describe five Assembly techniques utilized in this chapter.

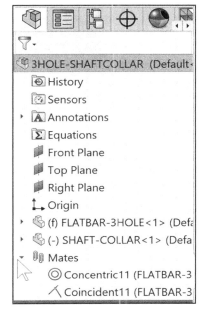

4. True or False. A fixed (f) component cannot move and is locked to the Origin.

5. Describe the purpose of an Assembly Layout diagram.

6. Describe the difference between a Pattern Driven Component Pattern and a Linear Component Pattern.

7. Describe the difference between copied geometry and mirrored geometry utilizing the Mirror Components tool.

8. True or False. An assembly contains one or more configurations.

9. Identify how you can incorporate design intent into an assembly.

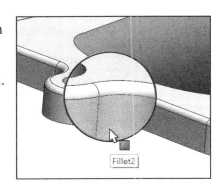

10. Review the Keyboard Short Cut keys in the Appendix. Identify the Short Cut keys you incorporated into this chapter. What is the g key used for?

Project assemblies below are created by students in my Freshman Engineering class.

Exercises

Exercise 6.1: Butterfly Valve Assembly Project

Copy the components from the Chapter 6 Homework\Butterfly Valve Assembly Project folder. View all components.

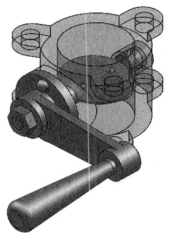

Create an ANSI IPS Butterfly Valve assembly document. Insert all needed components and mates to assemble the assembly. Simulate the proper movement. You are the designer. Address all tolerancing and dimension modifications. Use Standard, Advanced and Mechanical Mates. Create and insert any additional components if needed.

Create a C-ANSI Landscape - Third Angle Isometric Exploded Drawing document with Explode lines of the assembly using your knowledge of SOLIDWORKS. Insert a BOM with Balloons. Insert all needed General notes and Custom Properties in the Title Block.

Chapter 6 Homework

Name

- Welder Arm Assembly Project
- Shock Assembly Project
- Shaper Tool Head Assembly Project
- Radial Engine Assembly Project
- Quick Acting Clamp Assembly Project
- Pulley Assembly Project
- Pipe Vice Assembly Project
- Kant Twist Clamp Assembly Project
- Drill Guide Assembly Project
- Butterfly Valve Assembly Project
- Bench Vice Assembly Project

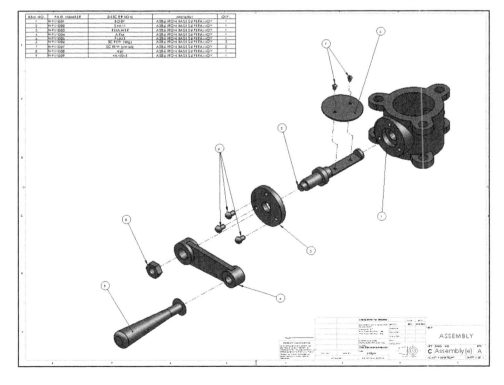

Exercise 6.2: Shock Assembly Project

Copy the components from the Chapter 6
Homework\Shock Assembly Project folder.

Create an ANSI IPS Shock assembly document. Insert all
needed components and mates to assemble the assembly.
Simulate the proper movement.

You are the designer. Address all tolerancing and
dimension modifications if needed. Be creative. Use
Standard, Advanced and Mechanical Mates. Create and
insert any additional components if needed.

Create a C-ANSI Landscape - Third Angle Isometric Exploded
Drawing document with Explode lines of the assembly using your
knowledge of SOLIDWORKS. Insert a BOM with Balloons. Insert
all needed General notes in the Title Block.

Chapter 6 Homework

Name

- Welder Arm Assembly Project
- Shock Assembly Project
- Shaper Tool Head Assembly Project
- Radial Engine Assembly Project
- Quick Acting Clamp Assembly Project
- Pulley Assembly Project
- Pipe Vice Assembly Project
- Kant Twist Clamp Assembly Project
- Drill Guide Assembly Project
- Butterfly Valve Assembly Project
- Bench Vice Assembly Project

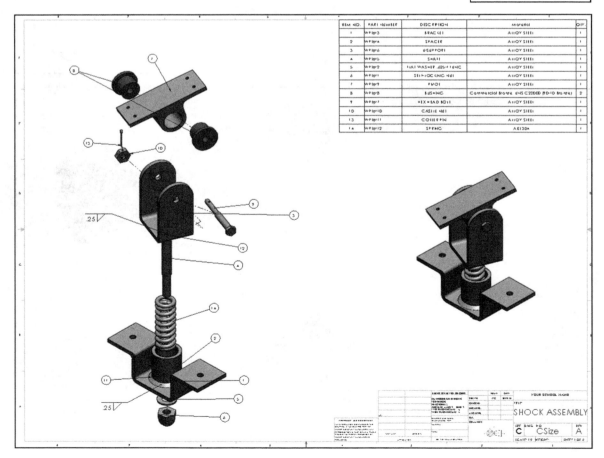

Exercise 6.3: Quick Acting Clamp Assembly Project

Copy the components from the Chapter 6 Homework\Quick Acting Clamp Assembly Project folder. View all components.

Create an ANSI IPS Quick Acting Clamp assembly document. Insert all needed components and mates to assemble the assembly. Simulate the proper movement.

You are the designer. Address all tolerancing and dimension modifications if needed. Be creative. Use Standard, Advanced and Mechanical Mates.

Create and insert any additional components if needed.

Create a C-ANSI Landscape - Third Angle Isometric Exploded Drawing document with Explode lines of the assembly using your knowledge of SOLIDWORKS. Insert a BOM with Balloons. Insert all needed General notes in the Title Block.

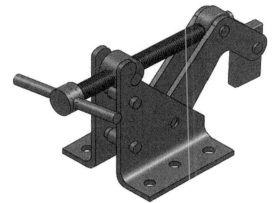

Chapter 6 Homework

Name

- Welder Arm Assembly Project
- Shock Assembly Project
- Shaper Tool Head Assembly Project
- Radial Engine Assembly Project
- Quick Acting Clamp Assembly Project
- Pulley Assembly Project
- Pipe Vice Assembly Project
- Kant Twist Clamp Assembly Project
- Drill Guide Assembly Project
- Butterfly Valve Assembly Project
- Bench Vice Assembly Project

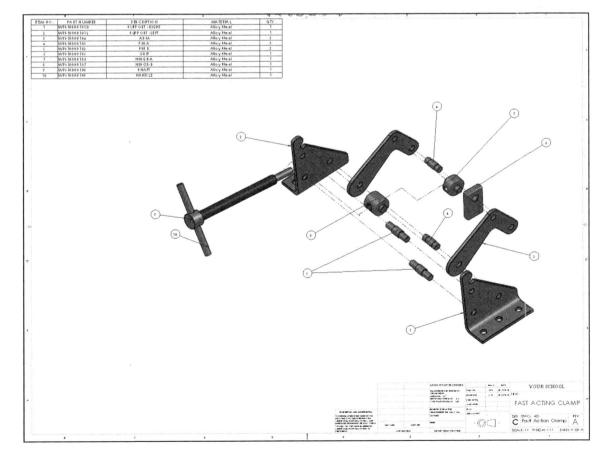

Exercise 6.4: Drill Guide Assembly Project

Copy the components from the Chapter 6
Homework\Drill Guide Assembly Project folder.
View all components.

Create an ANSI IPS Drill Guide assembly document.
Insert all needed components and mates to assemble
the assembly. Simulate the proper movement.

You are the designer. Address all tolerancing and
dimension modifications if needed. Be creative. Use
Standard, Advanced and Mechanical Mates.

Create and insert any additional components if
needed.

Create a C-ANSI Landscape - Third Angle Isometric
Exploded Drawing document with Explode lines of
the assembly using your knowledge of
SOLIDWORKS. Insert a BOM with Balloons. Insert all needed
General notes in the Title Block.

Chapter 6 Homework

Name

- Welder Arm Assembly Project
- Shock Assembly Project
- Shaper Tool Head Assembly Project
- Radial Engine Assembly Project
- Quick Acting Clamp Assembly Project
- Pulley Assembly Project
- Pipe Vice Assembly Project
- Kant Twist Clamp Assembly Project
- Drill Guide Assembly Project
- Butterfly Valve Assembly Project
- Bench Vice Assembly Project

ITEM NO.	PART NUMBER	DESCRIPTION	MATERIAL	QTY.
1	WPI-1000-01	BASE	2014 ALLOY	1
2	WPI-1000-02	ROTATOR	PLAIN CARBON STEEL	2
3	WPI-1000-03	ROD GUIDE	CAST ALLOY STEEL	2
4	WPI-1000-04	SLIDE	2014 ALLOY	1
5	WPI-1000-05	COLLAR	2014 ALLOY	2
6	WPI-1000-06	BUSHING	ALUMINUM BRONZE	2
7	WPI-1000-07	RETAINING RING B27.1 - NA2-45	ALLOY STEEL	2
8	WPI-1000-08	DRILL ADAPTOR	PLAIN CARBON STEEL	1
9	WPI-1000-09	THUMB SCREW .25-20x0.51 TYPE B, FLAT POINT-C	2014 ALLOY	4

Exercise 6.5: Pulley Assembly Project

Copy the components from the Chapter 6 Homework\Pulley Assembly Project folder. View all components.

Create an ANSI Pulley assembly document. Insert all needed components and mates to assemble the assembly. Simulate the proper movement.

You are the designer. Address all tolerancing and dimension modifications if needed. Be creative. Use Standard, Advanced and Mechanical Mates.

Create and insert any additional components if needed.

Create a C-ANSI Landscape - Third Angle Isometric Exploded Drawing document with Explode lines of the assembly using your knowledge of SOLIDWORKS. Insert a BOM with Balloons. Insert all needed General notes in the Title Block.

Chapter 6 Homework

Name

- Welder Arm Assembly Project
- Shock Assembly Project
- Shaper Tool Head Assembly Project
- Radial Engine Assembly Project
- Quick Acting Clamp Assembly Project
- Pulley Assembly Project
- Pipe Vice Assembly Project
- Kant Twist Clamp Assembly Project
- Drill Guide Assembly Project
- Butterfly Valve Assembly Project
- Bench Vice Assembly Project

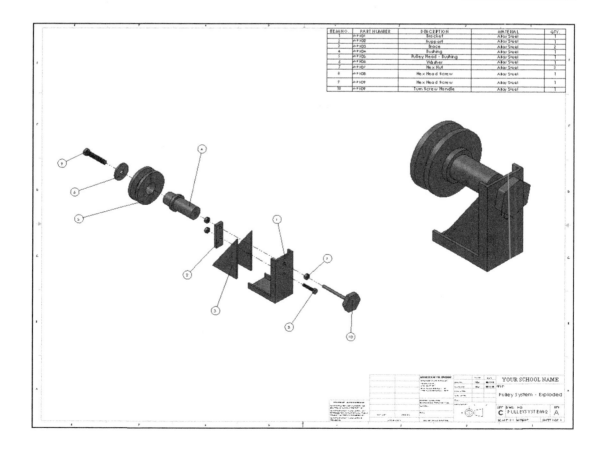

Exercise 6.6: Welder Arm Assembly Project

Copy the components from the Chapter 6 Homework\Welder Arm Assembly Project folder. View all components.

Create an ANSI Welder Arm assembly document. Insert all needed components and mates to assemble the assembly. Simulate the proper movement.

You are the designer. Address all tolerancing and dimension modifications if needed. Be creative. Use Standard, Advanced and Mechanical Mates. Create and insert any additional components if needed.

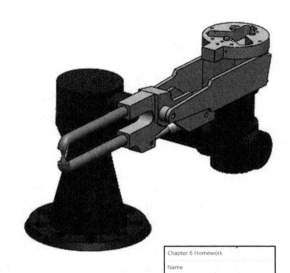

Chapter 6 Homework

Name

📁 Welder Arm Assembly Project
📁 Shock Assembly Project
📁 Shaper Tool Head Assembly Project
📁 Radial Engine Assembly Project
📁 Quick Acting Clamp Assembly Project
📁 Pulley Assembly Project
📁 Pipe Vice Assembly Project
📁 Kant Twist Clamp Assembly Project
📁 Drill Guide Assembly Project
📁 Butterfly Valve Assembly Project
📁 Bench Vice Assembly Project

Create a C-ANSI Landscape - Third Angle Isometric Exploded Drawing document with Explode lines of the assembly using your knowledge of SOLIDWORKS. Insert a BOM with Balloons. Insert all needed General notes in the Title Block.

ITEM NO	PART NUMBER	DESCRIPTION	MATERIAL	QTY
1	WM01-01-001	bw	Gav Cast Iron	1
2	WM01-01-002	baseplace	Cast Alloy Steel	1
3	WM01-01-003	support	Cast Alloy Steel	2
4	WM01-01-004	holder	Cast Alloy Steel	1
5	WM01-01-005	block	Cast Alloy Steel	1
6	WM01-01-006	thing	Iron	2
7	WM01-01-007	thrust	Cast Alloy Steel	1
8	WM01-01-008	B-4Hex0.125---0.5 1.4	Cast Alloy Steel	3
9	WM01-01-009	SPS 0.1875x0.75	Aluminum 6061 Alloy	1
10	WM01-01-010	SPS 0.1875x1.5	Aluminum 6061 Alloy	1
11	WM01-01011	SPS 0.125x0.5	Aluminum 6061 Alloy	1

YOUR SCHOOL NAME

Welder Arm Exp

C | WM01-03-001

Exercise 6.7: Radial Engine Assembly Project

Copy the components from the Chapter 6 Homework\Radial Engine Assembly Project folder. View all components.

Create an ANSI Radial Engine assembly document. Insert all needed components and mates to assemble the assembly. Simulate the proper movement.

You are the designer. Address all tolerancing and dimension modifications if needed. Be creative.

Use Standard, Advanced and Mechanical Mates.

Create and insert any additional components if needed.

Create a C-ANSI Landscape - Third Angle Isometric Exploded Drawing document with Explode lines of the assembly using your knowledge of SOLIDWORKS.

Insert a BOM with Balloons. Insert all needed General notes in the Title Block.

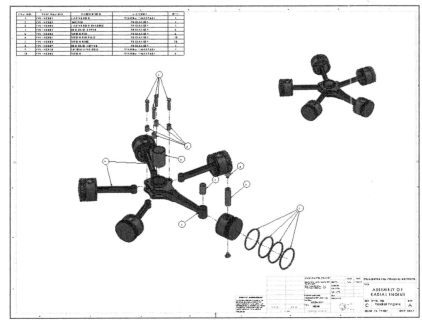

Exercise 6.8: Bench Vice Assembly Project

Copy the components from the Chapter 6 Homework\Bench Vice Assembly Project folder. View all components.

Create an ANSI IPS Bench Vice assembly document. Insert all needed components and mates to assemble the assembly. Simulate the proper movement.

You are the designer. Address all tolerancing and dimension modifications if needed. Use Standard, Advanced and Mechanical Mates.

Create and insert any additional components if needed.

Create a C-ANSI Landscape - Third Angle Isometric Exploded Drawing document with Explode lines of the assembly using your knowledge of SOLIDWORKS.

Insert a BOM with Balloons. Insert all needed General notes and Custom Properties in the Title Block.

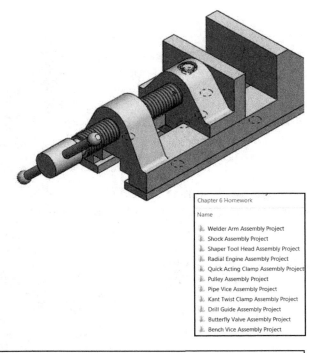

Chapter 6 Homework

Name

- Welder Arm Assembly Project
- Shock Assembly Project
- Shaper Tool Head Assembly Project
- Radial Engine Assembly Project
- Quick Acting Clamp Assembly Project
- Pulley Assembly Project
- Pipe Vice Assembly Project
- Kant Twist Clamp Assembly Project
- Drill Guide Assembly Project
- Butterfly Valve Assembly Project
- Bench Vice Assembly Project

ITEM NO.	PART NUMBER	DESCRIPTION	MATERIAL	QTY.
1	WPI-0001	Base	Alloy Steel	1
2	WPI-0002	Base Plate	Alloy Steel	2
3	WPI-0003	Vice Jaw	Alloy Steel	1
4	WPI-0004	Clamping Plate	Alloy Steel	1
5	WPI-0005	Jaw Screw	Alloy Steel	1
6	WPI-0006	Screw Bar	Alloy Steel	1
7	WPI-0007	Bar Globes	Alloy Steel	2
8	WPI-008	Hex Screw M6x1.0x14-14WN	Alloy Steel	4
9	WPI-0009	Bolt	Alloy Steell	1
10	WPI-00019	Blot1	Alloy Steel	2

MAIN MACHINE

TITLE: ARBOR PRESS

SIZE **A**   DWG. NO. 10-344   REV

SCALE: 1:3   WEIGHT: SHEET 1 OF 1

Exercise 6.9: Kant Twist Clamp Assembly Project

Copy the components from the Chapter 6 Homework\Kant Twist Clamp Assembly Project folder. View all components.

Create an ANSI IPS Kant Twist Clamp assembly document. Insert all needed components and mates to assemble the assembly. Simulate the proper movement.

You are the designer. Address all tolerancing and dimension modifications if needed.

Use Standard, Advanced and Mechanical Mates. Create and insert any additional components if needed.

Create a C-ANSI Landscape - Third Angle Isometric Exploded Drawing document with Explode lines of the assembly using your knowledge of SOLIDWORKS.

Insert a BOM with Balloons. Insert all needed General notes and Custom Properties in the Title Block.

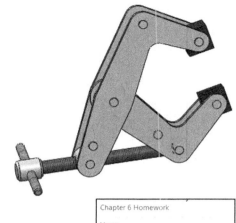

Chapter 6 Homework

Name

- Welder Arm Assembly Project
- Shock Assembly Project
- Shaper Tool Head Assembly Project
- Radial Engine Assembly Project
- Quick Acting Clamp Assembly Project
- Pulley Assembly Project
- Pipe Vice Assembly Project
- Kant Twist Clamp Assembly Project
- Drill Guide Assembly Project
- Butterfly Valve Assembly Project
- Bench Vice Assembly Project

Notes:

CHAPTER 7 - CSWA INTRODUCTION AND DRAFTING COMPETENCIES

Introduction

DS SOLIDWORKS Corp. offers various types of certification. Each stage represents increasing levels of expertise in 3D CAD: Certified SOLIDWORKS Associate CSWA, Certified SOLIDWORKS Professional CSWP and Certified SOLIDWORKS Expert CSWE along with specialty fields.

The CSWA Academic exam is provided either in a single 3 hour segment, or 2 - 90 minute segments.

Part 1 of the CSWA Academic exam is 90 minutes, minimum passing score is 80, with 6 questions. There are two questions in the Basic Part Creation and Modification category, two questions in the Intermediate Part Creation and Modification category and two questions in the Assembly Creation and Modification category.

Part 2 of the CSWA Academic exam is 90 minutes, minimum passing score is 80 with 8 questions. There are three questions on the CSWA Academic exam in the Drafting Competencies category, three questions in the Advanced Part Creation and Modification category and two questions in the Assembly Creation and Modification category.

The CSWA exam for industry is only provided in a single 3 hour segment. The exam consists of 14 questions in five categories worth a total of 240 points. All exams cover the same material.

The CSWA exam consists of 14 questions in the following five categories and subject areas:

- *Drafting Competencies*: (Three questions - multiple choice - 5 points each).

    - Questions on general drawing views: Projected, Section, Break, Crop, Detail, Alternate Position, etc.

> Drafting Competencies - To create drawing view 'B' it is necessary to select drawing view 'A' and insert which SolidWorks view type?

Screen shot from an exam

- *Basic Part Creation and Modification*:
  (Two questions - one multiple choice/one
  single answer - 15 points each).

  - Sketch Planes:

    - Front, Top, Right.

  - 2D Sketching:

    - Geometric Relations and
      Dimensioning.

  - Extruded Boss/Base Feature.

  - Extruded Cut feature.

  - Modification of Basic part.

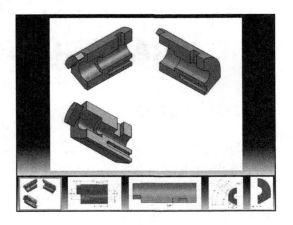

🔆 In the *Basic Part Creation and
Modification* category there is a dimension
modification question based on the first
(multiple choice) question. You should be
within 1% of the multiple choice answer before
you go on to the modification single answer section.

- *Intermediate Part Creation and Modification;* (Two
  questions - one multiple choice/one single answer -
  15 points each).

  - Sketch Planes:

    - Front, Top, Right.

  - 2D Sketching:

    - Geometric Relations and
      Dimensioning.

  - Extruded Boss/Base Feature.

  - Extruded Cut Feature.

  - Revolved Boss/Base Feature.

  - Mirror and Fillet Feature.

  - Circular and Linear Pattern Feature.

  - Plane Feature.

  - Modification of Intermediate Part:

    - Sketch, Feature, Pattern, etc.

  - Modification of Intermediate part.

Intermediate Part (Wheel) - Step 1
Build this part in SolidWorks.
(Save part after each question in a different file in
case it must be reviewed)

Unit system: MMGS (millimeter, gram, second)
Decimal places: 2
Part origin: Arbitrary
All holes through all unless shown otherwise.
Material: Aluminium 1060 Alloy

A = 134.00
B = 890.00

Note: All geometry is symmetrical about the plane
represented by the line labeled F'' in the M-M
Section View.

What is the overall mass of the part (grams)?

Hint: If you don't find an option within 1%
of your answer please re-check your
model(s).

Screen shots from an exam

In the *Intermediate Part Creation and Modification* category, there are two dimension modification questions based on the first (multiple choice) question. You should be within 1% of the multiple choice answer before you go on to the modification single answer section.

- *Advanced Part Creation and Modification; (Three questions - one multiple choice/two single answers - 15 points each).*

  - Sketch Planes:

    - Front, Top, Right, Face, Created Plane etc.

    - 2D Sketching or 3D Sketching.

  - Sketch Tools:

    - Offset Entities, Entities, etc.

  - Extruded Boss/Base Feature.

  - Extruded Cut Feature.

  - Revolved Boss/Base Feature.

  - Mirror and Fillet Feature.

  - Circular and Linear Pattern Feature.

  - Shell Feature.

  - Plane Feature.

  - More Difficult Geometry Modifications.

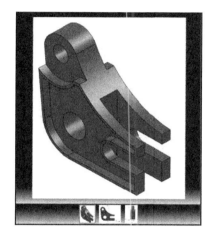

Advanced Part (Bracket) - Step 1
Build this part in SolidWorks.
(Save part after each question in a different file in case it must be reviewed)

Unit system: MMGS (millimeter, gram, second)
Decimal places: 2
Part origin: Arbitrary
All holes through all unless shown otherwise.
Material: AISI 1020 Steel
Density = 0.0079 g/mm^3

A = 64.00
B = 20.00
C = 26.50

What is the overall mass of the part (grams)?

Hint: If you don't find an option within 1% of your answer please re-check your model(s).

In the *Advanced Part Creation and Modification* category, there are two dimension modification questions based on the first (multiple choice) question. You should be within 1% of the multiple choice answer before you go on to the modification single answer section.

- *Assembly Creation and Modification; (Two different assemblies - four questions - two multiple choice/two single answers - 30 points each).*

  - Insert the first (fixed) component.

  - Insert all needed components.

  - Standard Mates.

  - Modification of key parameters in the assembly.

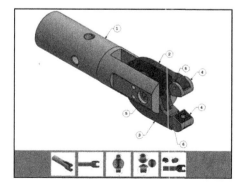

Screen shots from an exam

In the *Assembly Creation and Modification* category, expect to see five to seven components. There are two dimension modification questions based on the first (multiple choice) question. You should be within 1% of the multiple choice answer before you go on to the modification single answer section.

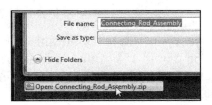

Download the needed components in a zip folder during the exam to create the assembly.

Do not use feature recognition when you open the downloaded components for the assembly. This is a timed exam. Additional model information is not needed in the exam.

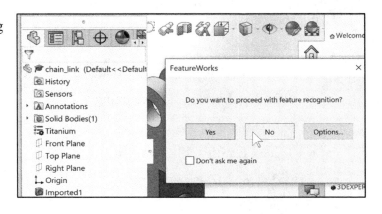

Illustrations may vary depending on your SOLIDWORKS version and system setup.

Use the view indicator to increase or decrease the active model in the view window.

Additional model views

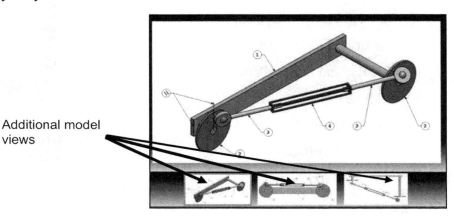

View indicator

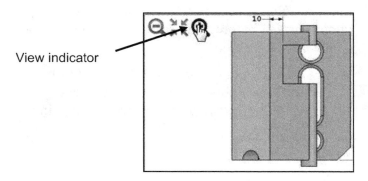

Screen shots from an exam

During the Exam

During the exam, SOLIDWORKS provides the ability to click on a detail view below (as illustrated) to obtain additional details and dimensions during the exam.

🔆 No Simulation questions are on the CSWA exam at this time.

🔆 No Sheetmetal questions are on the CSWA exam at this time.

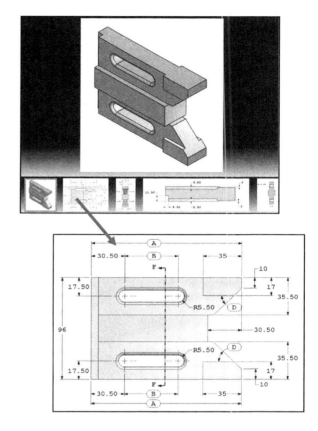

During the exam, use the control keys at the bottom of the screen to:

- *Show the Previous Question.*

- *Reset the Question.*

- *Show the Summary Screen.*

- *Move to the Next Question.*

When you are finished, press the End Examination button. The tester will ask you if you want to end the test. Click Yes.

If there are any unanswered questions, the tester will provide a warning message as illustrated.

If you do not pass the certification exam (either segment), you will need to wait 30 days until you can retake that segment of the exam.

To obtain additional CSWA exam information, visit the SOLIDWORKS VirtualTester Certification site at https://SOLIDWORKS.virtualtester.com/.

Use the clock in the tester to view the amount of time that you used and the amount of time that is left in the exam.

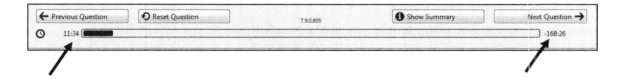

Objectives

Drafting Competencies is one of the five categories (Drafting Competencies, Basic Part Creation and Modification, Intermediate Part Creation and Modification, Advance Part Creation and Modification and Assembly Creation and Modification) on the CSWA exam.

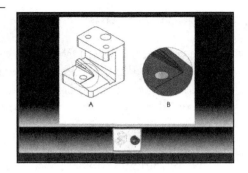

Screen shot from the exam

There are three questions (total) on the CSWA exam in the *Drafting Competencies* category. Each question is worth five (5) points. Drafting Competency questions are addressed in part 2 of the CSWA exam.

The three questions are in a multiple choice single answer format. You are allowed to answer the questions in any order you prefer. Use the Summary Screen during the exam to view the list of all questions you have or have not answered.

In the *Drafting Competencies* category of the exam, you are **not required** to create or perform an analysis on a part, assembly, or drawing but you are required to have general drafting/drawing knowledge and understanding of various drawing view methods.

On the completion of the chapter, you will be able to:

- Identify the procedure to create a named drawing view:

    - Projected view, Section view, Break view, Crop view, Detail, Alternate Position view, etc.

In the *Basic Part Creation and Modification, Intermediate Part Creation and Modification, Advanced Part Creation and Modification and Assembly Creation and Modification* categories, you are required to read and interpret various types of drawing views and understand various types of drawing annotations.

Download all needed model files (SW-TUTORIAL-2020 folder) from the SDC Publications website (www.sdcpublications.com) to a local hard drive. All SOLIDWORKS models (initial and final) are provided.

SOLIDWORKS CSWA Model Folder
Name
Chapter 7
Chapter 7 Final Solutions
Chapter 8
Chapter 8 Final Solutions
Chapter 9
Chapter 9 Final Solutions
Chapter 10
Chapter 10 Final Solutions

Procedure to Create a Named Drawing view

You need the ability to identify the procedure to create a named drawing view: *Standard 3 View, Model View, Projected View, Auxiliary View, Section View, Removed Section, Detail View, Broken-out Section, Break, Crop View and Alternate Position View.*

Create a Section view in a drawing by cutting the parent view with a section line. The Section view can be a straight cut section or an offset section defined by a stepped section line. The section line can also include concentric arcs.

Create an Aligned Section view in a drawing through a model, or portion of a model, that is aligned with a selected section line segment. The Aligned Section view is similar to a Section view, but the section line for an aligned section comprises two or more lines connected at an angle.

Create a Detail view in a drawing to show a portion of a view, usually at an enlarged scale. This detail may be of an Orthographic view, a Non-planar (isometric) view, Section view, Crop view an Exploded Assembly view, or another Detail view.

🔅 Crop any drawing view except a Detail view, a view from which a Detail view has been created, or an Exploded view. To create a Crop view, sketch a closed profile such as a circle or spline. The view outside the closed profile disappears as illustrated.

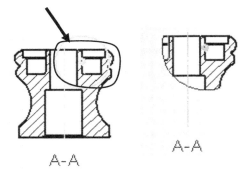

🔅 Create a Detail view in a drawing to display a portion of a view, usually at an enlarged scale. This detail may be of an orthographic view, a non-planar (isometric) view, a Section view, a Crop view, an Exploded assembly view, or another Detail view.

Tutorial: Drawing Named Procedure 7-1

Identify the drawing name view and understand the procedure to create the name view.

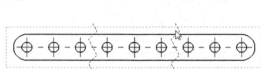

1. **View** the illustrated drawing views. The top drawing view is a Break view. The Break view is created by adding a break line to a selected view.

💡 Broken views make it possible to display the drawing view in a larger scale on a smaller size drawing sheet. Reference dimensions and model dimensions associated with the broken area reflect the actual model values.

💡 In views with multiple breaks, the Break line style must be the same.

Tutorial: Drawing Named Procedure 7-2

Identify the drawing name view and understand the procedure to create the name view.

1. **View** the illustrated drawing views. The top drawing view is a Section view of the bottom view. The Section view is created by cutting the parent view with a cutting section line.

💡 Create a Section view in a drawing by cutting the parent view with a section line. The section view can be a straight cut section or an offset section defined by a stepped section line. The section line can also include Concentric arcs.

Tutorial: Drawing Named Procedure 7-3

Identify the drawing name view and understand the procedure to create the name view.

1. **View** the illustrated drawing views. The view to the right is an Auxilary view of the Front view. Select a reference edge to create an Auxiliary view as illustrated.

💡 An Auxiliary view is similar to a Projected view, but it is unfolded normal to a reference edge in an existing view.

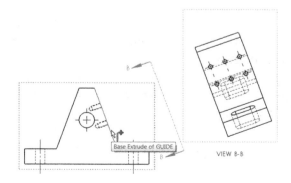

Tutorial: Drawing Named Procedure 7-4

Identify the drawing name view and understand the procedure to create the name view.

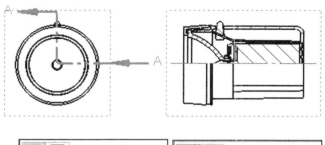

1. **View** the illustrated drawing views. The right drawing view is an Aligned half Section view of the view to the left. The Section view is created by using two lines connected at an angle. Create an Aligned half Section view in a drawing through a model, or portion of a model, that is aligned with a selected section line segment.

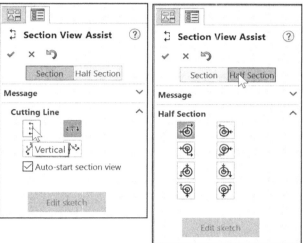

☀ The Aligned Section view is very similar to a Section View, with the exception that the section line for an aligned half section is comprised of two or more lines connected at an angle.

Tutorial: Drawing Named Procedure 7-5

Identify the drawing name view and understand the procedure to create the name view.

1. **View** the illustrated drawing views. The left drawing view is a Detail view of the Section view. The Detail view is created by sketching a circle with the Circle Sketch tool. Click and drag for the location.

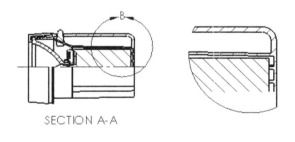

SECTION A-A

The Detail view ⓐ tool provides the ability to add a Detail view to display a portion of a view, usually at an enlarged scale.

☀ To create a profile other than a circle, sketch the profile before clicking the Detail view tool. Using a sketch entity tool, create a closed profile around the area to be detailed.

Tutorial: Drawing Named Procedure 7-6

Identify the drawing name view and understand the procedure to create the name view.

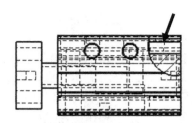

1. **View** the illustrated drawing views. The top drawing view is a Broken-out Section view of the bottom drawing view. The Broken-out Section View is part of an existing drawing view, not a separate view. Create the Broken-out Section view with a closed profile, usually by using the Spline Sketch tool. Material is removed to a specified depth to expose inner details.

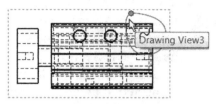

Tutorial: Drawing Named Procedure 7-7

Identify the drawing name view and understand the procedure to create the name view.

1. **View** the illustrated drawing view. The top drawing view is a Crop view. The Crop view is created by a closed sketch profile such as a circle, or spline as illustrated.

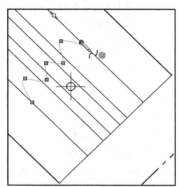

The Crop View provides the ability to crop an existing drawing view. You cannot use the Crop tool on a Detail view, a view from which a Detail view has been created, or an Exploded view.

Use the Crop tool to save steps. Example: instead of creating a Section View and then a Detail view, then hiding the unnecessary Section view, use the Crop tool to crop the Section view directly.

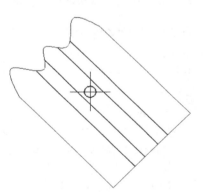

☀ In the exam, you are allowed to answer the questions in any order. Use the Summary Screen during the exam to view the list of all questions you have or have not answered.

Tutorial: Drawing Named Procedure 7-8

Identify the drawing name view and understand the procedure to create the name view.

1. **View** the illustrated drawing view. The drawing view is an Alternate Position View. The

   Alternate Position view tool ⊞ provides the ability to superimpose an existing drawing view precisely on another. The alternate position is displayed with phantom lines.

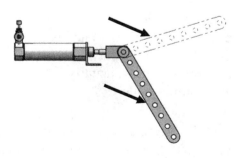

Intended Audience

The intended audience is anyone with a minimum of 6 - 9 months of SOLIDWORKS experience and basic knowledge of engineering fundamentals and practices.

SOLIDWORKS recommends that you review their SOLIDWORKS Tutorials on Parts, Assemblies and Drawings as a prerequisite and have at least 45 hours of classroom time learning SOLIDWORKS or using SOLIDWORKS with basic engineering design principles and practices.

To prepare for the CSWA exam, it is recommended that you first perform the following:

- Take a CSWA exam preparation class or review a text book written for the CSWA exam.

- Visit the SOLIDWORKS VirtualTester Certification site at https://SOLIDWORKS.virtualtester.com/. Download and open the CSWA sample exam folder. Follow the instructions to login and take a sample exam.

- Complete the SOLIDWORKS Tutorials.

- Practice creating models from the isometric working drawings sections of any Technical Drawing or Engineering Drawing Documentation text books.

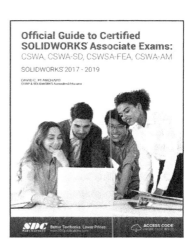

Additional references to help you prepare are as follows:

- **Official Guide to Certified SOLIDWORKS Associate Exams: CSWA, CSWA-SD, CSWSA-FEA, CSWA-AM Version 4; 2017-2019**, Version 3; 2015 - 2017, Version 2; 2015 - 2012, Version 1; 2012, 2013.

- **Engineering Drawing and Design**, Jensen & Helsel, Glencoe, 1990.

Questions

1. Identify the illustrated Drawing view.

- A: Projected
- B: Alternative Position
- C: Extended
- D: Aligned Section

2. Identify the illustrated Drawing view.

- A: Crop
- B: Break
- C: Broken-out Section
- D: Aligned Section

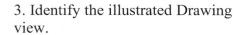

3. Identify the illustrated Drawing view.

- A: Section
- B: Crop
- C: Broken-out Section
- D: Aligned

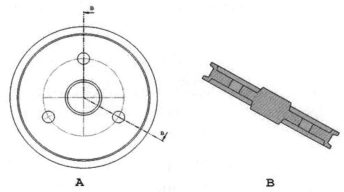

A   B

4. Identify the view procedure. To create the following view, you need to insert a:

- A: Rectangle Sketch tool
- B: Closed Profile: Spline
- C: Open Profile: Circle
- D: None of the above

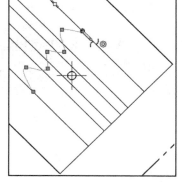

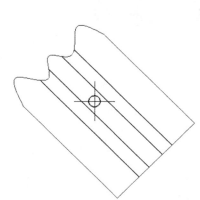

5. Identify the view procedure. To create the following view, you need to insert a:

- A: Open Spline
- B: Closed Spline
- C: 3 Point Arc
- D: None of the above

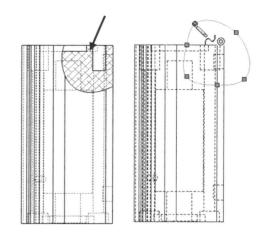

6. Identify the illustrated view type.

- A: Crop
- B: Section
- C: Projected
- D: Detail

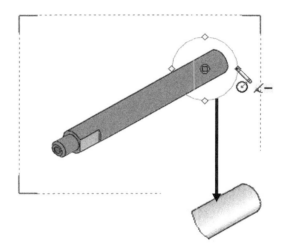

7. To create View B from Drawing View A insert which View type?

- A: Crop
- B: Section
- C: Aligned Section
- D: Projected

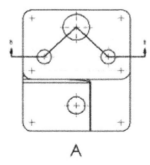 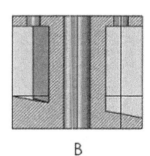

A   B

8. To create View B it is necessary to sketch a closed spline on View A and insert which View type?

- A: Broken out Section

- B: Detail

- C: Section

- D: Projected

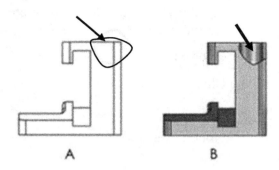

A   B

9. To create View B it is necessary to sketch a closed spline on View A and insert which View type?

- A: Horizontal Break

- B: Detail

- C: Section

- D: Broken out Section

A   B

Chapter 8 - Basic Part and Intermediate Part Creation and Modification

Objectives

Basic Part Creation and Modification and Intermediate Part Creation and Modification are two of the five categories on the CSWA exam. This chapter covers the knowledge to create and modify models in these categories from detailed dimensioned illustrations.

The main difference between the *Basic Part Creation and Modification* category and the *Intermediate Part Creation and Modification* or the *Advance Part Creation and Modification* category is the complexity of the sketches and the number of dimensions and geometric relations along with an increase in the number of features.

There are two questions on the CSWA Academic exam (part 1) in the *Basic Part Creation and Modification* category and two questions in the *Intermediate Part Creation and Modification* category.

The first question is in a multiple-choice single answer format. You should be within 1% of the multiple-choice answer before you move on to the modification single answer section, (fill in the blank format).

Each question is worth fifteen (15) points for a total of thirty (30) points. You are required to build a model with six or more features and to answer a question either on the overall mass, volume, or the location of the Center of mass for the created model relative to the default part Origin location. You are then requested to modify the part and answer a fill in the blank format question.

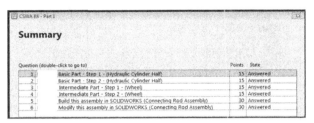

Question (double-click to go to)		Points	State
1	Basic Part - Step 1 - (Hydraulic Cylinder Half)	15	Answered
2	Basic Part - Step 2 - (Hydraulic Cylinder Half)	15	Answered
3	Intermediate Part - Step 1 - (Wheel)	15	Answered
4	Intermediate Part - Step 2 - (Wheel)	15	Answered
5	Build this assembly in SOLIDWORKS (Connecting Rod Assembly)	30	Answered
6	Modify this assembly in SOLIDWORKS (Connecting Rod Assembly)	30	Answered

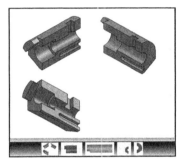

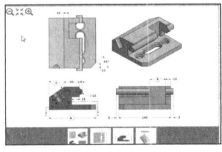

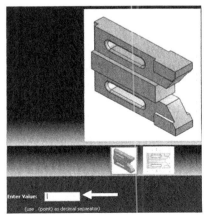

Screen shots from an exam

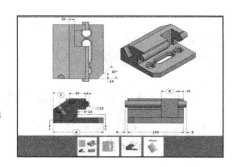

On the completion of the chapter, you will be able to:

- Read and understand an Engineering document used in the CSWA exam:

  - Identify the Sketch plane, part Origin location, part dimensions, geometric relations and design intent of the sketch and feature.

- Build a part from a detailed dimensioned illustration using the following SOLIDWORKS tools and features:

  - 2D & 3D sketch tools, Extruded Boss/Base, Extruded Cut, Fillet, Mirror, Revolved Base, Chamfer, Reference geometry, Plane, Axis, Calculate the overall mass and volume of the created part and Locate the Center of mass for the created part relative to the Origin.

The complexity of the models along with the features progressively increases throughout this chapter to simulate the final types of models that would be provided on the exam.

FeatureManager names were changed through various revisions of SOLIDWORKS. Example: Extrude1 vs. Boss-Extrude1. These changes do not affect the models or answers in this book.

Read and Understand an Engineering Document

A 2D drawing view is displayed in the *Basic Part Creation and Modification, Intermediate Part Creation and Modification, Advance Part Creation and Modification, and Assembly Creation and Modification* categories of the CSWA exam to clarify dimensions and details.

The ability to interpret a 2D drawing view is required.

- Example 1: *8X Ø.19 EQ. SP*. Eight holes with a .19in. diameter are required that are equally (.55in.) spaced.

- Example 2: *R2.50 TYP*. Typical radius of 2.50. The dimension has a two decimal place precision.

- Example 3: ⊽ . The Depth/Deep ⊽ symbol with a 1.50 dimension associated with the hole. The hole Ø.562 has a three decimal place precision.

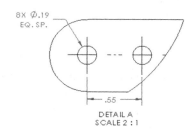

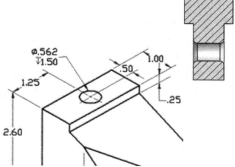

- Example 4: *A+40*. A is provided to you on the CSWA exam. A + 40mm.

🔆 N is a Detail view of the M-M Section view.

- Example 5: *ØB*. Diameter of B. B is provided to you on the exam.

- Example 6: ⃠. Parallelism.

- Example 7: ⊗ The faces are coincident.

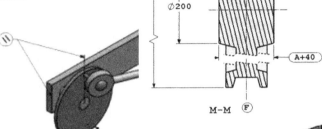

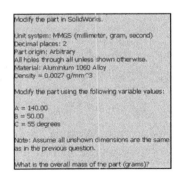

During the exam, each question will display an information table on the left side of the screen and drawing information on the right. Read the provided information and apply it to the drawing. Various values are provided on each question.

🔆 If you don't find your answer (within 1%) in the multiple-choice single answer format section - recheck your solid model for precision and accuracy.

Download all needed model files (SW-TUTORIAL-2020 folder) from the SDC Publications website (www.sdcpublications.com) to a local hard drive. All SOLIDWORKS models (initial and final) are provided.

Modify the part in SolidWorks.

Unit system: MMGS (millimeter, gram, second)
Decimal places: 2
Part origin: Arbitrary
All holes through all unless shown otherwise.
Material: Aluminium 1060 Alloy
Density = 0.0027 g/mm^3

Modify the part using the following variable values:

A = 140.00
B = 50.00
C = 55 degrees

Note: Assume all unshown dimensions are the same as in the previous question.

What is the overall mass of the part (grams)?

Screen shot from an exam

Build a Basic Part from a Detailed illustration

Tutorial: Volume/Center of Mass 8-1

Build this model in SOLIDWORKS. Calculate the volume of the part and locate the Center of mass with the provided information.

1. **Create** a New part in SOLIDWORKS.

2. **Build** the illustrated dimensioned model. The model displays all edges on perpendicular planes. Think about the steps to build the model. Insert two features: Extruded Base (Boss-Extrude1) and Extruded Cut (Cut-Extrude1). The part Origin is located in the front left corner of the model. Think about your Base Sketch plane. Keep your Base Sketch simple.

3. **Set** the document properties for the model.

4. Create **Sketch1**. Select the Front Plane as the Sketch plane. Sketch1 is the Base sketch. Sketch1 is the profile for the Extruded Base (Boss-Extrude1) feature. Insert the required geometric relations and dimensions.

5. Create the **Extruded Base** feature. Boss-Extrude1 is the Base feature. Blind is the default End Condition in Direction 1. Depth = 2.25in. Identify the extrude direction to maintain the location of the Origin.

6. Create **Sketch2**. Select the Top right face as the Sketch plane for the second feature. Sketch a square. Sketch2 is the profile for the Extruded Cut feature. Insert the required geometric relations and dimensions.

7. Create the **Extruded Cut** feature. Select Through All for End Condition in Direction 1.

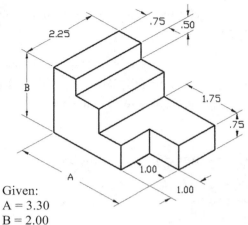

Given:
A = 3.30
B = 2.00
Material: 2014 Alloy
Density = .101 lb/in^3
Units: IPS
Decimal places = 2

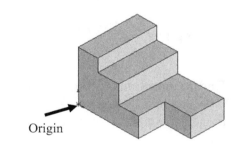

Origin

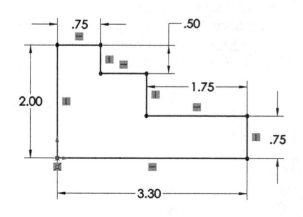

8. **Assign** 2014 Alloy material to the part. Material is required to locate the Center of mass.

9. **Calculate** the volume. The volume = 8.28 cubic inches.

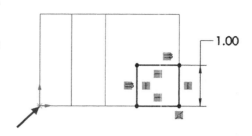

💡 There are numerous ways to build the models in this chapter. A goal is to display different design intents and techniques.

10. **Locate** the Center of mass. The location of the Center of mass is derived from the part Origin.

- X: 1.14 inch

- Y: 0.75 inch

- Z: -1.18 inch

11. **Save** the part and name it Volume-Center of mass 8-1.

12. **Close** the model.

Mass = 0.84 pounds

Volume = 8.28 cubic inches

Surface area = 29.88 square inches

Center of mass: (inches)
   X = 1.14
   Y = 0.75
   Z = -1.18

💡 The principal axes and Center of mass are displayed graphically on the model in the Graphics window.

Tutorial: Volume/Center of Mass 8-2

Build this model. Calculate the volume of the part and locate the Center of mass with the provided information.

1. **Create** a New part in SOLIDWORKS.

2. **Build** the illustrated dimensioned model. The model displays all edges on perpendicular planes. Think about the steps that are required to build this model. Remember, there are numerous ways to create the models in this chapter.

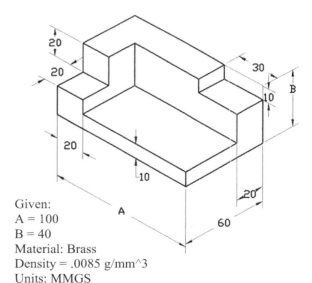

Given:
A = 100
B = 40
Material: Brass
Density = .0085 g/mm^3
Units: MMGS

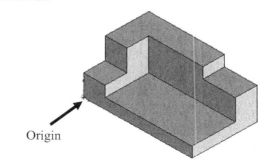

Origin

The CSWA exam is timed. Work efficiently.

View the provided Part FeatureManagers. Both FeatureManagers create the same illustrated model. In Option1, there are four sketches and four features (Extruded Base and three Extruded Cuts) that are used to build the model.

In Option2, there are three sketches and three features (Extruded Boss/Base) that are used to build the model. Which FeatureManager is better? In a timed exam, optimize your time and use the least amount of features through mirror, pattern, symmetry, etc.

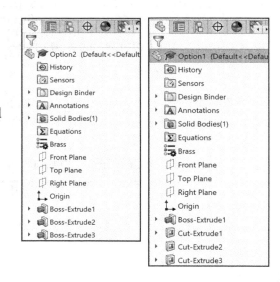

Use Centerlines to create symmetrical sketch elements and revolved features, or as construction geometry.

Create the model using the Option2 Part FeatureManager.

3. **Set** the document properties for the model.

4. Create **Sketch1**. Select the Top Plane as the Sketch plane. Sketch a rectangle. Insert the required relations and dimensions.

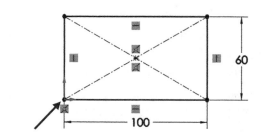

5. Create the **Extruded** feature. Boss-Extrude1 is the Base feature. Blind is the default End Condition in Direction 1. Depth = 10mm.

6. Create **Sketch2**. Select the back face of Boss-Extrude1.

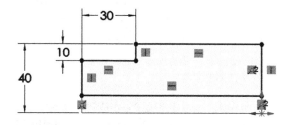

7. Select **Normal To** view. Sketch2 is the profile for the second Extruded feature. Insert the required geometric relations and dimensions as illustrated.

8. Create the second Extruded feature (**Boss-Extrude2**). Blind is the default End Condition in Direction 1. Depth = 20mm. Note the direction of the extrude towards the front of the model.

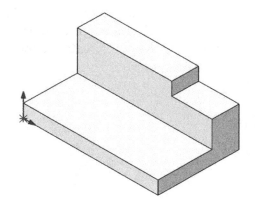

9. Create **Sketch3**. Select the left face of Boss-Extrude1 as the Sketch plane. Sketch3 is the profile for the third Extrude feature. Insert the required geometric relations and dimensions.

10. Create the third Extruded feature (**Boss-Extrude3**). Blind is the default End Condition in Direction 1. Depth = 20mm.

11. **Assign** brass material to the part.

12. **Calculate** the volume of the model. The volume = 130,000.00 cubic millimeters.

13. **Locate** the Center of mass. The location of the Center of mass is derived from the part Origin.

- X: 43.46 millimeters

- Y: 15.00 millimeters

- Z: -37.69 millimeters

14. **Save** the part and name it Volume-Center of mass 8-2.

15. **Calculate** the volume of the model using the IPS unit system. The volume = 7.93 cubic inches.

16. **Locate** the Center of mass using the IPS unit system. The location of the Center of mass is derived from the part Origin.

- X: 1.71 inches

- Y: 0.59 inches

- Z: -1.48 inches

17. **Save** the part and name it Volume-Center of mass 8-2-IPS.

18. **Close** the model.

There are numerous ways to create the models in this chapter. A goal is to display different design intents and techniques.

All SOLIDWORKS models for the next few chapters (initial and final) are provided in the SOLIDWORKS CSWA model folder.

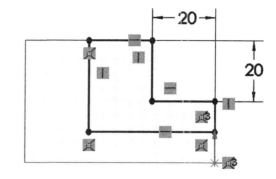

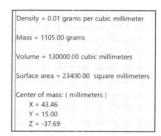

Density = 0.01 grams per cubic millimeter

Mass = 1105.00 grams

Volume = 130000.00 cubic millimeters

Surface area = 23400.00 square millimeters

Center of mass: (millimeters)
  X = 43.46
  Y = 15.00
  Z = -37.69

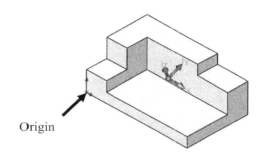

Origin

Mass = 2.44 pounds

Volume = 7.93 cubic inches

Surface area = 36.27 square inches

Center of mass: (inches)
  X = 1.71
  Y = 0.59
  Z = -1.48

SOLIDWORKS CSWA Model Folder

Name

- Chapter 7
- Chapter 7 Final Solutions
- Chapter 8
- Chapter 8 Final Solutions
- Chapter 9
- Chapter 9 Final Solutions
- Chapter 10
- Chapter 10 Final Solutions

Tutorial: Mass-Volume 8-3

Build this model. Calculate the overall mass of the illustrated model with the provided information.

1. **Create** a New part in SOLIDWORKS.

2. **Build** the illustrated model. The model displays all edges on perpendicular planes. Think about the steps required to build the model. Apply the Mirror Sketch tool to the Base sketch. Insert an Extruded Base (Boss-Extrude1) and Extruded-Cut (Cut-Extrude1) feature.

3. **Set** the document properties for the model.

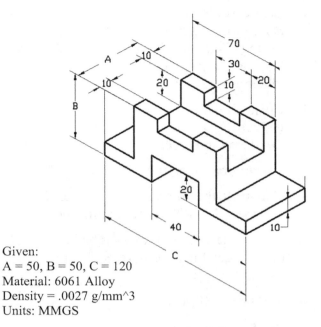

Given:
A = 50, B = 50, C = 120
Material: 6061 Alloy
Density = .0027 g/mm^3
Units: MMGS

💡 To activate the Mirror Sketch tool, click **Tools**, **Sketch Tools**, **Mirror** from the Menu bar menu. The Mirror PropertyManager is displayed.

4. Create **Sketch1**. Select the Front Plane as the Sketch plane. Apply the Mirror Sketch tool. Select the construction geometry to mirror about as illustrated. Select the Entities to mirror. Insert the required geometric relations and dimensions.

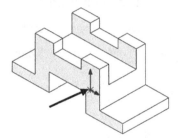

Construction geometry is ignored when the sketch is used to create a feature. Construction geometry uses the same line style as centerlines.

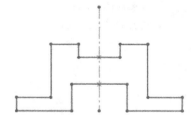

💡 When you create a new part or assembly, the three default Planes (Front, Right and Top) are aligned with specific views.

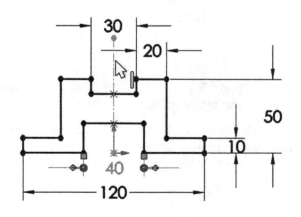

5. Create the **Boss-Extrude1** feature. Boss-Extrude1 is the Base feature. Apply the Mid Plane End Condition in Direction 1 for symmetry. Depth = 50mm.

6. Create **Sketch2**. Select the right face for the Sketch plane. Sketch2 is the profile for the Extruded Cut feature. Insert the required geometric relations and dimensions. Apply construction geometry.

7. Create the **Extruded Cut** feature. Through All is the selected End Condition in Direction 1.

8. **Assign** 6061 Alloy material to the part.

9. **Calculate** the overall mass. The overall mass = 302.40 grams.

10. **Save** the part and name it Mass-Volume 8-3.

11. **Close** the model.

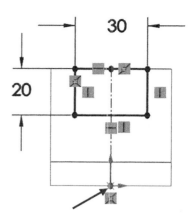

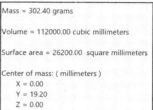

Mass = 302.40 grams

Volume = 112000.00 cubic millimeters

Surface area = 26200.00 square millimeters

Center of mass: (millimeters)
  X = 0.00
  Y = 19.20
  Z = 0.00

Tutorial: Mass-Volume 8-4

Build this model. Calculate the overall mass of the part and locate the Center of mass with the provided information.

1. **Create** a New part in SOLIDWORKS.

2. **Build** the illustrated model. All edges of the model are not located on Perpendicular planes. Think about the steps required to build the model. Insert two features: Extruded Base (Boss-Extrude1) and Extruded Cut (Cut-Extrude1).

3. **Set** the document properties for the model.

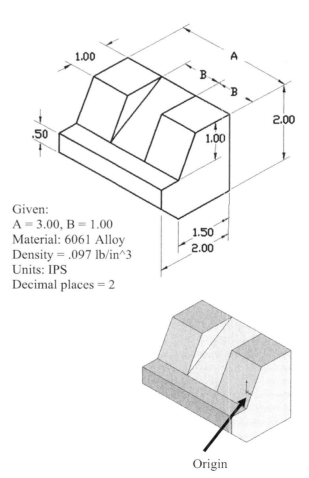

Given:
A = 3.00, B = 1.00
Material: 6061 Alloy
Density = .097 lb/in^3
Units: IPS
Decimal places = 2

Origin

4. Create **Sketch1**. Select the Right Plane as the Sketch plane. Apply construction geometry. Insert the required geometric relations and dimensions.

5. Create the **Extruded Base** feature. Boss-Extrude1 is the Base feature. Apply symmetry. Select Mid Plane as the End Condition in Direction 1. Depth = 3.00in.

6. Create **Sketch2**. Select the Right Plane as the Sketch plane. Select the Line Sketch tool. Insert the required geometric relations. Sketch2 is the profile for the Extruded Cut feature.

7. Create the **Extruded Cut (Cut-Extrude1)** feature. Apply symmetry. Select Mid Plane as the End Condition in Direction 1. Depth = 1.00in.

8. **Assign** 6061 Alloy material to the part.

9. **Calculate** overall mass. The overall mass = 0.87 pounds.

10. **Locate** the Center of mass. The location of the Center of mass is derived from the part Origin.

- X: 0.00 inches

- Y: 0.86 inches

- Z: 0.82 inches

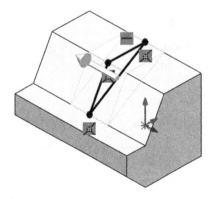

Mass = 0.87 pounds

Volume = 8.88 cubic inches

Surface area = 28.91 square inches

Center of mass: (inches)
   X = 0.00
   Y = 0.86
   Z = 0.82

In this category an exam question could read "Build this model. Locate the Center of mass with respect to the part Origin."

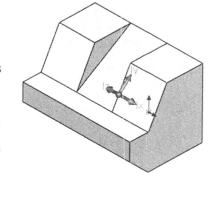

- A: X = 0.10 inches, Y = -0.86 inches, Z = -0.82 inches

- B: X = 0.00 inches, Y = 0.86 inches, Z = 0.82 inches

- C: X = 0.15 inches, Y = -0.96 inches, Z = -0.02 inches

- D: X = 1.00 inches, Y = -0.89 inches, Z = -1.82 inches

The correct answer is B.

11. **Save** the part and name it Mass-Volume 8-4.

12. **Close** the model.

As an exercise, modify the Mass-Volume 8-4 part using the MMGS unit system. Assign Nickel as the material. Calculate the overall mass. The overall mass of the part = 1236.20 grams. Save the part and name it Mass-Volume 8-4-MMGS.

Tutorial: Mass-Volume 8-5

Build this model. Calculate the overall mass of the part and locate the Center of mass with the provided information.

1. **Create** a New part in SOLIDWORKS.

2. **Build** the illustrated model. Insert five sketches and five features to build the model: Extruded Base, three Extruded Cut features and a Mirror feature.

🔅 There are numerous ways to build the models in this chapter. A goal is to display different design intents and techniques.

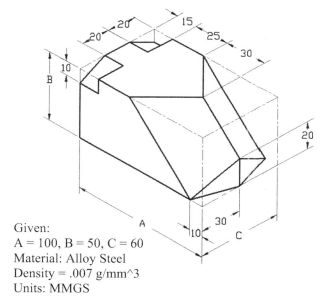

Given:
A = 100, B = 50, C = 60
Material: Alloy Steel
Density = .007 g/mm^3
Units: MMGS

3. **Set** the document properties for the model.

4. Create **Sketch1**. Select the Front Plane as the Sketch plane. Sketch a rectangle. Insert the required relations and dimensions. The part Origin is located in the lower left corner of the sketch.

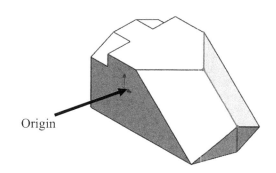

Origin

5. Create the **Extruded Base (Boss-Extrude1)** feature. Apply symmetry. Select the Mid Plane End Condition for Direction 1. Depth = 60mm.

6. Create **Sketch2**. Select the left face of Boss-Extrude1 as the Sketch plane. Insert the required geometric relations and dimensions.

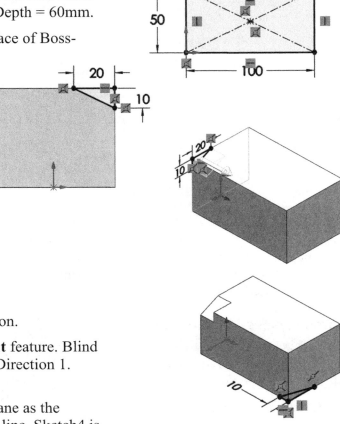

7. Create the first **Extruded Cut** feature. Blind is the default End Condition in Direction 1. Depth = 15mm. Note the direction of the extrude feature.

8. Create **Sketch3**. Select the bottom face of Boss-Extrude1 for the Sketch plane. Insert the required geometric relations and dimension.

9. Create the second **Extruded Cut** feature. Blind is the default End Condition in Direction 1. Depth = 20mm.

10. Create **Sketch4**. Select Front Plane as the Sketch plane. Sketch a diagonal line. Sketch4 is the direction of extrusion for the third Extruded Cut feature. Insert the required dimension.

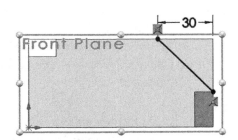

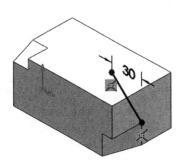

11. Create **Sketch5**. Select the top face of Boss-Extrude1 as the Sketch plane. Sketch5 is the sketch profile for the third Extruded Cut feature. Apply construction geometry. Insert the required geometric relations and dimensions.

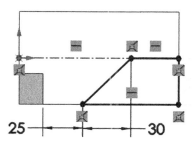

12. Create the third **Extruded Cut** feature. Select Through All for End Condition in Direction 1.

13. Select **Sketch4** in the Graphics window for Direction of Extrusion. Line1@Sketch4 is displayed in the Cut-Extrude PropertyManager.

14. Create the **Mirror** feature. Mirror the three Extruded Cut features about the Front Plane. Use the fly-out FeatureManager.

15. **Assign** Alloy Steel material to the part.

16. **Calculate** the overall mass. The overall mass = 1794.10 grams.

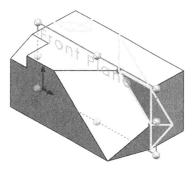

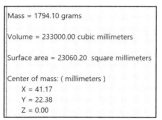

Mass = 1794.10 grams

Volume = 233000.00 cubic millimeters

Surface area = 23060.20 square millimeters

Center of mass: (millimeters)
  X = 41.17
  Y = 22.38
  Z = 0.00

17. **Locate** the Center of mass. The location of the Center of mass is derived from the part Origin.

- X = 41.17 millimeters

- Y = 22.38 millimeters

- Z = 0.00 millimeters

View the triad location of the Center of mass for the part.

18. **Save** the part and name it Mass-Volume 8-5.

19. **Close** the model.

Set document precision from the Document Properties dialog box or from the Dimension PropertyManager. You can also address Callout value, Tolerance type, and Dimension Text symbols in the Dimension PropertyManager.

☼ You are allowed to answer the questions in any order you prefer. Use the Summary Screen during the CSWA exam to view the list of all questions you have or have not answered.

There are no Surfacing or Boundary feature questions on the CSWA exam at this time.

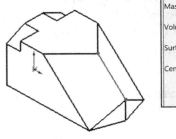

Mass = 1794.10 grams

Volume = 233000.00 cubic millimeters

Surface area = 23060.20 square millimeters

Center of mass: (millimeters)
 X = 41.17
 Y = 22.38
 Z = 0.00

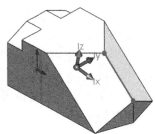

System Options	Document Properties

Drafting Standard
⊞ Annotations
⊞ Dimensions
  Virtual Sharps
⊞ Tables
⊞ DimXpert
Detailing
Grid/Snap
Units
Model Display
Material Properties
Image Quality
Sheet Metal MBD
Sheet Metal
Weldments
Plane Display
Configurations

Unit system
○ MKS (meter, kilogram, second)
○ CGS (centimeter, gram, second)
○ MMGS (millimeter, gram, second)
◉ IPS (inch, pound, second)
○ Custom

Type	Unit	Decimals
Basic Units		
Length	inches	.12
Dual Dimension Length	inches	.12
Angle	degrees	.12

Build Additional Basic Parts

Tutorial: Mass-Volume 8-6

Build this model. Calculate the overall mass of the part and locate the Center of mass with the provided information.

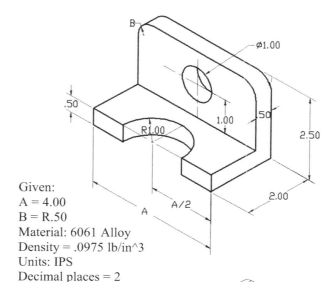

1. **Create** a New part in SOLIDWORKS.

2. **Build** the illustrated model. Think about the required steps to build this part. Insert four features: Extruded Base, two Extruded Cuts, and a Fillet. All Thru Holes.

Given:
A = 4.00
B = R.50
Material: 6061 Alloy
Density = .0975 lb/in^3
Units: IPS
Decimal places = 2

🔆 There are numerous ways to build the models in this chapter. A goal is to display different design intents and techniques.

3. **Set** the document properties for the model.

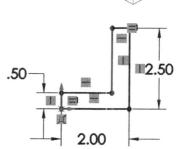

4. Create **Sketch1**. Select the Right Plane as the Sketch plane. The part Origin is located in the lower left corner of the sketch. Insert the required geometric relations and dimensions.

5. Create the **Extruded Base (Boss-Extrude1)** feature. Apply symmetry. Select the Mid Plane End Condition for Direction 1. Depth = 4.00in.

6. Create **Sketch2**. Select the top flat face of Boss-Extrude1 as the Sketch plane. Sketch a circle. The center of the circle is located at the part Origin. Insert the required dimension.

7. Create the first **Extruded Cut** feature. Select Through All for End Condition in Direction 1.

8. Create **Sketch3**. Select the front vertical face of Extrude1 as the Sketch plane. Sketch a circle. Insert the required geometric relations and dimensions.

9. Create the second **Extruded Cut** feature. Select Through All for End Condition in Direction 1.

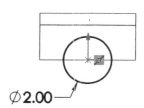

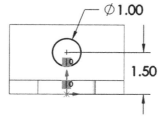

10. Create the **Fillet** feature. Constant radius is selected by default. Fillet the top two edges as illustrated. Radius = .50in.

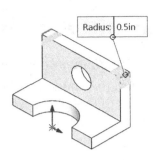

☀ A Fillet feature removes material. Selecting the correct radius value is important to obtain the correct mass and volume answer in the exam.

11. **Assign** the defined material to the part.

12. **Calculate** the overall mass. The overall mass = 0.66 pounds.

13. **Locate** the Center of mass. The location of the Center of mass is derived from the part Origin.

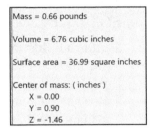

- X: 0.00 inches

- Y: 0.90 inches

- Z: -1.46 inches

In this category an exam question could read "Build this model. Locate the Center of mass relative to the part Origin."

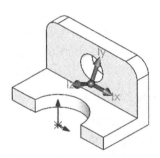

- A: X = -2.63 inches, Y = 4.01 inches, Z = -0.04 inches

- B: X = 4.00 inches, Y = 1.90 inches, Z = -1.64 inches

- C: X = 0.00 inches, Y = 0.90 inches, Z = -1.46 inches

- D: X = -1.69 inches, Y = 1.00 inches, Z = 0.10 inches

The correct answer is C. Note: Tangent edges and Origin is displayed for educational purposes.

14. **Save** the part and name it Mass-Volume 8-6.

15. **Close** the model.

As an exercise, calculate the overall mass of the part using the MMGS unit system, and assign 2014 Alloy material to the part.

The overall mass of the part = 310.17 grams.

Save the part and name it Mass-Volume 8-6-MMGS.

Tutorial: Mass-Volume 8-7

Build this model. Calculate the overall mass of the part and locate the Center of mass with the provided information.

1. **Create** a New part in SOLIDWORKS.

2. **Build** the illustrated model. Insert two features: Extruded Base (Boss-Extrude1) and Revolved Boss.

3. **Set** the document properties for the model.

☀ Tangent edges and the Origin are displayed for educational purposes.

4. Create **Sketch1**. Select the Top Plane as the Sketch plane. Apply construction geometry. Apply the Tangent Arc and Line Sketch tool. Insert the required geometric relations and dimensions.

5. Create the **Extruded Base** feature. Blind is the default End Condition. Depth = 8mm.

6. Create **Sketch2**. Select the Front Plane as the Sketch plane. Apply construction geometry for the Revolved Boss feature. Insert the required geometric relations and dimension.

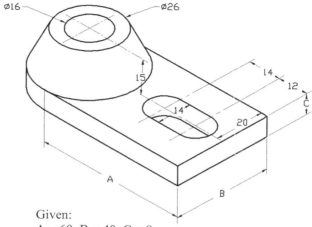

Given:
A = 60, B = 40, C = 8
Material: Cast Alloy Steel
Density = .0073 g/mm^3
Units: MMGS

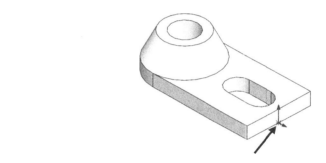

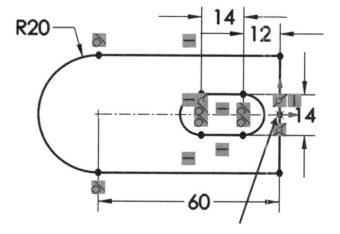

7. Create the **Revolved Boss** feature. The default angle is 360deg. Select the centerline for Axis of Revolution.

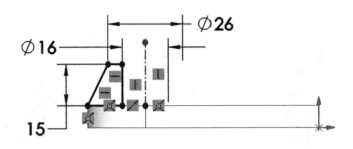

8. **Assign** the defined material to the part.

9. **Calculate** the overall mass. The overall mass = 229.46 grams.

10. **Locate** the Center of mass. The location of the Center of mass is derived from the part Origin.

- X = -46.68 millimeters

- Y = 7.23 millimeters

- Z = 0.00 millimeters

Mass = 229.46 grams

Volume = 31433.02 cubic millimeters

Surface area = 9459.63 square millimeters

Center of mass: (millimeters)
   X = -46.68
   Y = 7.23
   Z = 0.00

In this category an exam question could read "Build this model. What is the overall mass of the part?"

- A: 229.46 grams

- B: 249.50 grams

- C: 240.33 grams

- D: 120.34 grams

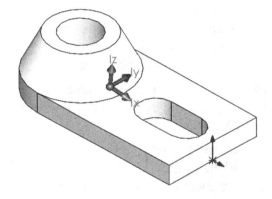

The correct answer is A.

11. **Save** the part and name it Mass-Volume 8-7.

12. **Close** the model.

Tangent edges and Origin are displayed for educational purposes.

☀ When you create a new part or assembly, the three default Planes (Front, Right and Top) are aligned with specific views. The Plane you select for the Base sketch determines the orientation of the part.

Tutorial: Basic/Intermediate Part 8-1

Build this model. Calculate the overall mass of the part and locate the Center of mass with the provided information.

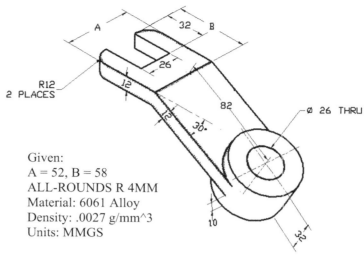

Given:
A = 52, B = 58
ALL-ROUNDS R 4MM
Material: 6061 Alloy
Density: .0027 g/mm^3
Units: MMGS

1. **Create** a New part in SOLIDWORKS.

2. **Build** the illustrated model. Think about the various features that create the part. Insert seven features and a plane to build this part: Extruded-Thin1, Boss-Extrude1, Cut-Extrude1, Cut-Extrude2 and three Fillets. Apply reference construction planes to build the circular features. All Thru Holes.

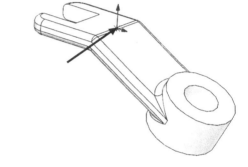

3. **Set** the document properties for the model.

4. Create **Sketch1**. Select the Front Plane as the Sketch plane. Apply construction geometry as the reference line for the 30deg angle. Insert the required geometric relations and dimensions. Note the location of the Origin.

5. Create the **Extrude-Thin1** feature. This is the Base feature. Apply symmetry. Select Mid Plane for End Condition in Direction 1 to maintain the location of the Origin. Depth = 52mm. Thickness = 12mm.

☼ Use the Thin Feature option to control the extrude thickness, not the Depth.

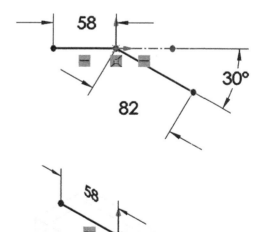

6. Create **Plane1**. Plane1 is the Sketch plane for the Extruded Boss (Boss-Extrude1) feature. Select the midpoint and the top face as illustrated. Plane1 is located in the middle of the top and bottom faces. Select Parallel Plane at Point for option.

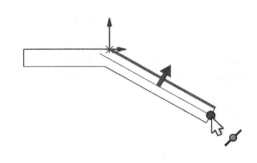

 Plane1 uses the Depth dimension of 32mm.

7. Create **Sketch2**. Select Plane1 as the Sketch plane. Use the Normal To view tool. Sketch a circle to create the Extruded Boss feature. Insert the required geometric relations.

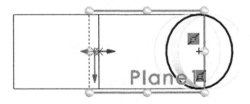

 The Normal To view tool rotates and zooms the model to the view orientation normal to the selected plane, planar face, or feature.

8. Create the **Extruded Boss** feature. Apply Symmetry. Select Mid Plane for End Condition in Direction 1. Depth = 32mm.

9. Create **Sketch3**. Select the top circular face of Boss-Extrude1 as the Sketch plane. Sketch a circle. Insert the required geometric relation and dimension.

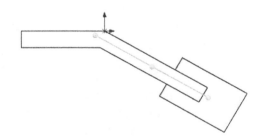

There are numerous ways to create the models in this chapter. A goal is to display different design intents and techniques.

10. Create the first **Extruded Cut** feature. Select Through All for End Condition in Direction 1.

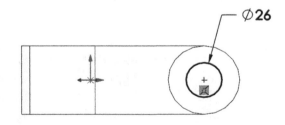

11. Create **Sketch4**. Select the top face of Extrude-Thin1 as the Sketch plane. Apply construction geometry. Insert the required geometric relations and dimensions.

12. Create the second **Extruded Cut** feature. Select Through All for End Condition in Direction 1.

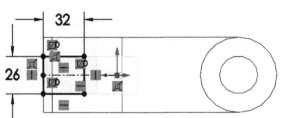

13. Create the **Fillet1** feature. Fillet
the left and right edges of Extrude-
Thin1 as illustrated.
Radius = 12mm.

14. Create the **Fillet2** feature. Fillet
the top and bottom edges of
Extrude-Thin1 as illustrated.
Radius = 4mm.

15. Create the **Fillet3** feature. Fillet
the rest of the model, six edges as
illustrated. Radius = 4mm.

16. **Assign** the defined material to the
part.

17. **Calculate** the overall mass of the part. The overall
mass = 300.65 grams.

18. **Locate** the Center of mass. The location of the
Center of mass is derived from the part Origin.

- X: 34.26 millimeters

- Y: -29.38 millimeters

- Z: 0.00 millimeters

19. **Save** the part and name it Part-Modeling 8-1.

20. **Close** the model.

As an exercise, modify the Fillet2 and Fillet3 radius
from 4mm to 2mm. Modify the Fillet1 radius from 12m
to 10mm. Modify the material from 6061 Alloy to ABS.

Modify the Sketch1 angle from 30deg to 45deg. Modify
the Extrude depth from 32mm to 38mm. Recalculate the
location of the Center of mass with respect to the part
Origin.

- X = 27.62 millimeters

- Y = -40.44 millimeters

- Z = 0.00 millimeters

21. **Save** the part and name it Part-Modeling 8-1-Modify.

 In the exam, you can answer the questions in any
order. Use the Summary Screen during the exam to view
the list of all questions you have or have not answered.

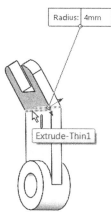

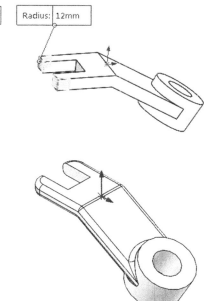

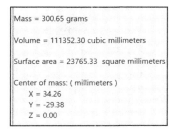

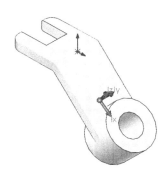

Mass = 300.65 grams

Volume = 111352.30 cubic millimeters

Surface area = 23765.33 square millimeters

Center of mass: (millimeters)
 X = 34.26
 Y = -29.38
 Z = 0.00

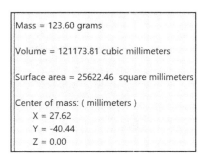

Mass = 123.60 grams

Volume = 121173.81 cubic millimeters

Surface area = 25622.46 square millimeters

Center of mass: (millimeters)
 X = 27.62
 Y = -40.44
 Z = 0.00

Tutorial: Basic/Intermediate-Part 8-2

Build this model. Calculate the volume of the part and locate the Center of mass with the provided information.

1. **Create** a New part in SOLIDWORKS.

2. **Build** the illustrated model. Think about the various features that create this model. Insert five features and a plane to build this part: Extruded Base, two Extruded Bosses, Extruded Cut and a Rib. Insert a reference plane to create the Boss-Extrude2 feature. All Thru Holes.

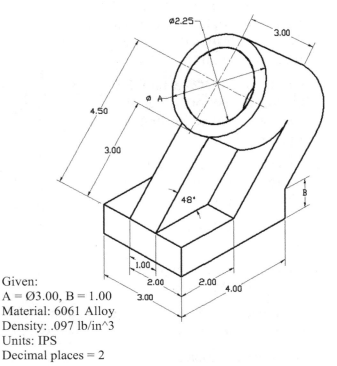

Given:
A = Ø3.00, B = 1.00
Material: 6061 Alloy
Density: .097 lb/in^3
Units: IPS
Decimal places = 2

3. **Set** the document properties for the model.

4. Create **Sketch1**. Select the Top Plane as the Sketch plane. Sketch a center rectangle. Use the horizontal construction line as the Plane1 reference. Insert the required relations and dimensions.

5. Create the **Extruded Base** feature. Blind is the default End Condition in Direction 1. Depth = 1.00in. Note the extrude direction is downward.

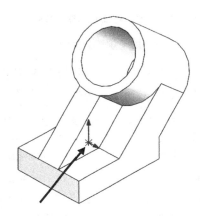

🔅 Create planes to aid in the modeling for the exam. Use planes to sketch, to create a section view, for a neutral plane in a draft feature, and so on.

🔅 The created plane is displayed 5% larger than the geometry on which the plane is created, or 5% larger than the bounding box. This helps reduce selection problems when planes are created directly on faces or from orthogonal geometry.

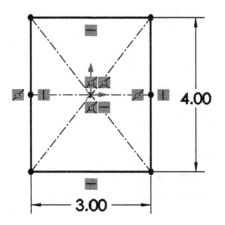

6. Create **Plane1**. Plane1 is the Sketch plane for the Extruded Boss feature. Show Sketch1. Select the horizontal construction line in Sketch1 and the top face of Boss-Extrude1. Angle = 48deg.

🔅 Click **View**, **Hide/Show**, **Sketches** from the Menu bar menu to displayed sketches in the Graphics window.

🔅 The Normal To view tool rotates and zooms the model to the view orientation normal to the selected plane, planar face, or feature.

7. Create **Sketch2**. Select Plane1 as the Sketch plane. Create the Extruded Boss profile. Insert the required geometric relations and dimension. Note: Dimension to the front **top edge** of Boss-Extrude1 as illustrated.

8. Create the first **Extruded Boss** feature. Select the Up To Vertex End Condition in Direction 1. Select the back top right vertex point as illustrated.

9. Create **Sketch3**. Select the back angled face of Boss-Extrude2 as the Sketch plane. Sketch a circle. Insert the required geometric relations.

10. Create the third **Extruded Boss** feature. Blind is the default End Condition in Direction 1. Depth = 3.00in.

11. Create **Sketch4**. Select the front face of Boss-Extrude3 as the Sketch plane. Sketch a circle. Sketch4 is the profile for the Extruded Cut feature. Insert the required geometric relation and dimension.

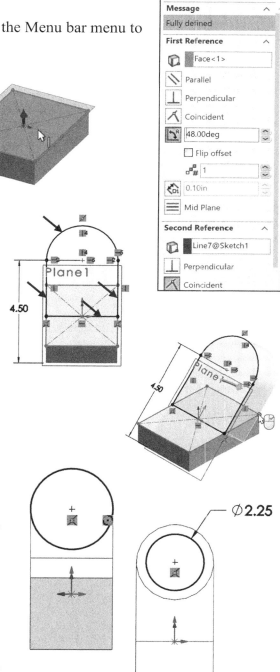

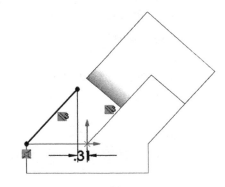

The part Origin is displayed in blue.

12. Create the **Extruded Cut** feature. Select Through All for End Condition in Direction 1.

13. Create **Sketch5**. Select the Right Plane as the Sketch plane. Insert a Parallel relation to partially define Sketch5. Sketch5 is the profile for the Rib feature. Sketch5 does not need to be fully defined. Sketch5 locates the end conditions based on existing geometry.

14. Create the **Rib** feature. Thickness = 1.00in.

The Rib feature is a special type of extruded feature created from open or closed sketched contours. The Rib feature adds material of a specified thickness in a specified direction between the contour and an existing part. You can create a rib feature using single or multiple sketches.

15. **Assign** 6061 Alloy material to the part.

16. **Calculate** the volume. The volume = 30.65 cubic inches.

17. **Locate** the Center of mass. The location of the Center of mass is derived from the part Origin.

- X: 0.00 inches

- Y: 0.73 inches

- Z: -0.86 inches

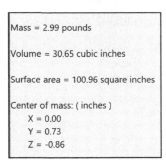

Mass = 2.99 pounds

Volume = 30.65 cubic inches

Surface area = 100.96 square inches

Center of mass: (inches)
   X = 0.00
   Y = 0.73
   Z = -0.86

18. **Save** the part and name it Part-Modeling 8-2.

19. **Close** the model.

As an exercise, modify the Rib1 feature from 1.00in to 1.25in. Modify the Extrude depth from 3.00in to 3.25in. Modify the material from 6061 Alloy to Copper.

Modify the Plane1 angle from 48deg to 30deg. Recalculate the volume of the part. The new volume = 26.94 cubic inches.

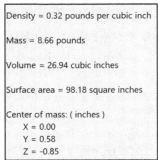

Density = 0.32 pounds per cubic inch

Mass = 8.66 pounds

Volume = 26.94 cubic inches

Surface area = 98.18 square inches

Center of mass: (inches)
   X = 0.00
   Y = 0.58
   Z = -0.85

20. **Save** the part and name it Part-Modeling 8-2-Modify.

Summary

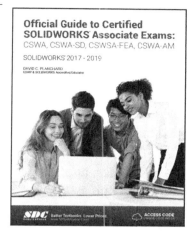

Basic Part Creation and Modification and *Intermediate Part Creation and Modification* are two of the five categories on the CSWA exam.

There are two questions on the CSWA Academic exam (Part 1) in the *Basic Part Creation and Modification* category. One question is in a multiple-choice single answer format and the other question (Modification of the model) is in the fill in the blank format. Each question is worth fifteen (15) points for a total of thirty (30) points.

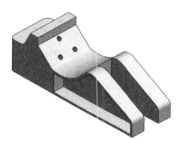

The main difference between the *Basic Part Creation and Modification* category and the *Intermediate Part Creation and Modification* or the *Advance Part Creation and Modification* category is the complexity of the sketches and the number of dimensions and geometric relations along with an increase in the number of features.

During the CSWA exam, SOLIDWORKS provides various model views. Click on the additional view icons during the exam to better understand the part and provided information. Read each question carefully. Identify the dimensions, center of mass, units and location of the Origin. Apply needed material.

To obtain additional CSWA exam information, visit the SOLIDWORKS VirtualTester Certification site at https://SOLIDWORKS.virtualtester.com/.

Questions

Question 1:

Build the model from the provided information. Set document properties. Identify the correct Sketch planes, apply the correct Sketch and Feature tools and apply material.

Calculate the overall mass of the part, volume, and locate the Center of mass with the provided illustrated information.

- Material: 6061 Alloy.

- Units: MMGS.

- All Thru Holes.

When you create a new part or assembly, the three default planes (Front, Right and Top) are aligned with specific views. The plane you select for the Base sketch determines the orientation of the part.

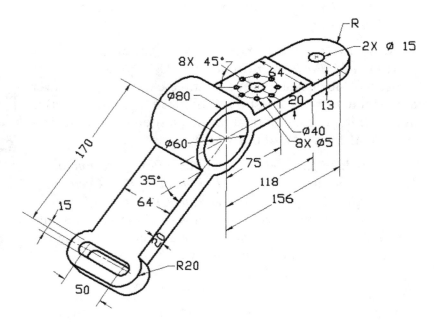

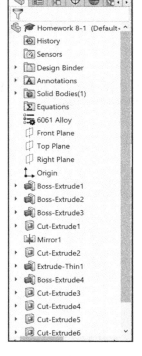

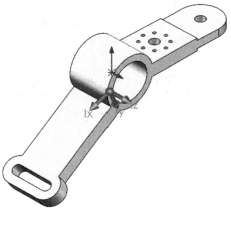

Question 2:

Build the model from the provided information. Set document properties. Apply material.

Calculate the overall mass of the part, volume, and locate the Center of mass with the provided information.

- Material: 6061 Alloy.
- Units: MMGS.

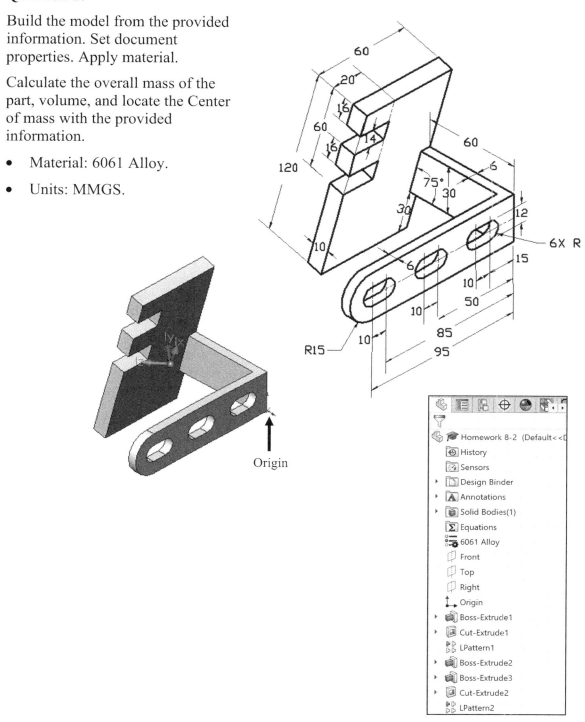

Origin

Question 3:

Build the model from the provided information. Set document properties. Apply material.

Calculate the overall mass of the part, volume, and locate the Center of mass with the provided information.

- Material: 6061.

- Units: IPS.

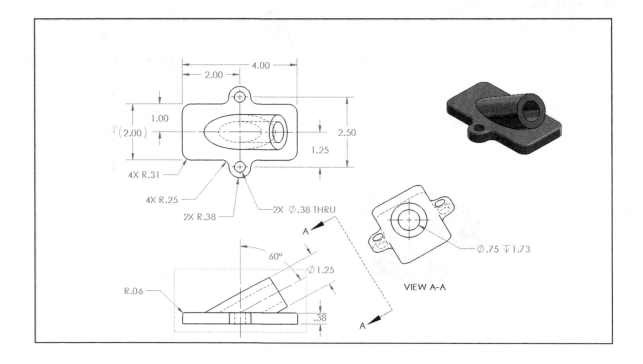

Question 4:

Build the model from the provided information.

Set document properties. Apply material.

Calculate the overall mass of the part, volume, and locate the Center of mass with the provided information.

- Material: 6061.

- Units: IPS.

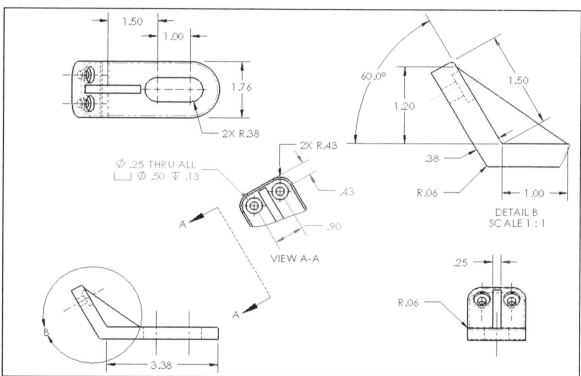

Question 5:

Build the model from the provided information.

Set document properties.

Apply material.

Calculate the overall mass and volume of the part.

- Precision for linear dimensions = **2**.

- Material: **AISI 304**.

- Units: **MMGS**.

- All Holes ⫪ **25**mm.

- All Rounds **5**mm.

- All Holes Ø**4**mm.

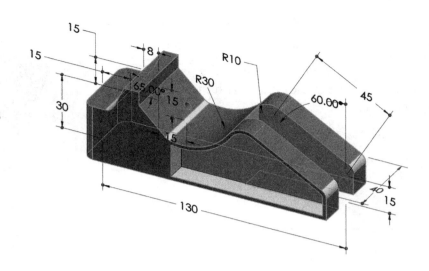

As a general rule, add fillets at the end of the modeling procedure.

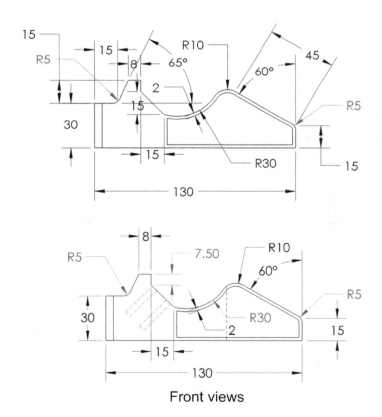

Front views

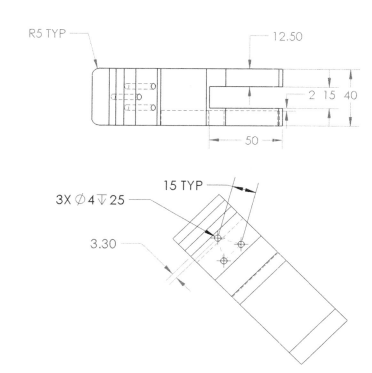

Top and Auxiliary view

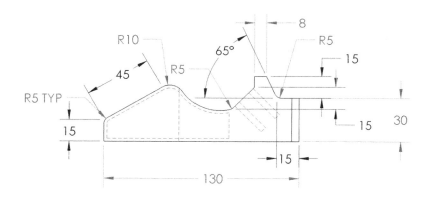

Calculate the mass:

A = **888.48grams**

B = 990.50grams

C = 788.48grams

D = 809.57grams

Back view

💡 If you don't find your answer (within 1%) in the multiple-choice single answer format section, recheck your solid model for precision and accuracy. It could be as simple as missing a few fillets.

Calculate the volume:

A = 122259.43 cubic millimeters
B = 133359.47 cubic millimeters
C = 111059.43 cubic millimeters
D = 145059.49 cubic millimeters

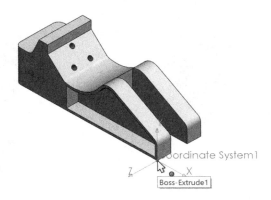

Question 5A:

Create a new coordinate system.

Center a new coordinate system with the provided illustration. The new coordinate system location is at the front right bottom point (vertex) of the model.

Enter the Center of Mass:

X = -80.39 millimeters

Y = 15.93 millimeters

Z = -22.65 millimeters

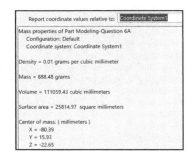

Question 5B:

Modify the model. Calculate the overall mass and volume of the part.

- Modify all fillets (rounds) to 7mm.

- Modify the overall length to 140mm.

- Modify material to 1060 alloy.

Enter the mass:

309.75 grams

Enter the volume:

114721.22 cubic millimeters

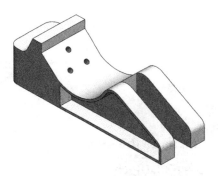

☀ If you don't find your answer (within 1%) in the multiple-choice single answer format section, recheck your solid model for precision and accuracy. It could be as simple as missing a few fillets.

Question 6:

Build the illustrated model.

Set document properties.

Apply material.

Calculate the overall mass and volume.

- Precision for linear dimensions = **2**.

- Material: **1060 Alloy**.

- Units: **MMGS**.

- TYP $\emptyset 12$.

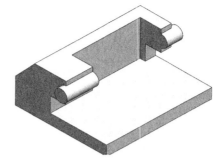

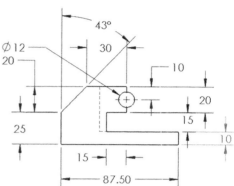

Front view

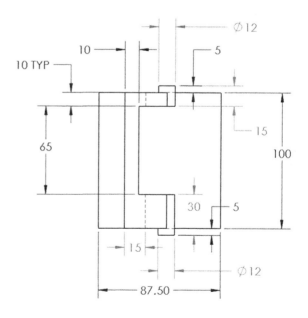

Top view

Calculate the mass:

A = 610.92 grams

B = 509.92 grams

C = 701.93 grams

D = 619.34 grams

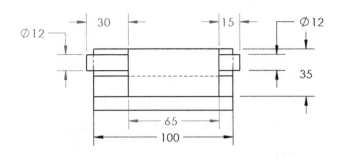

Right view

Calculate the volume:

A = 188860.93 cubic millimeters
B = 206660.93 cubic millimeters
C = 198880.65 cubic millimeters
D = 230021.67 cubic millimeters

Mass properties of Homework 8-6
   Configuration: Default
   Coordinate system: -- default --

Density = 0.00 grams per cubic millimeter

Mass = 509.92 grams

Volume = 188860.93 cubic millimeters

Surface area = 32545.06 square millimeters

Center of mass: (millimeters)
   X = 31.39
   Y = 16.55
   Z = 1.37

Question 6A:

Modify the model.

Calculate the overall mass and volume of the model.

- Modify material to Plain Carbon Steel.

- Modify TYP Hole diameter from TYP 12 to TPY 10.

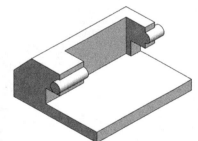

Enter the mass:

1465.70 grams

Enter the volume:

187910.60 cubic millimeters

   Configuration: modified
   Coordinate system: -- default --

Density = 0.01 grams per cubic millimeter

Mass = 1465.70 grams

Volume = 187910.60 cubic millimeters

Surface area = 32373.16 square millimeters

Center of mass: (millimeters)
   X = 31.29
   Y = 16.45
   Z = 1.33

Question 6B:

Create a new coordinate system.

The new coordinate system location is at the front left bottom point (vertex) of the model.

Enter the Center of Mass:

X = 31.39 millimeters

Y = 16.45 millimeters

Z = -48.67 millimeters

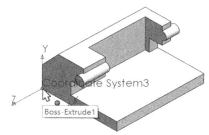

Question 7:

Build the illustrated model.

Calculate the overall mass and volume of the part with the provided information.

- Precision for linear dimensions = **2**.

- Material: **Plain Carbon Steel**.

- Units: **MMGS**.

- The part is **symmetrical** about the Front Plane.

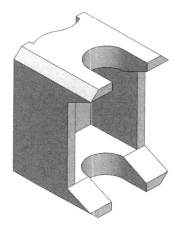

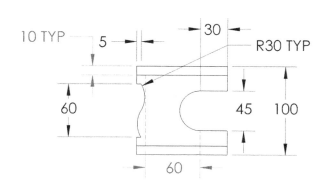

Top view

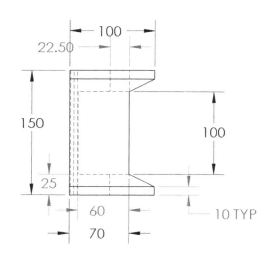

Front view

Calculate the mass:

A = 4311.50 grams

B = 4079.32 grams

C = 4234.30 grams

D = 5322.00 grams

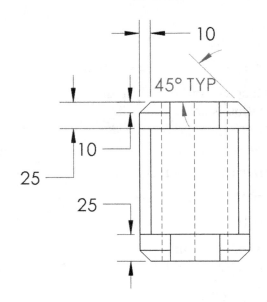

Right view

Calculate the volume:

A = 522989.22 cubic millimeters
B = 555655.11 cubic millimeters
C = 591233.34 cubic millimeters
D = 655444.00 cubic millimeters

Question 7A:

Create a new coordinate system.

Center a new coordinate system with the provided illustration.

The new coordinate system location is at the back right bottom point (vertex) of the model.

Enter the Center of Mass:

X = -64.09 millimeters

Y = 75.00 millimeters

Z = 40.70 millimeters

Mass properties of Homework 8-7
    Configuration: Default
    Coordinate system: -- default --

Density = 0.01 grams per cubic millimeter

Mass = 4079.32 grams

Volume = 522989.22 cubic millimeters

Surface area = 92824.57 square millimeters

Center of mass: (millimeters)
    X = 35.91
    Y = 0.00
    Z = 0.70

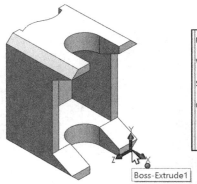

Mass = 4079.32 grams

Volume = 522989.22 cubic millimeters

Surface area = 92824.57 square millimeters

Center of mass: (millimeters)
    X = -64.09
    Y = 75.00
    Z = 40.70

Question 8:

Build the provided model.

Set document properties.

Calculate the overall mass and volume of the part. There are no illustrated dimensions.

Open Homework problem 8-8 from the Homework chapter exercises. Use the rollback bar to obtain features and dimensions. Think about the various ways that this model can be built.

- Precision for linear dimensions = **2**.

- Material: **1060 Alloy**.

- Units: **MMGS**.

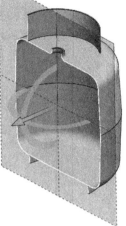

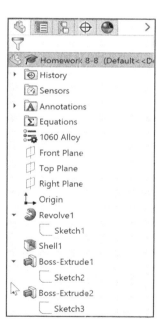

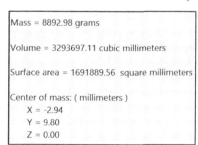

Mass = 8892.98 grams

Volume = 3293697.11 cubic millimeters

Surface area = 1691889.56 square millimeters

Center of mass: (millimeters)
   X = -2.94
   Y = 9.80
   Z = 0.00

Sample screen shots from an older CSWA exam for a simple part. Click on the additional views to understand the part and to provide information. Read each question carefully. Understand the dimensions, center of mass and units. Apply needed materials.

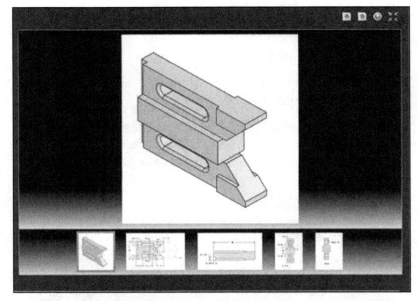

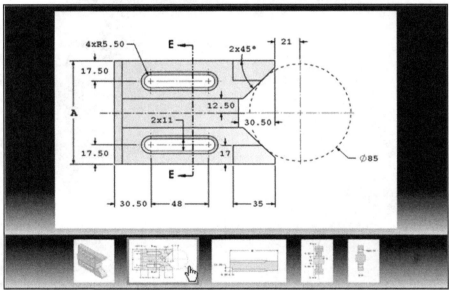

Screen shots from an exam

Zoom in on the part or view if needed.

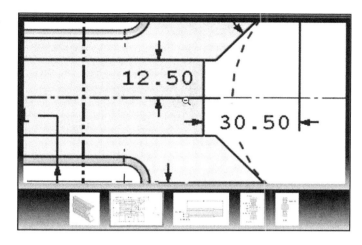

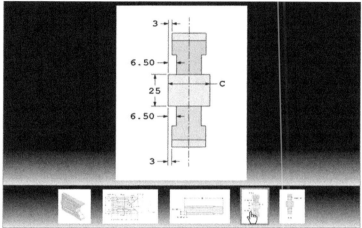

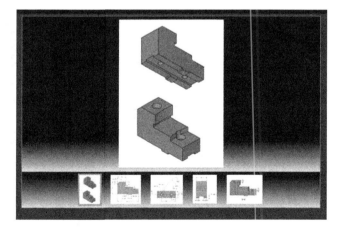

Screen shots from an exam

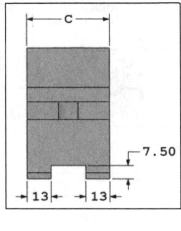

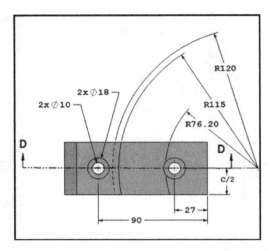

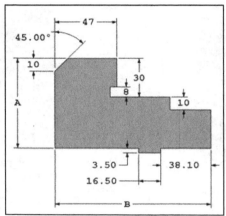

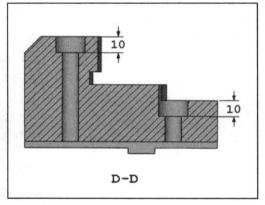

D-D

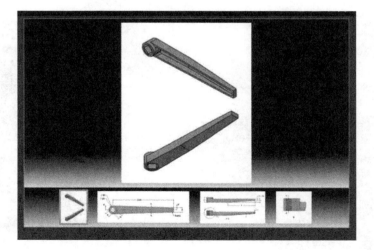

Screen shots from an exam

Chapter 9 - CSWA Advanced Part Creation and Modification

Objectives

Advanced Part Creation and Modification is one of the five categories on the CSWA exam. The main difference between the *Advanced Part Creation and Modification* and the *Basic Part Creation and Modification* or the *Intermediate Part Creation and Modification* category is the complexity of the sketches and the number of dimensions and geometric relations along with an increased number of features.

There are three questions on the CSWA Academic exam (part 2) in this category. The first question is in a multiple-choice single answer format and the other two questions (Modification of the model) are in the fill in the blank format. Each question is worth fifteen (15) points for a total of forty five (45) points.

You are required to build a model with six or more features and to answer a question either on the overall mass, volume, or the location of the Center of mass for the created model relative to the default part Origin location. You are then requested to modify the model and answer fill in the blank format questions.

Download all needed model files (SW-TUTORIAL-2020 folder) from the SDC Publications website (www.sdcpublications.com). All SOLIDWORKS models (initial and final) are provided.

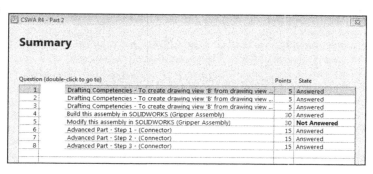

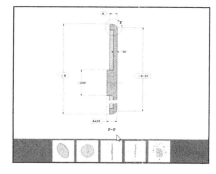

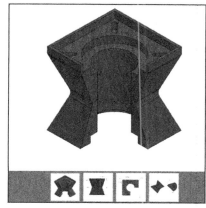

Screen shots from an exam

On the completion of the chapter, you will be able to:

- Specify Document Properties.

- Interpret Engineering terminology.

- Build an advanced part from a detailed dimensioned illustration using the following tools and features:

  - 2D & 3D Sketch tools, Extruded Boss/Base, Extruded Cut, Fillet, Mirror, Revolved Boss/Base, Linear & Circular Pattern, Chamfer and Revolved Cut.

- Locate the Center of mass relative to the part Origin.

- Create a coordinate system location.

- Locate the Center of mass relative to a created Coordinate system.

Build an Advanced Part from a Detailed Dimensioned Illustration

Tutorial: Advanced Part 9-1

An exam question in this category could read "Build this part. Calculate the overall mass and locate the Center of mass of the illustrated model."

1. **Create** a New part in SOLIDWORKS.

2. **Build** the illustrated model. Insert seven features: Extruded Base, two Extruded Bosses, two Extruded Cuts, a Chamfer and a Fillet.

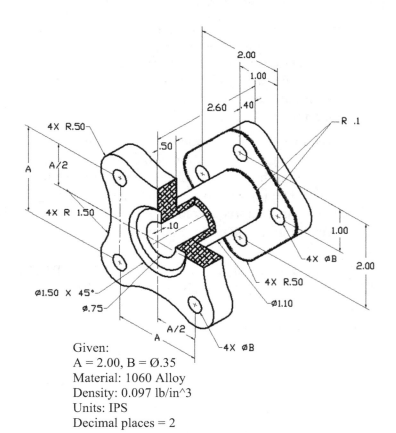

Given:
A = 2.00, B = Ø.35
Material: 1060 Alloy
Density: 0.097 lb/in^3
Units: IPS
Decimal places = 2

Think about the steps that you would take to build the illustrated part. Identify the location of the part Origin. Start with the back base flange. Review the provided dimensions and annotations in the illustration.

The main difference between the *Advanced Part Creation and Modification* and the *Basic Part Creation and Modification* or the *Intermediate Part Creation and Modification* category is the complexity of the sketches and the number of dimensions and geometric relations along with an increased number of features.

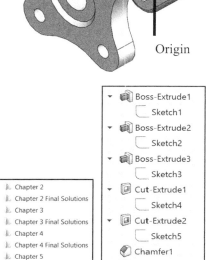

Origin

All SOLIDWORKS models for the next few chapters (initial and final) are provided in the SOLIDWORKS CSWA model folder.

3. **Set** the document properties for the model.

4. Create **Sketch1**. Sketch1 is the Base sketch. Select the Front Plane as the Sketch plane. Apply construction geometry. Sketch a horizontal and vertical centerline. Sketch four circles. Insert an Equal relation. Insert a Symmetric relation about the vertical and horizontal centerlines. Sketch two top angled lines and a tangent arc. Apply the Mirror Sketch tool. Complete the sketch. Insert the required geometric relations and dimensions.

In a Symmetric relation, the selected items remain equidistant from the centerline, on a line perpendicular to the centerline. Sketch entities to select: a centerline and two points, lines, arcs or ellipses.

- Boss-Extrude1
    - Sketch1
- Boss-Extrude2
    - Sketch2
- Boss-Extrude3
    - Sketch3
- Cut-Extrude1
    - Sketch4
- Cut-Extrude2
    - Sketch5
- Chamfer1
- Fillet1

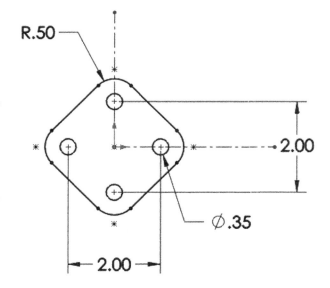

R.50

2.00

Ø.35

2.00

In the exam, you are allowed to answer the questions in any order. Use the Summary Screen during the exam to view the list of all questions you have or have not answered.

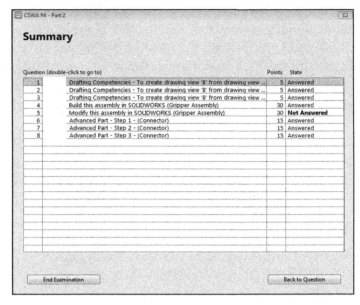

💡 The Sketch Fillet tool rounds the selected corner at the intersection of two sketch entities, creating a tangent arc.

5. Create the **Extruded Base** feature. Boss-Extrude1 is the Base feature. Blind is the default End Condition in Direction 1. Depth = .40in.

6. Create **Sketch2**. Select the front face of Boss-Extrude1 as the Sketch plane. Sketch a circle. Insert the required geometric relation and dimension.

7. Create the first **Extruded Boss** feature. Blind is the default End Condition in Direction 1. The Extrude feature is the tube between the two flanges. Depth = 1.70in. Note: 1.70in = 2.60in - (.50in + .40in).

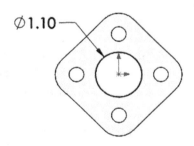

The complexity of the models along with the features progressively increases throughout this chapter to simulate the final types of parts that could be provided on the CSWA exam.

When you create a new part or assembly, the three default Planes (Front, Right and Top) are aligned with specific views. The Plane you select for the Base sketch determines the orientation of the part.

💡 There are no Surfacing or Boundary feature questions on the CSWA exam at this time.

The book is designed to expose the new user to many tools, techniques and procedures. It may not always use the most direct tool or process.

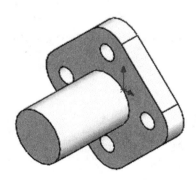

8. Create **Sketch3**. Select the front circular face of Boss-Extrude2 as the Sketch plane. Sketch a horizontal and vertical centerline. Sketch the top two circles. Insert an Equal and Symmetric relation between the two circles. Mirror the top two circles about the horizontal centerline. Insert dimensions to locate the circles from the Origin. Apply either the 3 Point Arc or the Centerpoint Arc Sketch tool. The center point of the Tangent Arc is aligned with a Vertical relation to the Origin. Complete the sketch.

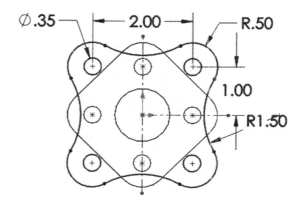

🔆 Use the Centerpoint Arc Sketch tool to create an arc from a centerpoint, a start point, and an end point.

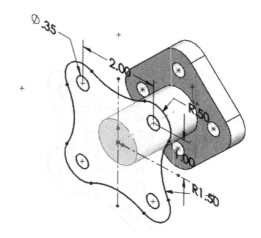

🔆 Apply the Tangent Arc Sketch tool to create an arc, tangent to a sketch entity.

The Arc PropertyManager controls the properties of a sketched Centerpoint Arc, Tangent Arc, and 3 Point Arc.

9. Create the second **Extruded Boss** feature. Blind is the default End Condition in Direction 1. Depth = .50in.

10. Create **Sketch4**. Select the front face of the Extrude feature as the Sketch plane. Sketch a circle. Insert the required geometric relation and dimension.

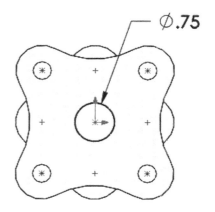

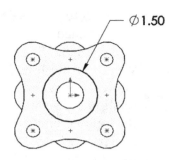

11. Create the first **Extruded Cut** feature. Select the Through All End Condition for Direction 1.

12. Create **Sketch5**. Select the front face of the Extrude feature as the Sketch plane. Sketch a circle. Insert the required geometric relation and dimension.

13. Create the second **Extruded Cut** feature. Blind is the default End Condition for Direction1. Depth = .10in.

14. Create the **Chamfer** feature. In order to have the outside circle 1.50in, select the inside edge of the sketched circle. Create an Angle distance chamfer. Distance = .10in. Angle = 45deg.

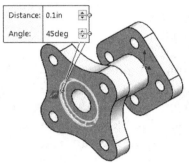

The Chamfer feature creates a beveled feature on selected edges, faces or a vertex.

15. Create the **Fillet** feature. Fillet the two edges as illustrated. Radius = .10in.

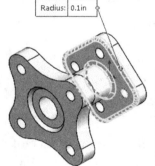

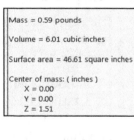

16. **Assign** 1060 Alloy material to the part. Material is required to calculate the overall mass of the part.

17. **Calculate** the overall mass. The overall mass = 0.59 pounds.

18. **Locate** the Center of mass. The location of the Center of mass is relative to the part Origin.

- X: 0.00

- Y: 0.00

- Z: 1.51

19. **Save** the part and name it Advanced Part 9-1.

20. **Close** the model.

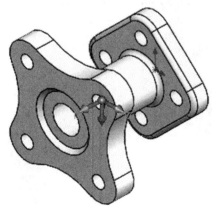

Tutorial: Advanced Part 9-2

An exam question in this category could read "Build this part in SOLIDWORKS. Calculate the overall mass and locate the Center of mass of the illustrated model."

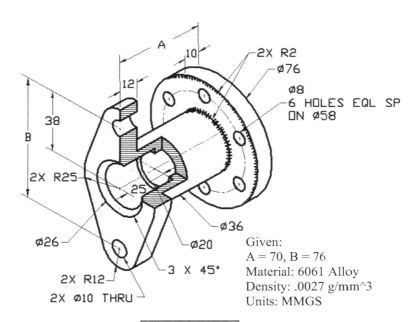

1. **Create** a New part in SOLIDWORKS.

2. **Build** the illustrated dimensioned model. Insert eight features: Extruded Base, Extruded Cut, Circular Pattern, two Extruded Bosses, Extruded Cut, Chamfer and Fillet.

Given:
A = 70, B = 76
Material: 6061 Alloy
Density: .0027 g/mm^3
Units: MMGS

Think about the steps that you would take to build the illustrated part. Review the provided information. Start with the six hole flange.

💡 Tangent Edges are displayed for educational purposes.

3. **Set** the document properties for the model.

4. Create **Sketch1**. Sketch1 is the Base sketch. Select the Front Plane as the Sketch plane. Sketch two circles. Insert the required geometric relations and dimensions.

5. Create the **Extruded Base** feature. Blind is the default End Condition in Direction 1. Depth = 10mm. Note the direction of the extrude feature to maintain the Origin location.

6. Create **Sketch2**. Select the front face of Boss-Extrude1 as the Sketch plane. Sketch2 is the profile for first Extruded Cut feature. The Extruded Cut feature is the seed feature for the Circular Pattern. Apply construction reference geometry. Insert the required geometric relations and dimensions.

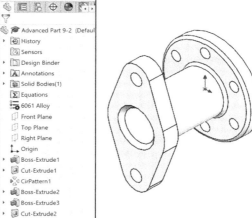

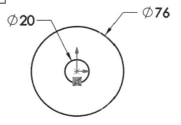

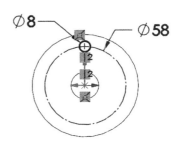

7. Create the **Extruded Cut** feature. Cut-Extrude1 is the first bolt hole. Select Through All for End Condition in Direction 1.

8. Create the **Circular Pattern** feature. Default Angle = 360deg. Number of instances = 6. Select the center axis for the Pattern Axis box.

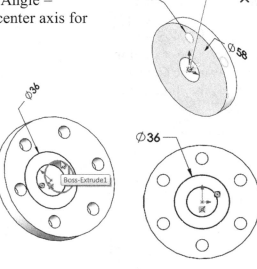

🔅 The Circular Pattern PropertyManager is displayed when you pattern one or more features about an axis.

9. Create **Sketch3**. Select the front face of the Extrude feature as the Sketch plane. Sketch two circles. Insert a Coradial relation on the inside circle. The two circles share the same center point and radius. Insert the required dimension.

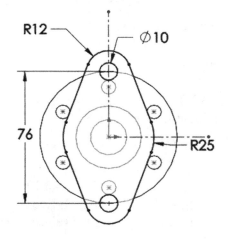

10. Create the first **Extruded Boss (Boss-Extrude2)** feature. The Boss-Extrude2 feature is the connecting tube between the two flanges. Blind is the default End Condition in Direction 1. Depth = 48mm.

11. Create **Sketch4**. Select the front circular face of Extrude3 as the Sketch plane. Sketch a horizontal and vertical centerline from the Origin. Sketch the top and bottom circles symmetric about the horizontal centerline. Dimension the distance between the two circles and their diameter. Create the top center point arc with the center point Coincident to the top circle. The start point and the end point of the arc are horizontal. Sketch the two top angled lines symmetric about the vertical centerline. Apply symmetry. Mirror the two lines and the center point arc about the horizontal centerline. Insert the left and right tangent arcs with a center point Coincident with the Origin. Complete the sketch.

12. Create the second **Extruded Boss** (Boss-Extrude3) feature. Blind is the default End Condition in Direction 1. Depth = 12mm.

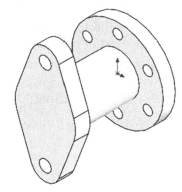

13. Create **Sketch5**. Select the front face of the Extrude feature as the Sketch plane. Sketch a circle. The part Origin is located in the center of the model. Insert the required dimension.

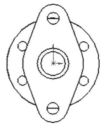
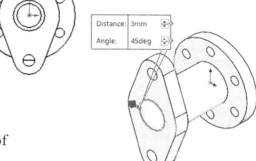

14. Create the second **Extruded Cut** feature. Blind is the default End Condition in Direction 1. Depth = 25mm.

15. Create the **Chamfer** feature. Chamfer1 is an Angle distance chamfer. Chamfer the inside edge of the Extrude feature as illustrated. Distance = 3mm. Angle = 45deg.

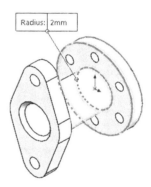

16. Create the **Fillet** feature. Fillet the two edges of Extrude1. Radius = 2mm.

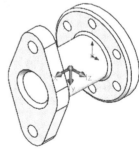

17. **Assign** 6061 Alloy material to the part.

18. **Calculate** the overall mass of the part. The overall mass = 276.97 grams.

19. **Locate** the Center of mass. The location of the Center of mass is relative to the part Origin.

- X: 0.00 millimeters

- Y: 0.00 millimeters

- Z: 21.95 millimeters

20. **Save** the part and name it Advanced Part 9-2.

21. **Close** the model.

In the Advanced Part Modeling category, an exam question could read "Build this model. Locate the Center of mass with respect to the part Origin."

- A: X = 0.00 millimeters, Y = 0.00 millimeters, Z = 21.95 millimeters

- B: X = 21.95 millimeters, Y = 10.00 millimeters, Z = 0.00 millimeters

- C: X = 0.00 millimeters, Y = 0.00 millimeters, Z = -27.02 millimeters

- D: X= 1.00 millimeters, Y = -1.01 millimeters, Z = -0.04 millimeters

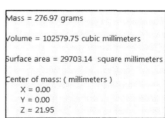

Mass = 276.97 grams

Volume = 102579.75 cubic millimeters

Surface area = 29703.14 square millimeters

Center of mass: (millimeters)
 X = 0.00
 Y = 0.00
 Z = 21.95

The correct answer is A.

Calculate the Center of Mass Relative to a Created Coordinate System Location

In the Simple Part Modeling chapter, you located the Center of mass relative to the default part Origin. In the Advanced Part Modeling category, you may need to locate the Center of mass relative to a created coordinate system location. The exam model may display a created coordinate system location. Example:

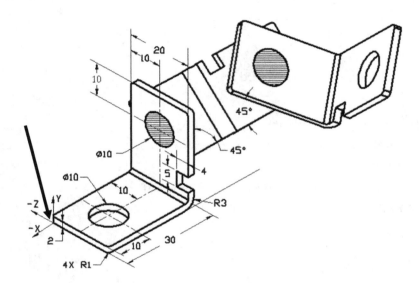

💡 The SOLIDWORKS software displays positive values for (X, Y, Z) coordinates for a reference coordinate system. The CSWA exam displays either a positive or negative sign in front of the (X, Y, Z) coordinates to indicate direction as illustrated (-X, +Y, -Z).

The following section reviews creating a Coordinate System location for a part.

Tutorial: Coordinate location 9-1

Use the Mass Properties tool to calculate the Center of mass for a part located at a new coordinate location through a point.

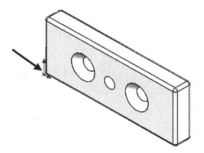

1. **Open** the Plate-3-Point part from the SOLIDWORKS CSWA Folder\Chapter 9 location. View the location of the part Origin.

2. **Locate** the Center of mass. The location of the Center of mass is relative to the part Origin.

- X = 28 millimeters

- Y = 11 millimeters

- Z = -3 millimeters

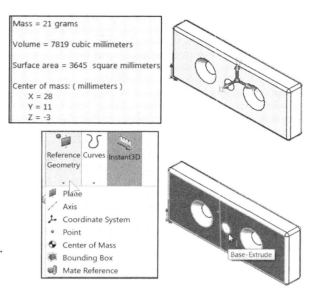

Create a new coordinate system location. Locate the new coordinate system location at the center of the center hole as illustrated.

3. **Exit** the Mass Properties dialog box.

4. Right-click the **front face** of Base-Extrude.

5. Click **Sketch** from the Context toolbar.

6. Click the **edge** of the center hole as illustrated.

7. Click **Convert Entities** from the Sketch toolbar. The center point for the new coordinate location is displayed.

8. **Exit** the sketch. Sketch4 is displayed.

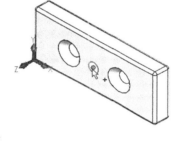

9. Click the **Coordinate System** tool from the Consolidated Reference Geometry toolbar. The Coordinate System PropertyManager is displayed.

10. Click the **center point** of the center hole in the Graphics window. Point2@Sketch4 is displayed in the Selections box as the Origin.

11. Click **OK** from the Coordinate System PropertyManager. Coordinate System1 is displayed.

12. **View** the new coordinate location at the center of the center hole.

View the Mass Properties of the part with the new coordinate location.

13. Click the **Mass Properties** tool from the Evaluate tab.

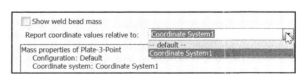

14. Select **Coordinate System1** from the Output box. The Center of mass relative to the new location is located at the following coordinates: X = 0 millimeters, Y = 0 millimeters, Z = -3 millimeters

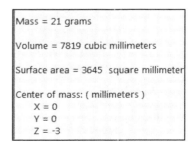

15. **Reverse** the direction of the axes as illustrated. On the CSWA exam, the coordinate system axes could be represented by (+X, -Y, -Z).

16. **Close** the model.

 To reverse the direction of an axis, click its **Reverse Axis Direction** button in the Coordinate System PropertyManager.

Tutorial: Coordinate location 9-2

Create a new coordinate system location. Locate the new coordinate system at the top back point as illustrated.

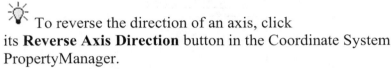

1. **Open** the Plate-X-Y-Z part from the SOLIDWORKS CSWA Folder\Chapter 9 location.

2. **View** the location of the part Origin.

3. Drag the **Rollback bar** under the Base-Extrude feature in the FeatureManager.

4. Click the **Coordinate System** tool from the Consolidated Reference Geometry toolbar. The Coordinate System PropertyManager is displayed.

5. Click the **back left vertex** as illustrated. Address direction if needed.

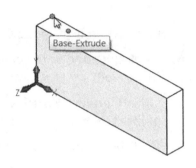

6. Click the **top back horizontal** edge as illustrated. Do not select the midpoint. Address direction if needed.

7. Click the **back left vertical** edge as illustrated. Address direction if needed.

8. Click **OK** from the Coordinate System PropertyManager. Coordinate System1 is displayed in the FeatureManager and in the Graphics window.

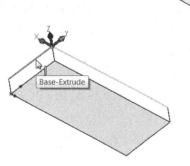

9. Drag the **Rollback bar** to the bottom of the FeatureManager.

10. **Calculate** the Center of mass relative to the new coordinate system.

11. Select **Coordinate System1**. The Center of mass relative to the new location is located at the following coordinates:

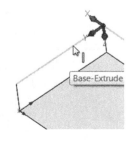

X = -28 millimeters

Y = 11 millimeters

Z = 4 millimeters

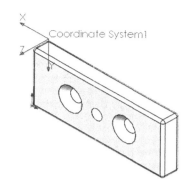

12. **Close** the model.

🔆 Define a Coordinate system for a part or assembly. Apply a Coordinate system with the Measure and Mass Properties tool.

Tutorial: Advanced Part 9-3

An exam question in this category could read "Build this part in SOLIDWORKS. Calculate the overall mass and locate the Center of mass of the illustrated model."

1. **Create** a New part in SOLIDWORKS.

2. **Build** the illustrated dimensioned model. Insert thirteen features: Extrude-Thin1, Fillet, two Extruded Cuts, Circular Pattern, two Extruded Cuts, Mirror, Chamfer, Extruded Cut, Mirror, Extruded Cut and Mirror.

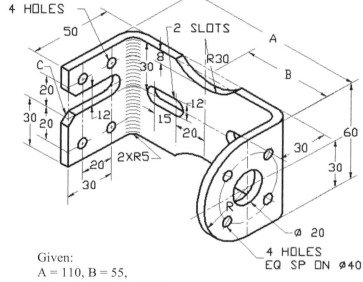

Given:
A = 110, B = 55,
C = 5 X 45Ø CHAMFER
Material: 5MM, 6061 Alloy
Density: .0027 g/mm^3
Units: MMGS
ALL HOLES 6MM

Think about the steps that you would take to build the illustrated part. Review the provided information. The depth of the left side is 50mm. The depth of the right side is 60mm.

🔆 There are numerous ways to build the models in this chapter. A goal is to display different design intents and techniques.

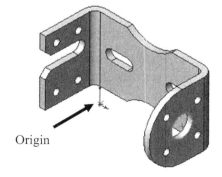

Origin

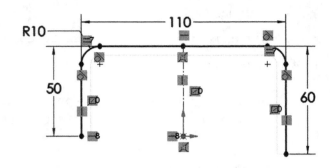

💡 If the inside radius = 5mm and the material thickness = 5mm, then the outside radius = 10mm.

3. **Set** the document properties for the model.

4. Create **Sketch1**. Sketch1 is the Base sketch. Select the Top Plane as the Sketch plane. Apply the Line and Sketch Fillet Sketch tools. Apply construction geometry. Insert the required geometric relations and dimensions.

5. Create the **Extrude-Thin1** feature. Extrude-Thin1 is the Base feature. Apply symmetry in Direction 1. Depth = 60mm. Thickness = 5mm. Check the Auto-fillet corners box. Radius = 5mm.

💡 The Auto-fillet corners option creates a round at each edge where lines meet at an angle.

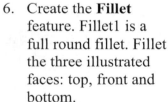

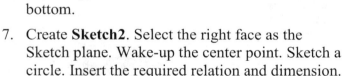

6. Create the **Fillet** feature. Fillet1 is a full round fillet. Fillet the three illustrated faces: top, front and bottom.

7. Create **Sketch2**. Select the right face as the Sketch plane. Wake-up the center point. Sketch a circle. Insert the required relation and dimension.

8. Create the first **Extruded Cut** feature. Select Up To Next for the End Condition in Direction 1.

💡 The Up To Next End Condition extends the feature from the sketch plane to the next surface that intercepts the entire profile. The intercepting surface must be on the same part.

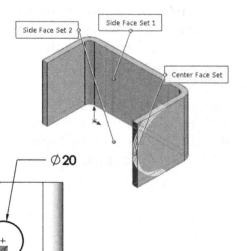

9. Create **Sketch3**. Select the right face as the Sketch plane. Create the profile for the second Extruded Cut feature. This is the seed feature for CirPattern1. Apply construction geometry to locate the center point of Sketch3. Insert the required relations and dimensions.

10. Create the second **Extruded Cut** feature. Select Up To Next for the End Condition in Direction 1.

11. Create the **Circular Pattern** feature. Number of Instances = 4. Default angle = 360deg.

12. Create **Sketch4**. Select the left outside face of Extrude-Thin1 as the Sketch plane. Apply the Line and Tangent Arc Sketch tool to create Sketch4. Insert the required geometric relations and dimensions.

13. Create the third **Extruded Cut** feature. Select Up To Next for End Condition in Direction 1. The Slot on the left side of Extrude-Thin1 is created.

14. Create **Sketch5**. Select the left outside face of Extrude-Thin1 as the Sketch plane. Sketch two circles. Insert the required geometric relations and dimensions.

15. Create the fourth **Extruded Cut** feature. Select Up To Next for End Condition in Direction 1.

☼ There are numerous ways to create the models in this chapter. A goal is to display different design intents and techniques.

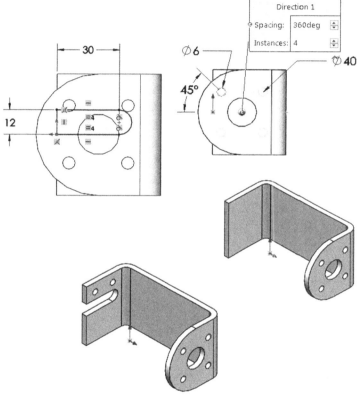

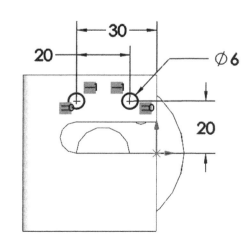

16. Create the first **Mirror** feature. Mirror the top two holes about the Top Plane.

17. Create the **Chamfer** feature. Create an Angle distance chamfer. Chamfer the selected edges as illustrated. Distance = 5mm. Angle = 45deg.

18. Create **Sketch6**. Select the front face of Extrude-Thin1 as the Sketch plane. Insert the required geometric relations and dimensions.

19. Create the fifth **Extruded Cut** feature. Select Through All for End Condition in Direction 1.

20. Create the second **Mirror** feature. Mirror Extrude5 about the Right Plane.

21. Create **Sketch7**. Select the front face of Extrude-Thin1 as the Sketch plane. Apply the 3 Point Arc Sketch tool. Apply the min First Arc Condition option. Insert the required geometric relations and dimensions.

22. Create the last **Extruded Cut** feature. Through All is the End Condition in Direction 1 and Direction 2.

23. Create the third **Mirror** feature. Mirror the Extrude feature about the Top Plane as illustrated.

24. **Assign** the material to the part.

25. **Calculate** the overall mass of the part. The overall mass = 132.45 grams.

26. **Locate** the Center of mass relative to the part Origin:

- X: 1.83 millimeters

- Y: -0.27 millimeters

- Z: -35.38 millimeters

27. **Save** the part and name it Advanced Part 9-3.

28. **Close** the model.

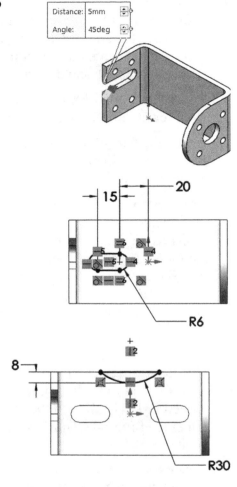

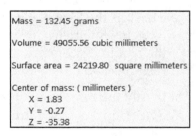

Mass = 132.45 grams

Volume = 49055.56 cubic millimeters

Surface area = 24219.80 square millimeters

Center of mass: (millimeters)
  X = 1.83
  Y = -0.27
  Z = -35.38

☼ Tangent edges and Origin are displayed for educational purposes.

All questions on the exam are in a multiple choice single answer or fill in the blank format. In the Advanced Part Modeling category, an exam question could read "Build this model. Calculate the overall mass of the part with the provided information."

- A: 139.34 grams

- B: 155.19 grams

- C: 132.45 grams

- D: 143.91 grams

The correct answer is C.

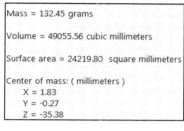

Mass = 132.45 grams

Volume = 49055.56 cubic millimeters

Surface area = 24219.80 square millimeters

Center of mass: (millimeters)
   X = 1.83
   Y = -0.27
   Z = -35.38

Use the Options button in the Mass Properties dialog box to apply custom settings to units.

Tutorial: Advanced Part 9-3A

An exam question in this category could read "Build this part in SOLIDWORKS. Locate the Center of mass." Note the coordinate system location of the model as illustrated.

Where do you start? Build the model as you did in the Tutorial: Advanced Part 9-3. Create Coordinate System1 to locate the Center of mass.

1. **Open** Advanced Part 9-3 that you created in the previous exercise.

Create the illustrated coordinate system location.

2. Show **Sketch2** from the FeatureManager design tree.

3. Click the **center point** of Sketch2 in the Graphics window as illustrated.

Click on the additional views during the CSWA exam to better understand the part and provided information.

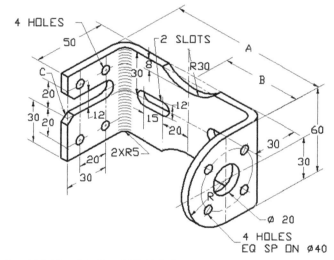

A = 110, B = 55, C = 5 X 45Ø CHAMFER
Material: 5MM, 6061 Alloy
Density: .0027 g/mm^3
Units: MMGS
ALL HOLES 6MM

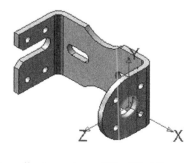

Coordinate system: +X, +Y, +Z

4. Click the **Coordinate System** tool from the Consolidated Reference Geometry toolbar. The Coordinate System PropertyManager is displayed. Point2@Sketch2 is displayed in the Origin box.

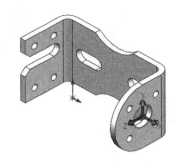

5. Click **OK** from the Coordinate System PropertyManager. Coordinate System1 is displayed.

6. **Locate** the Center of mass based on the location of the illustrated coordinate system. Select Coordinate System1.

- X: -53.17 millimeters

- Y: -0.27 millimeters

- Z: -15.38 millimeters

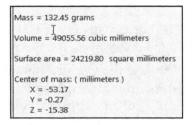

Mass = 132.45 grams

Volume = 49055.56 cubic millimeters

Surface area = 24219.80 square millimeters

Center of mass: (millimeters)
   X = -53.17
   Y = -0.27
   Z = -15.38

7. **Save** the part and name it Advanced Part 9-3A.

8. **Close** the model.

Tutorial: Advanced Part 9-3B

Build this part in SOLIDWORKS. Locate the Center of mass. View the location of the coordinate system. The coordinate system is located at the left front point of the model.

Build the illustrated model as you did in the Tutorial: Advanced Part 9-3. Create Coordinate System1 to locate the Center of mass for the model.

1. **Open** Advanced Part 9-3 that you created in the previous exercise.

Create the illustrated coordinate system.

2. Click the **vertex** as illustrated for the Origin location.

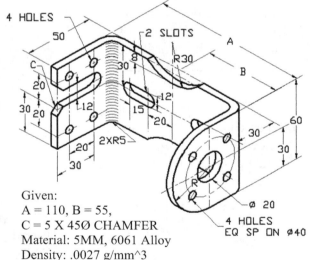

Given:
A = 110, B = 55,
C = 5 X 45Ø CHAMFER
Material: 5MM, 6061 Alloy
Density: .0027 g/mm^3
Units: MMGS
ALL HOLES 6MM

To reverse the direction of an axis, click the **Reverse Axis Direction** button in the Coordinate System PropertyManager.

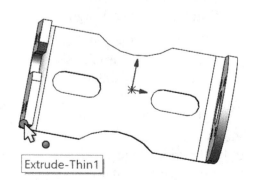

Extrude-Thin1

3. Click the **Coordinate System** tool from the Consolidated Reference Geometry toolbar. The Coordinate System PropertyManager is displayed. Vertex<1> is displayed in the Origin box.

4. Click the **bottom horizontal edge** as illustrated. Edge<1> is displayed in the X Axis Direction box.

5. Click the **left back vertical edge** as illustrated. Edge<2> is displayed in the Y Axis Direction box.

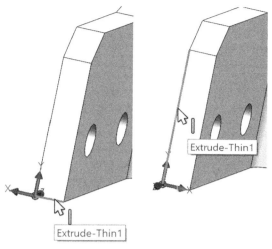

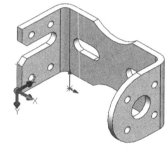

6. Click **OK** from the Coordinate System PropertyManager. Coordinate System1 is displayed.

9. **Locate** the Center of mass based on the location of the illustrated coordinate system. Select Coordinate System1.

- X: 1.83 millimeters

- Y: -0.27 millimeters

- Z: -35.38 millimeters

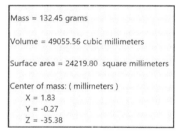

10. **Save** the part and name it Advanced Part 9-3B.

11. **Close** the model.

In the Advanced Part Modeling category, an exam question could read "Build this model. Locate the Center of mass."

- A: X = -1.83 millimeters, Y = -0.27 millimeters, Z = -35.38 millimeters

- B: X = 2.80 millimeters, Y = -1.27 millimeters, Z = -45.54 millimeters

- C: X = -59.20 millimeters, Y = -0.27 millimeters, Z = -15.54 millimeters

- D: X= -1.80 millimeters, Y = 3.05 millimeters, Z = -0.14 millimeters

The correct answer is A.

Tutorial: Advanced Part 9-4

An exam question in this category could read "Build this part. Calculate the overall mass and locate the Center of mass of the illustrated model."

1. **Create** a new part in SOLIDWORKS.

2. **Build** the illustrated dimensioned model. Insert twelve features and a Reference plane: Extrude-Thin1, two Extruded Bosses, Extruded Cut, Extruded Boss, Extruded Cut, Plane1, Mirror and five Extruded Cuts.

Think about the steps that you would take to build the illustrated part. Create an Extrude-Thin1 feature as the Base feature.

3. **Set** the document properties for the model. Review the given information.

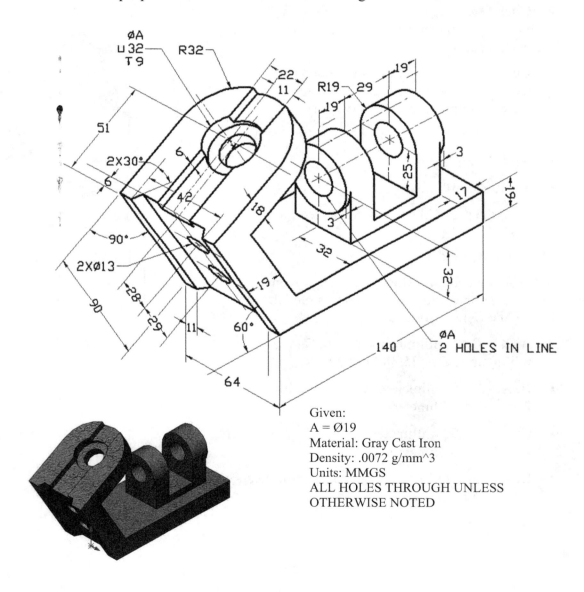

Given:
A = Ø19
Material: Gray Cast Iron
Density: .0072 g/mm^3
Units: MMGS
ALL HOLES THROUGH UNLESS
OTHERWISE NOTED

4. Create **Sketch1**. Sketch1 is the Base sketch. Select the Right Plane as the Sketch plane. Apply construction geometry. Insert the required geometric relations and dimensions. Sketch1 is the profile for Extrude-Thin1. Note the location of the Origin.

5. Create the **Extrude-Thin1** feature. Apply symmetry. Select Mid Plane as the End Condition in Direction 1. Depth = 64mm. Thickness = 19mm.

6. Create **Sketch2**. Select the top narrow face of Extrude-Thin1 as the Sketch plane. Sketch three lines: two vertical and one horizontal and a tangent arc. Insert the required geometric relations and dimensions.

7. Create the **Boss-Extrude1** feature. Blind is the default End Condition in Direction 1. Depth = 18mm.

8. Create **Sketch3**. Select the Right Plane as the Sketch plane. Sketch a rectangle. Insert the required geometric relations and dimensions. Note: 61mm = (19mm - 3mm) x 2 + 29mm.

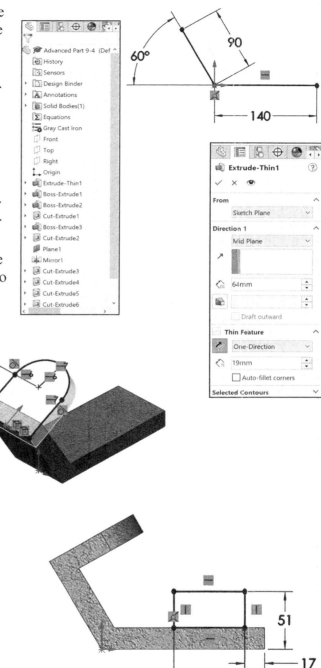

9. Create the **Boss-Extrude2** feature. Select Mid Plane for End Condition in Direction 1. Depth = 38mm. Note: 2 x R19.

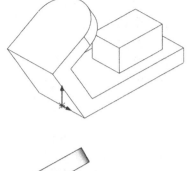

10. Create **Sketch4**. Select the Right Plane as the Sketch plane. Sketch a vertical centerline from the top midpoint of the sketch. The centerline is required for Plane1. Plane1 is a Reference plane. Sketch a rectangle symmetric about the centerline. Insert the required relations and dimensions. Sketch4 is the profile for Extrude3.

11. Create the first **Extruded Cut** feature. Extrude in both directions. Select Through All for End Condition in Direction 1 and Direction 2.

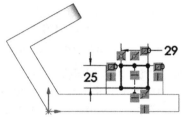

12. Create **Sketch5**. Select the inside face of the Extrude feature for the Sketch plane. Sketch a circle from the top midpoint. Sketch a construction circle. Construction geometry is required for future features. Complete the sketch.

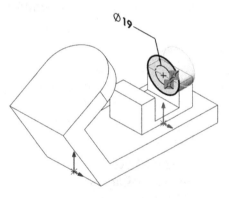

13. Create the **Extruded Boss** (Boss-Extrude3) feature. Blind is the default End Condition. Depth = 19mm.

14. Create **Sketch6**. Select the inside face for the Sketch plane. Show Sketch5. Select the construction circle in Sketch5. Apply the Convert Entities Sketch tool.

15. Create the second **Extruded Cut** feature. Select the Up To Next End Condition in Direction 1.

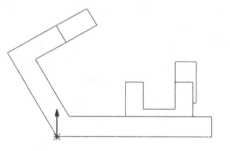

⚡ There are numerous ways to create the models in this chapter. A goal is to display different design intents and techniques.

⚡ Tangent edges and Origin are displayed for educational purposes.

16. Create **Plane1**. Apply symmetry. Create Plane1 to mirror Cut-Extrude2 and Boss-Extrude3. Create a Parallel Plane at Point. Select the midpoint of Sketch4 and Face<1> as illustrated. Point1@Sketch4 and Face<1> is displayed in the Selections box.

17. Create the **Mirror** feature. Mirror Cut-Extrude2 and Boss-Extrude3 about Plane1.

🔅 The Mirror feature creates a copy of a feature (or multiple features) mirrored about a face or a plane. You can select the feature or you can select the faces that comprise the feature.

18. Create **Sketch7**. Select the top front angled face of Extrude-Thin1 as the Sketch plane. Apply the Centerline Sketch tool. Insert the required geometric relations and dimensions.

19. Create the third **Extruded Cut** feature. Select Through All for End Condition in Direction 1. Select the edge for the vector to extrude.

🔅 Click on the additional views during the CSWA exam to better understand the part and provided information. Read each question carefully. Identify the dimensions, center of mass and units. Apply needed material.

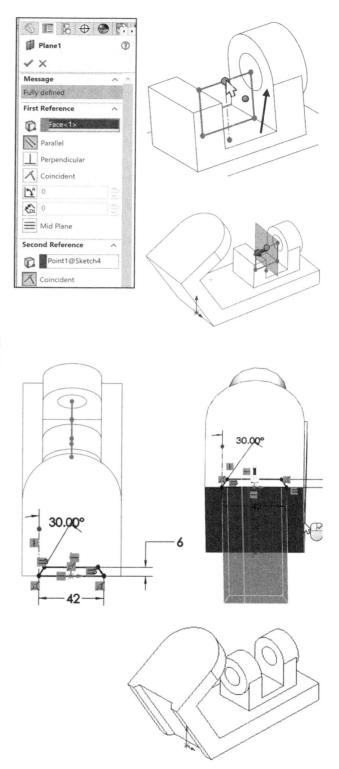

20. Create **Sketch8**. Select the top front angled face of Extrude-Thin1 as the Sketch plane. Sketch a centerline. Sketch two vertical lines and a horizontal line. Select the top arc edge. Apply the Convert Entities Sketch tool. Apply the Trim Sketch tool to remove the unwanted arc geometry. Insert the required geometric relations and dimension.

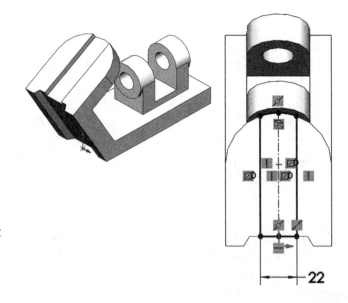

21. Create the fourth **Extruded Cut** feature. Blind is the default End Condition in Direction 1. Depth = 6mm.

22. Create **Sketch9**. Create a Cbore with Sketch9 and Sketch10. Select the top front angled face of Extrude-Thin1 as illustrated for the Sketch plane. Extrude8 is the center hole in the Extrude-Thin1 feature. Sketch a circle. Insert the required geometric relations and dimension.

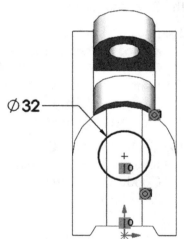

23. Create the fifth **Extrude Cut** feature. Blind is the default End Condition. Depth = 9mm. Note: This is the first feature for the Cbore.

24. Create **Sketch10**. Select the top front angled face of Extrude-Thin1 as the Sketch plane. Sketch a circle. Insert the required geometric relation and dimension. Note: A = Ø19.

25. Create the sixth **Extruded Cut** feature. Select the Up To Next End Condition in Direction 1. The Cbore is complete.

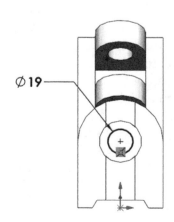

☀ In the exam, you can answer the questions in any order. Use the Summary Screen during the exam to view the list of all questions you have or have not answered.

26. Create **Sketch11**. Select the front angle face of the Extrude feature for the Sketch plane. Sketch two circles. Insert the required geometric relations and dimensions.

27. Create the last **Extruded Cut** feature. Select the Up To Next End Condition in Direction 1.

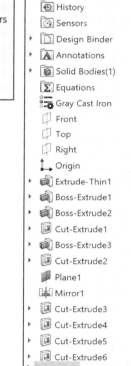

The FilletXpert automatically calls the FeatureXpert when it has trouble placing a fillet on the specified geometry.

28. **Assign** material to the part.

29. **Calculate** the overall mass of the part. The overall mass = 2536.59 grams.

Mass = 2536.59 grams

Volume = 352304.50 cubic millimeters

Surface area = 61252.90 square millimeters

Center of mass: (millimeters)
   X = 0.00
   Y = 34.97
   Z = -46.67

30. **Locate** the Center of mass relative to the part Origin:

- X: 0.00 millimeters

- Y: 34.97 millimeters

- Z: -46.67 millimeters

31. **Save** the part and name it Advanced Part 9-4.

The book is designed to expose the new user to many tools, techniques and procedures. It may not always use the most direct tool or process.

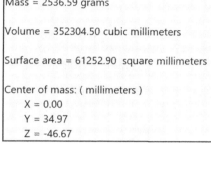

Advanced Part 9-4 (Def
- History
- Sensors
- Design Binder
- Annotations
- Solid Bodies(1)
- Equations
- Gray Cast Iron
- Front
- Top
- Right
- Origin
- Extrude-Thin1
- Boss-Extrude1
- Boss-Extrude2
- Cut-Extrude1
- Boss-Extrude3
- Cut-Extrude2
- Plane1
- Mirror1
- Cut-Extrude3
- Cut-Extrude4
- Cut-Extrude5
- Cut-Extrude6

Tutorial: Advanced Part 9-4A

An exam question in this category could read "Build this part. Locate the Center of mass for the illustrated coordinate system."

Where do you start? Build the illustrated model as you did in the Tutorial: Advanced Part 9-3. Create Coordinate System1 to locate the Center of mass for the model.

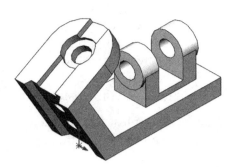

1. **Open** Advanced Part 9-4 that you created in the previous exercise.

Create the illustrated Coordinate system.

2. Click the **Coordinate System** tool from the Consolidated Reference Geometry toolbar. The Coordinate System PropertyManager is displayed.

3. Click the **bottom midpoint** of Extrude-Thin1 as illustrated. Point<1> is displayed in the Origin box.

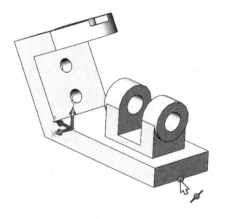

4. Click **OK** from the Coordinate System PropertyManager. Coordinate System1 is displayed.

5. **Locate** the Center of mass based on the location of the illustrated coordinate system. Select Coordinate System1.

*   X: 0.00 millimeters

*   Y: 34.97 millimeters

*   Z: 93.33 millimeters

6. **Save** the part and name it Advanced Part 9-4A.

7. **View** the Center of mass with the default coordinate system.

8. **Close** the model.

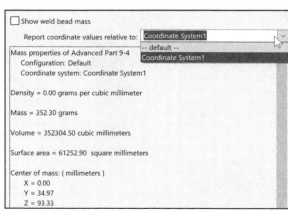
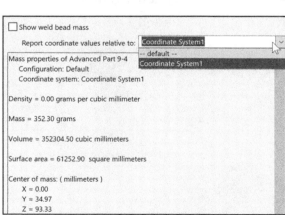

Show weld bead mass

Report coordinate values relative to: Coordinate System1
-- default --
Coordinate System1

Mass properties of Advanced Part 9-4
    Configuration: Default
    Coordinate system: Coordinate System1

Density = 0.00 grams per cubic millimeter

Mass = 352.30 grams

Volume = 352304.50 cubic millimeters

Surface area = 61252.90 square millimeters

Center of mass: (millimeters)
    X = 0.00
    Y = 34.97
    Z = 93.33

Summary

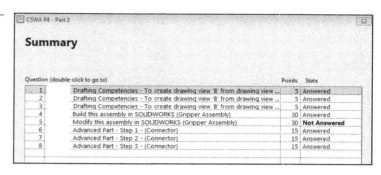

Advanced Part Creation and Modification is one of the five categories on the CSWA exam. The main difference between the *Advanced Part Creation and Modification* and the *Basic Part Creation and Modification* or the *Intermediate Part Creation and Modification* category is the complexity of the sketches and the number of dimensions and geometric relations along with an increased number of features.

There are three questions on the CSWA exam in this category. One question is in a multiple choice single answer format and the other two questions (Modification of the model) are in the fill in the blank format.

Each question is worth fifteen (15) points for a total of forty-five (45) points. You are required to build a model with six or more features and to answer a question either on the overall mass, volume, or the location of the Center of mass for the created model relative to the default part Origin location. You are then requested to modify the model and answer a fill in the blank format question.

Assembly Creation and Modification (Bottom-up) is the next chapter in this book. Up to this point, a Basic Part, Intermediate Part or an Advanced part was the focus. The *Assembly Creation and Modification* category addresses an assembly with numerous sub-components. All sub-components are provided to you in the exam.

The next chapter covers the general concepts and terminology used in the *Assembly Creation and Modification* category and then addresses the core elements. Knowledge of Standard mates is required in this category.

There are four questions. Multiple choice/single answer in the category. Each question is worth 30 points.

The CSWA Academic exam is provided either in a single 3 hour segment, or 2 - 90 minute segments. The CSWA exam for industry is only provided in a single 3 hour segment. All exams cover the same material.

This book addresses the CSWA Academic exam provided in 2 - 90 minute segments. Part 1 of the CSWA Academic exam is 90 minutes, minimum passing score is 80, with 6 questions. Part 2 of the CSWA Academic exam is 90 minutes, minimum passing score is 80 with 8 questions.

To obtain additional CSWA exam information, visit the SOLIDWORKS VirtualTester Certification site at https://SOLIDWORKS.virtualtester.com/.

Questions

Question 1:

In Tutorial: Advanced Part 9-1 you created the illustrated part. Modify the Base flange thickness from .40in to .50in. Modify the Chamfer feature angle from 45deg to 33deg. Modify the Fillet feature radius from .10in to .125in. Modify the material from 1060 Alloy to Nickel.

Calculate the overall mass of the part, volume, and locate the Center of mass with the provided information.

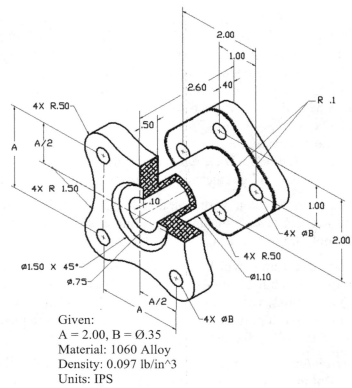

Given:
A = 2.00, B = Ø.35
Material: 1060 Alloy
Density: 0.097 lb/in^3
Units: IPS
Decimal places = 2

Question 2:

In the tutorial: Advanced Part 9-2 you created the illustrated part. Modify the CirPattern1 feature. Modify the number of instances from 6 to 8. Modify the seed feature from an 8mm diameter to a 6mm diameter.

Calculate the overall mass, volume, and the location of the Center of mass relative to the part Origin.

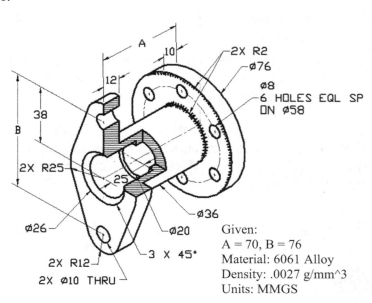

Given:
A = 70, B = 76
Material: 6061 Alloy
Density: .0027 g/mm^3
Units: MMGS

Question 3:

Build the illustrated model in SOLIDWORKS. Set document properties, identify the correct Sketch planes, apply the correct Sketch and Feature tools, and apply material.

Calculate the overall mass of the part, volume and locate the Center of mass with the provided information.

- Material: 6061 Alloy

- Units: MMGS

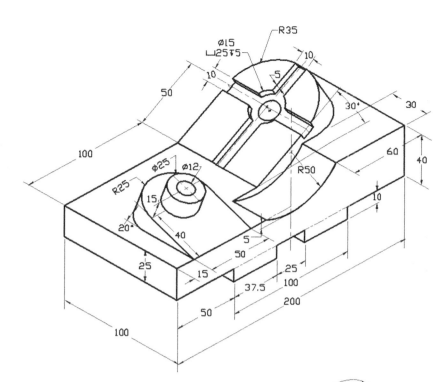

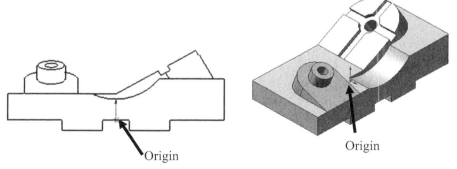

Origin   Origin

Question 4:

Build the illustrated model in SOLIDWORKS. Calculate the overall mass of the part, volume and locate the Center of mass with the provided information. Where do you start? Build the model as you did in the above exercise. Create Coordinate System1 to locate the Center of mass for the model.

- Material: 6061 Alloy

- Units: MMGS

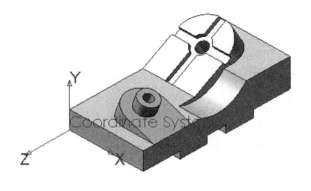

Question 5:

Build the illustrated model in SOLIDWORKS. Calculate the overall mass of the part, volume, and locate the Center of mass with the provided information.

- Material: 6061 Alloy

- Units: MMGS

Question 6:

Build the illustrated model in SOLIDWORKS. Calculate the overall mass of the part, volume and locate the Center of mass with the provided information.

Create Coordinate System1 to locate the Center of mass for the model.

- Material: 6061 Alloy

- Units: MMGS

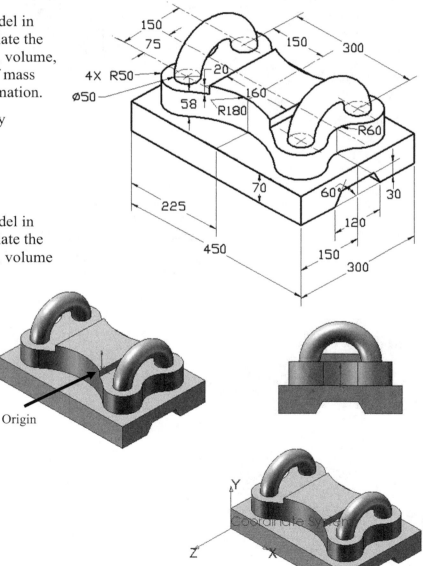

Question 7:

Build the provided model using SOLIDWORKS.

The model is provided in the SOILDWORKS CSWA Folder\Chapter 9.

Open Homework Part 9-7.

Utilize the rollback bar to view the sketches and features. Note: This is only one way to create the model.

Calculate the overall mass and volume of part with the provided information.

- Precision for linear dimensions = **2**.

- Material: **Case Stainless Steel**.

- Units: **MMGS**.

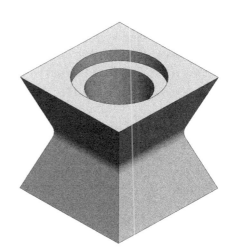

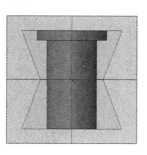

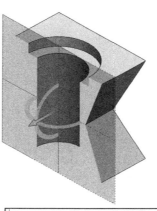

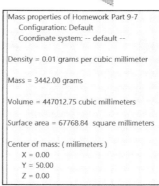

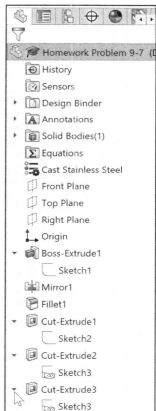

Sample screen shots from an older CSWA exam for an Advanced Modeling part. Click on the additional views to understand the part and provided information. Read each question carefully. Understand the dimensions, center of mass and units. Apply needed materials.

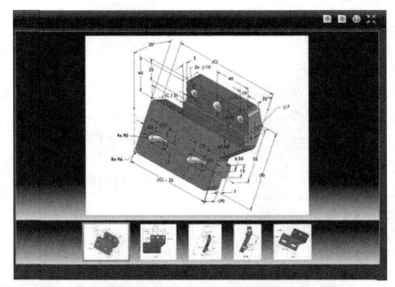

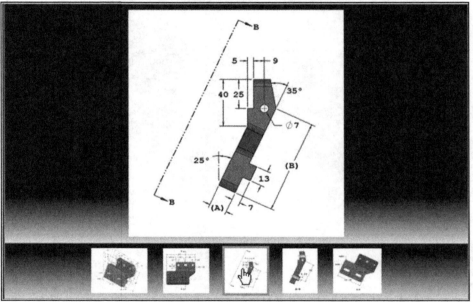

Screen shots from an exam

Zoom in on the part if needed.

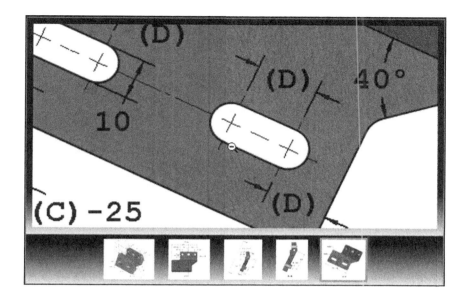

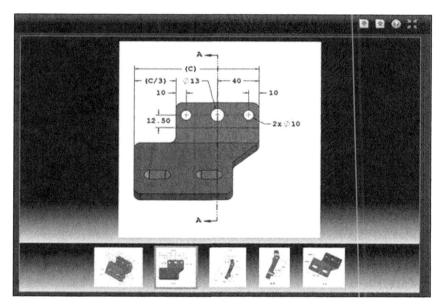

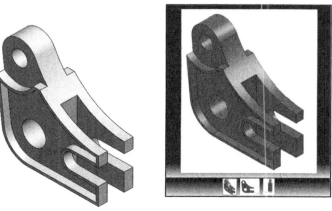

Notes:

Chapter 10 - CSWA - Assembly Creation and Modification

Objectives

Assembly Creation and Modification is one of the five categories on the CSWA exam. In the last two chapters, a Basic, Intermediate or Advanced model was the focus. The Assembly Creation and Modification (Bottom-up) category addresses an assembly with numerous components.

This chapter covers the general concepts and terminology used in Assembly modeling and then addresses the core elements that are aligned to the CSWA exam. Knowledge to insert mates and to create a new Coordinate system location is required in this category.

There are four questions on the CSWA Academic exam (2 questions in part 1, 2 questions in part 2) in the Assembly Creation and Modification category: (2) different assemblies - (4) questions - (2) multiple choice/(2) single answer - 30 points each.

The first question is in a multiple-choice single answer format. You should be within 1% of the multiple-choice answer before you move on to the modification single answer section, (fill in the blank format).

You are required to download the needed components from a provided zip file and insert them correctly to create the assembly.

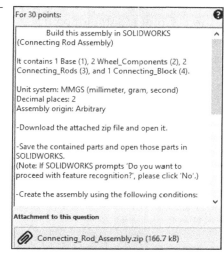

For 30 points:

> Build this assembly in SOLIDWORKS (Connecting Rod Assembly)
>
> It contains 1 Base (1), 2 Wheel_Components (2), 2 Connecting_Rods (3), and 1 Connecting_Block (4).
>
> Unit system: MMGS (millimeter, gram, second)
> Decimal places: 2
> Assembly origin: Arbitrary
>
> -Download the attached zip file and open it.
>
> -Save the contained parts and open those parts in SOLIDWORKS.
> (Note: If SOLIDWORKS prompts 'Do you want to proceed with feature recognition?', please click 'No'.)
>
> -Create the assembly using the following conditions:

Attachment to this question

🔗 Connecting_Rod_Assembly.zip (166.7 kB)

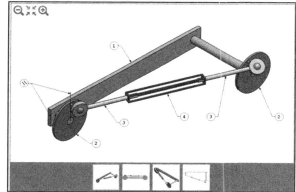

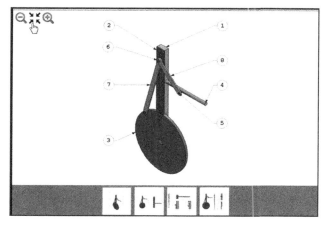

Screen shots from a CSWA exam

On the completion of the chapter, you will be able to:

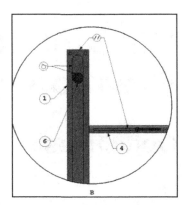

- Specify Document Properties.

- Identify the first fixed component in an assembly.

- Create a Bottom-up assembly with the following Standard mates:

  - Coincident, Concentric, Perpendicular, Parallel, Tangent, Distance, Angle, and Aligned, Anti-Aligned options.

- Apply the Mirror Component tool.

- Locate the Center of mass relative to the assembly Origin.

- Create a Coordinate system location.

- Locate the Center of mass relative to a created Coordinate system.

 All SOLIDWORKS models (initial and final) are provided in the SOLIDWORKS CSWA Model Folder.

Assembly Modeling

There are two key Assembly Modeling techniques:

- Top-down, "In-Context" assembly modeling.

- Bottom-up assembly modeling.

In Top-down assembly modeling, one or more features of a part are defined by something in an assembly, such as a sketch or the geometry of another component. The design intent comes from the top, and moves down into the individual components, hence the name Top-down assembly modeling.

 Mate the first component with respect to the assembly reference planes.

Bottom-up assembly modeling is a traditional method that combines individual components. Based on design criteria, the components are developed independently. The three major steps in a Bottom-up design approach are:

1. Create each part independent of any other component in the assembly.

2. Insert the parts into the assembly.

3. Mate the components in the assembly as they relate to the physical constraints of your design.

Build an Assembly from a Detailed Dimensioned illustration

An exam question in this category could read, "Build this assembly in SOLIDWORKS. Locate the Center of mass of the model with respect to the illustrated coordinate system. Set decimal place to 2."

The assembly contains the following: (1) Clevis component, (3) Axle components, (2) 5 Hole Link components, (2) 3 Hole Link components, and (6) Collar components. All holes Ø.190 THRU unless otherwise noted. Angle A = 150deg. Angle B = 120deg. Unit system: IPS.

Note: The location of the illustrated coordinate system (+X, +Y, +Z).

In the exam, download the zip file of the components. Unzip the components. Do not use feature recognition when you open the downloaded components. This is a timed exam. You do not need the additional feature information.

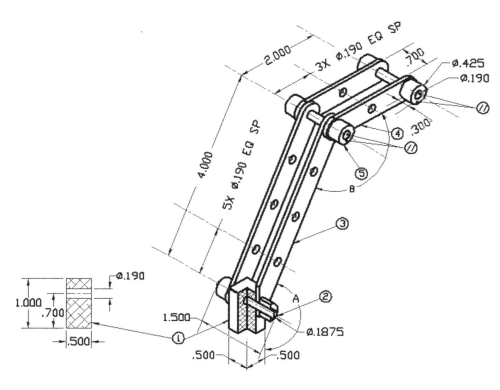

- Clevis, (Item 1): Material: 6061 Alloy. The two (5) Hole Link components are positioned with equal Angle mates, (150deg) to the Clevis component.

- Axle, (Item 2): Material: AISI 304. The first Axle component is mated Concentric and Coincident to the Clevis. The second and third Axle components are mated Concentric and Coincident to the 5 Hole Link and the 3 Hole Link components respectively.

- 5 Hole Link, (Item 3): Material: 6061 Alloy. Material thickness = .100in. Radius = .250in. Five holes located 1in. on center. The 5 Hole Link components are positioned with equal Angle mates, (120deg) to the 3 Hole Link components.

- 3 Hole Link, (Item 4): Material: 6061 Alloy. Material thickness = .100in. Radius = .250in. Three holes located 1in. on center. The 3 Hole Link components are positioned with equal Angle mates, (120deg) to the 5 Hole Link components.

- Collar, (Item 5): Material: 6061 Alloy. The Collar components are mated Concentric and Coincident to the Axle and the 5 Hole Link and 3 Hole Link components respectively.

Think about the steps that you would take to build the illustrated assembly. Identify the first fixed component. Insert the required Standard mates.

Locate the Center of mass of the model with respect to the illustrated coordinate system. In this example, start with the Clevis component.

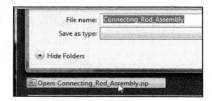

Do not use feature recognition when you open the downloaded components for the assembly in the CSWA exam. This is a timed exam. Manage your time. You do not need the additional feature information.

View the .pdf in the SOLIDWORKS CSWA Model Folder for a sample CSWA exam.

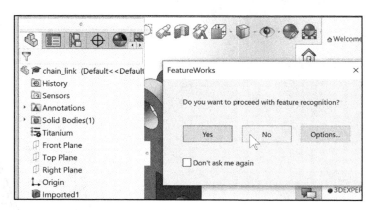

Tutorial: Assembly Model 10-1

Build the assembly in SOLIDWORKS.

1. **Download** the needed components from the SOLIDWORKS CSWA Model Folder\Chapter 10\Assembly Model 10-1 folder.

2. **Create** a new IPS assembly in SOLIDWORKS.

3. Click **Cancel** in the Open dialog box.

4. Click **Cancel** ✖ from the Begin Assembly PropertyManager. Assem1 is the default document name. Assembly documents end with the extension.sldasm.

5. **Set** the document properties for the model.

6. **Insert** the Clevis part.

7. **Fix** the component to the assembly Origin. Click OK from the Insert Component PropertyManager. The Clevis is displayed in the Assembly FeatureManager and in the Graphics window.

The first component or sub-assembly should be fixed **(f)** to the origin, fully defined to the assembly document or mated to an axis about the assembly origin.

🔅 Only insert the required mates (timed exam) to obtain the needed Mass properties information.

8. **Insert** the Axle part above the Clevis component as illustrated.

9. **Insert** a Concentric mate between the inside cylindrical face of the Clevis and the outside cylindrical face of the Axle.

10. **Insert** a Coincident mate between the Right Plane of the Clevis and the Right Plane of the Axle.

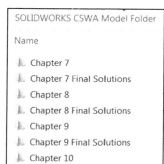

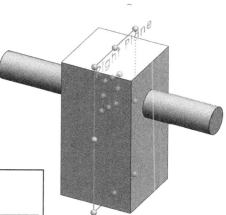

11. **Insert** the 5 Hole Link part.

12. **Rotate** the component as illustrated.

13. **Insert** a Concentric mate between the outside cylindrical face of the Axle and the inside cylindrical face of the 5 Hole Link. Concentric2 is created.

14. **Insert** a Coincident mate between the right face of the Clevis and the left face of the 5 Hole Link. Coincident2 is created.

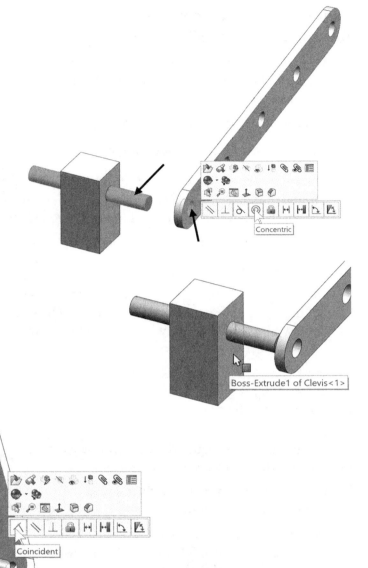

15. **Insert** an Angle mate between the bottom face of the 5 Hole Link and the back face of the Clevis. Angle = 30deg. If needed, click the Flip direction box and or the Aligned box.

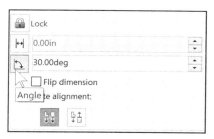

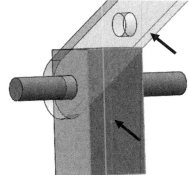

☼ Depending on the component orientation, select the Flip direction option and/or enter the supplement of the angle.

16. **Insert** the second Axle part. Locate the second Axle component near the end of the 5 Hole Link as illustrated.

17. **Insert** a Concentric mate between the inside cylindrical face of the 5 Hole Link and the outside cylindrical face of the Axle. Concentric3 is created.

18. **Insert** a Coincident mate between the Right Plane of the assembly and the Right Plane of the Axle. Coincident3 is created.

19. **Insert** the 3 Hole Link part.

20. **Rotate** the component as illustrated.

21. **Insert** a Concentric mate between the outside cylindrical face of the Axle and the inside cylindrical face of the 3 Hole Link. Concentric4 is created.

22. **Insert** a Coincident mate between the right face of the 5 Hole Link and the left face of the 3 Hole Link.

23. **Insert** an Angle mate between the bottom face of the 5 Hole Link and the bottom face of the 3 Hole Link. Angle = 60deg. Angle2 is created.

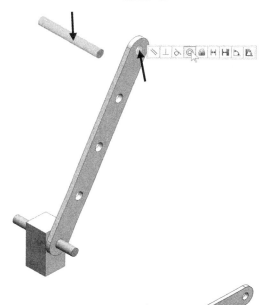

☼ Depending on the component orientation, select the Flip direction option and/or enter the supplement of the angle when needed.

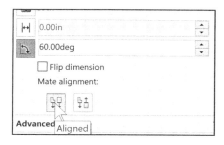

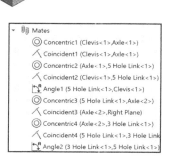

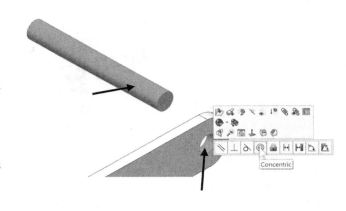

🔅 Apply the Measure tool to check the angle.

24. **Insert** the third Axle part.

25. **Insert** a Concentric mate between the inside cylindrical face of the 3 Hole Link and the outside cylindrical face of the Axle.

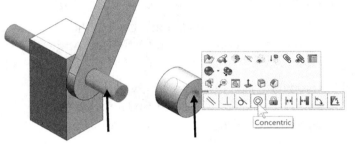

26. **Insert** a Coincident mate between the Right Plane of the assembly and the Right Plane of the Axle.

27. **Insert** the Collar part. Locate the Collar near the first Axle component.

28. **Insert** a Concentric mate between the inside cylindrical face of the Collar and the outside cylindrical face of the first Axle.

29. **Insert** a Coincident mate between the right face of the 5 Hole Link and the left face of the Collar.

30. **Insert** the second Collar part. Locate the Collar near the second Axle component.

31. **Insert** a Concentric mate between the inside circular face of the second Collar and the outside circular face of the second Axle.

32. **Insert** a Coincident mate between the right face of the 3 Hole Link and the left face of the second Collar.

33. **Insert** the third Collar part. Locate the Collar near the third Axle component.

▸ 🔩 Mates
  ◎ Concentric1 (Clevis<1>,Axle<1>)
  ⋀ Coincident1 (Clevis<1>,Axle<1>)
  ◎ Concentric2 (Axle<1>,5 Hole Link<1>)
  ⋀ Coincident2 (Clevis<1>,5 Hole Link<1>)
  📐 Angle1 (5 Hole Link<1>,Clevis<1>)
  ◎ Concentric3 (5 Hole Link<1>,Axle<2>)
  ⋀ Coincident3 (Axle<2>,Right Plane)
  ◎ Concentric4 (Axle<2>,3 Hole Link<1>)
  ⋀ Coincident4 (5 Hole Link<1>,3 Hole Link·
  📐 Angle2 (3 Hole Link<1>,5 Hole Link<1>)
  ◎ Concentric5 (3 Hole Link<1>,Axle<3>)
  ⋀ Coincident5 (Axle<3>,Right Plane)
  ◎ Concentric6 (Axle<1>,Collar<1>)
  ⋀ Coincident6 (5 Hole Link<1>,Collar<1>)
  ◎ Concentric7 (Axle<2>,Collar<2>)
  ⋀ Coincident7 (3 Hole Link<1>,Collar<2>)

34. **Insert** a Concentric mate between the inside cylindrical face of the Collar and the outside cylindrical face of the third Axle.

35. **Insert** a Coincident mate between the right face of the 3 Hole Link and the left face of the third Collar.

36. **Mirror** the components. Mirror the three Collars, 5 Hole Link and 3 Hole Link about the Right Plane. If using an older version of SOLIDWORKS, check the Recreate mates to new components box. Click Next in the Mirror Components PropertyManager. View your options.

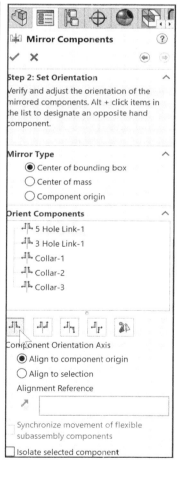

Click **Insert**, **Mirror Components** from the Menu bar menu or click the **Mirror Components** tool from the Linear Component Pattern Consolidated toolbar.

If using an older release of SOLIDWORKS, no check mark in the Components to Mirror box indicates that the components are copied. The geometry of a copied component is unchanged from the original, only the orientation of the component is different.

Create the coordinate system location for the assembly.

37. Select the front right **vertex** of the Clevis component as illustrated.

38. Click the **Coordinate System** tool from the Reference Geometry Consolidated toolbar. The Coordinate System PropertyManager is displayed.

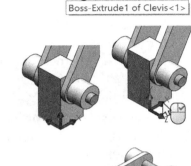

Boss-Extrude1 of Clevis<1>

39. Click the **right bottom edge** of the Clevis component.

40. Click the **front bottom edge** of the Clevis component as illustrated.

41. Address the **direction** for X, Y, Z as illustrated.

42. Click **OK** from the Coordinate System PropertyManager. Coordinate System1 is displayed.

43. **Locate** the Center of mass based on the location of the illustrated coordinate system. Select Coordinate System1.

- X: 1.79 inches

- Y: 0.25 inches

- Z: 2.61 inches

Mass properties of Assem1
Configuration: Default
Coordinate system: Coordinate System1
Mass = 0.14 pounds
Volume = 1.20 cubic inches
Surface area = 27.04 square inches
Center of mass: (inches)
X = 1.79
Y = 0.25
Z = 2.61

44. **Save** the part and name it Assembly Modeling 10-1.

45. **Close** the model.

Coordinate System1

There are numerous ways to create the models in this chapter. A goal in this text is to display different design intents and techniques.

💡 If you don't find an option within 1% of your answer on the exam re-check your assembly.

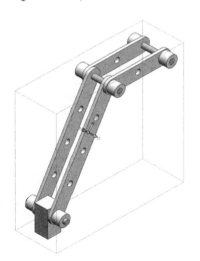

Tutorial: Assembly Model 10-2

An exam question in this category could read, "Build this assembly in SOLIDWORKS. Locate the Center of mass of the model with the illustrated coordinate system. Set decimal place to 2. Unit system: MMGS."

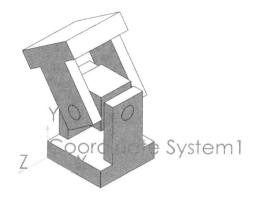

The assembly contains the following: (2) U-Bracket components, (4) Pin components and (1) Square block component.

- U-Bracket, (Item 1): Material: AISI 304. Two U-Bracket components are combined together Concentric to opposite holes of the Square block component. The second U-Bracket component is positioned with an Angle mate to the right face of the first U-Bracket and a Parallel mate between the top face of the first U-Bracket and the top face of the Square block component. Angle A = 125deg.

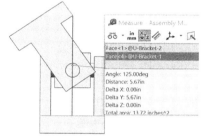

- Square block, (Item 2): Material: AISI 304. The Pin components are mated Concentric and Coincident to the 4 holes in the Square block (no clearance). The depth of each hole = 10mm.

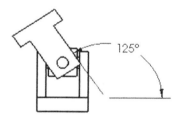

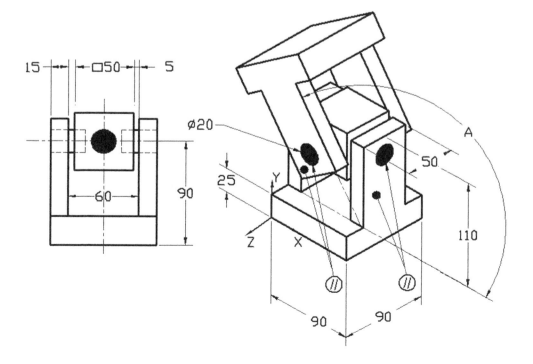

- Pin, (Item 3): Material: AISI 304. The Pin components are mated Concentric to the hole (no clearance). The end face of the Pin components are Coincident to the outer face of the U-Bracket components. The Pin component has a 5mm spacing between the Square block component and the two U-Bracket components.

Think about the steps that you would take to build the illustrated assembly. Identify the first fixed component. This is the Base component of the assembly. Insert the required Standard mates. Locate the Center of mass of the model with respect to the illustrated coordinate system. In this example, start with the U-Bracket part.

Create the assembly.

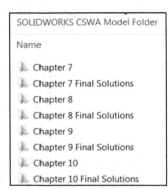

1. **Download** the needed components from the SOLIDWORKS CSWA Model Folder\Chapter 10\Assembly Model 10-2 folder.

2. **Create** a new assembly in SOLIDWORKS. The created models are displayed in the Open documents box.

3. Click **Cancel** in the Open dialog box.

4. Click **Cancel** ✖ from the Begin Assembly PropertyManager.

5. **Set** the document properties for the model.

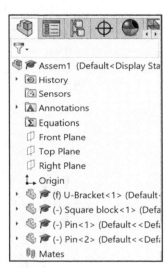

6. **Insert** the first U-Bracket component into the assembly document.

7. **Fix** the component to the assembly Origin. Click OK from the PropertyManager. The U-Bracket is displayed in the Assembly FeatureManager and in the Graphics window.

8. **Insert** the Square block above the U-Bracket component as illustrated.

9. **Insert** the first Pin part. Locate the first Pin to the front of the Square block.

10. **Insert** the second Pin part. Locate the second Pin to the back of the Square block.

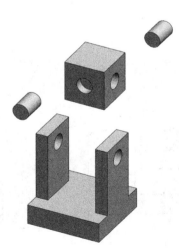

11. **Insert** the third Pin part. Locate the third Pin to the left side of the Square block.

12. **Rotate** the Pin as illustrated.

13. **Insert** the fourth Pin part. Locate the fourth Pin to the right side of the Square block.

14. **Rotate** the Pin as illustrated.

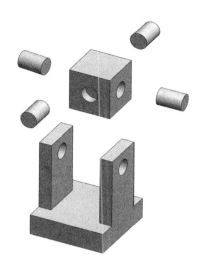

15. **Insert** a Concentric mate between the inside cylindrical face of the Square block and the outside cylindrical face of the first Pin. Concentric1 is created.

16. **Insert** a Coincident mate between the inside back circular face of the Square block and the flat back face of the first Pin. Coincident1 mate is created.

17. **Insert** a Concentric mate between the inside cylindrical face of the Square block and the outside cylindrical face of the second Pin. Concentric2 is created.

18. **Insert** a Coincident mate between the inside back circular face of the Square block and the front flat face of the second Pin. Coincident2 mate is created.

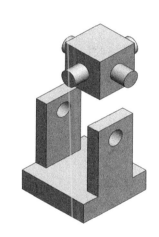

19. **Insert** a Concentric mate between the inside cylindrical face of the Square block and the outside cylindrical face of the third Pin. Concentric3 is created.

20. **Insert** a Coincident mate between the inside back circular face of the Square block and the right flat face of the third Pin. Coincident3 mate is created.

21. **Insert** a Concentric mate between the inside circular face of the Square block and the outside cylindrical face of the fourth Pin. Concentric4 is created.

22. **Insert** a Coincident mate between the inside back circular face of the Square block and the left flat face of the fourth Pin. Coincident4 mate is created.

> ▸ 🔩🎓 (f) U-Bracket<1> (Default<<Default>_
> ▸ 🔩🎓 (-) Square block<1> (Default<<Defau
> ▸ 🔩🎓 (-) Pin<1> (Default<<Default>_Appea
> ▸ 🔩🎓 (-) Pin<2> (Default<<Default>_Appea
> ▸ 🔩🎓 (-) Pin<3> (Default<<Default>_Appea
> ▸ 🔩🎓 (-) Pin<4> (Default<<Default>_Appea
> ▾ 🔗 Mates
>   ◎ Concentric1 (Square block<1>,Pin
>   ⟋ Coincident1 (Square block<1>,Pin<1>
>   ◎ Concentric2 (Square block<1>,Pin<2>
>   ⟋ Coincident2 (Square block<1>,Pin<2>
>   ◎ Concentric3 (Square block<1>,Pin<3>
>   ⟋ Coincident3 (Square block<1>,Pin<3>
>   ◎ Concentric4 (Square block<1>,Pin<4>
>   ⟋ Coincident4 (Square block<1>,Pin<4>

23. **Insert** a Concentric mate between the inside right cylindrical face of the Cut-Extrude feature on the U-Bracket and the outside cylindrical face of the right Pin. Concentric5 is created.

24. **Insert** a Coincident mate between the Right Plane of the Square block and the Right Plane of the assembly. Coincident5 is created.

25. **Insert** the second U-Bracket part above the assembly. Position the U-Bracket as illustrated.

26. **Insert** a Concentric mate between the inside cylindrical face of the second U-Bracket component and the outside cylindrical face of the second Pin. The mate is created.

27. **Insert** a Coincident mate between the outside circular edge of the second U-Bracket and the back flat face of the second Pin. The mate is created.

There are numerous ways to mate the models in this chapter. A goal is to display different design intents and techniques.

28. **Insert** an Angle mate between the top flat face of the first U-Bracket component and the front narrow face of the second U-Bracket component as illustrated. Angle1 is created. An Angle mate is required to obtain the correct Center of mass.

29. **Insert** a Parallel mate between the top flat face of the first U-Bracket and the top flat face of the Square block component.

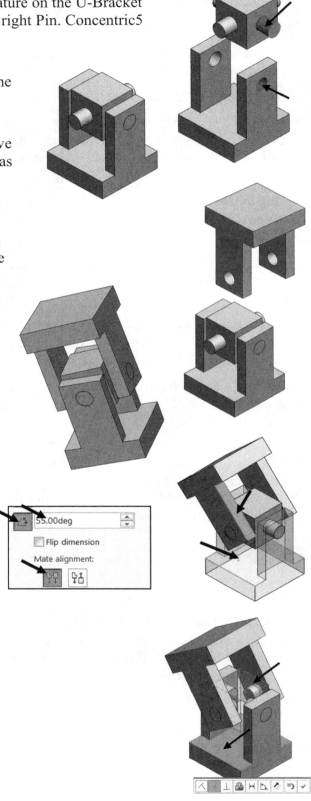

30. **Expand** the Mates folder and the components from the FeatureManager. View the created mates.

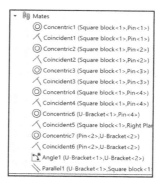

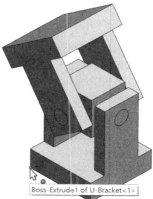

Create the coordinate location for the assembly.

31. Select the front **bottom left vertex** of the first U-Bracket component as illustrated.

32. Click the **Coordinate System** tool from the Reference Geometry Consolidated toolbar. The Coordinate System PropertyManager is displayed.

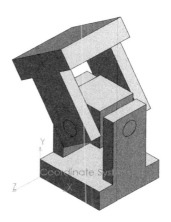

33. Click **OK** from the Coordinate System PropertyManager. Coordinate System1 is displayed.

34. **Locate** the Center of mass based on the location of the illustrated coordinate system. Select Coordinate System1.

- X: 31.54 millimeters

- Y: 85.76 millimeters

- Z: -45.00 millimeters

35. **Save** the part and name it Assembly Modeling 10-2.

36. **Close** the model.

☀ If you don't find an option within 1% of your answer on the exam re-check your assembly.

☀ Click on the additional views during the CSWA exam to better understand the assembly/component. Read each question carefully. Identify the dimensions, center of mass and units. Apply needed material.

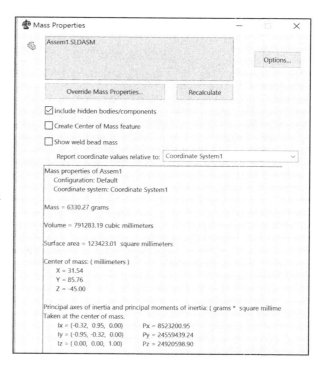

Tutorial: Assembly Model 10-3

An exam question in this category could read, "Build this assembly in SOLIDWORKS. Locate the Center of mass using the illustrated coordinate system. Set decimal place to 2. Unit system: MMGS."

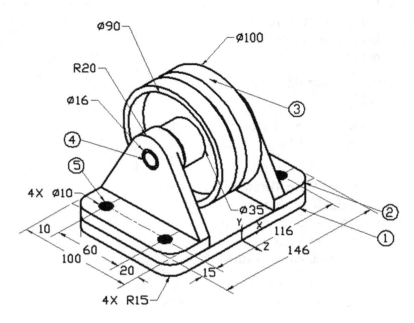

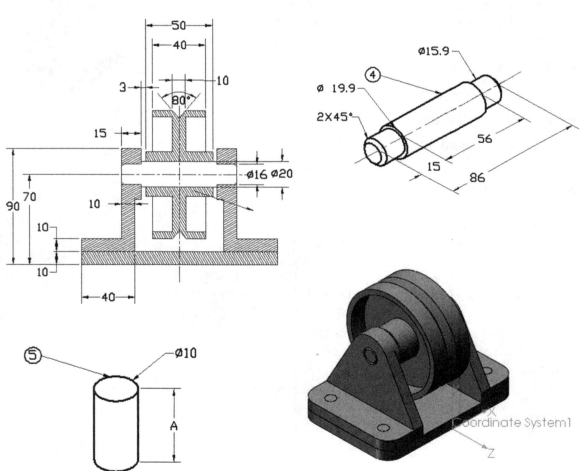

The assembly contains the following: (1) WheelPlate component, (2) Bracket100 components, (1) Axle40 component, (1) Wheel1 component and (4) Pin-4 components.

- WheelPlate, (Item 1): Material: AISI 304. The WheelPlate contains 4-Ø10 holes. The holes are aligned to the left Bracket100 and the right Bracket100 components. All holes are THRU ALL. The thickness of the WheelPlate = 10 mm.

- Bracket100, (Item 2): Material: AISI 304. The Bracket100 component contains 2-Ø10 holes and 1- Ø16 hole. All holes are through-all.

- Wheel1, (Item 3): Material AISI 304: The center hole of the Wheel1 component is Concentric with the Axle40 component. There is a 3mm gap between the inside faces of the Bracket100 components and the end faces of the Wheel hub.

- Axle40, (Item 4): Material AISI 304: The end faces of the Axle40 are Coincident with the outside faces of the Bracket100 components.

- Pin-4, (Item 5): Material AISI 304: The Pin-4 components are mated Concentric to the holes of the Bracket100 components (no clearance). The end faces are Coincident to the WheelPlate bottom face and the Bracket100 top face.

Think about the steps that you would take to build the illustrated assembly. Identify the first fixed component. This is the Base component of the assembly. Insert the required Standard mates.

Locate the Center of mass of the illustrated model with respect to the referenced coordinate system.

The referenced coordinate system is located at the bottom, right, midpoint of the Wheelplate. In this example, start with the WheelPlate part.

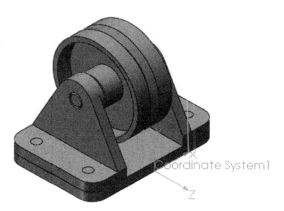

1. **Download** the needed components from the SOLIDWORKS CSWA Model Folder\Chapter 10\Assembly Model 10-3 folder.

2. **Create** a new assembly in SOLIDWORKS.

3. Click **Cancel** in the Open dialog box.

4. Click **Cancel** ✖ from the Begin Assembly PropertyManager.

5. **Set** the document properties for the assembly.

SOLIDWORKS CSWA Model Folder

Name

📁 Chapter 7
📁 Chapter 7 Final Solutions
📁 Chapter 8
📁 Chapter 8 Final Solutions
📁 Chapter 9
📁 Chapter 9 Final Solutions
📁 Chapter 10
📁 Chapter 10 Final Solutions

6. **Insert** the first component. Insert the WheelPlate. Fix the component to the assembly Origin. The WheelPlate is displayed in the Assembly FeatureManager and in the Graphics window. The WheelPlate component is fixed.

7. **Insert** the first Bracket100 part above the WheelPlate component as illustrated.

8. **Insert** a Concentric mate between the inside front left cylindrical face of the Bracket100 component and the inside front left cylindrical face of the WheelPlate. Concentric1 is created.

9. **Insert** a Concentric mate between the inside front right cylindrical face of the Bracket100 component and the inside front right cylindrical face of the WheelPlate. Concentric2 is created.

10. **Insert** a Coincident mate between the bottom flat face of the Bracket100 component and the top flat face of the WheelPlate component. Coincident1 is created.

11. **Insert** the Axle40 part above the first Bracket100 component as illustrated.

12. **Insert** a Concentric mate between the outside cylindrical face of the Axle40 component and the inside cylindrical face of the Bracket100 component. Concentric3 is created.

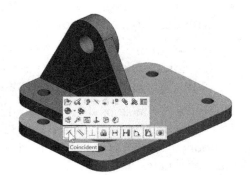

13. **Insert** a Coincident mate between the flat face of the Axle40 component and the front outside edge of the first Bracket100 component. Coincident2 is created.

To verify that the distance between holes of mating components is equal, utilize Concentric mates between pairs of cylindrical hole faces.

14. **Insert** the first Pin-4 part above the Bracket100 component.

15. **Insert** the second Pin-4 part above the Bracket100 component.

16. **Insert** a Concentric mate between the outside cylindrical face of the first Pin-4 component and the inside front left cylindrical face of the Bracket100 component. Concentric4 is created.

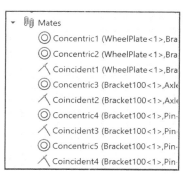

17. **Insert** a Coincident mate between the flat top face of the first Pin-4 component and the top face of the first Bracket100 component. Coincident3 is created.

18. **Insert** a Concentric mate between the outside cylindrical face of the second Pin-4 component and the inside front right cylindrical face of the Bracket100 component. Concentric5 is created.

19. **Insert** a Coincident mate between the flat top face of the second Pin-4 component and the top face of the first Bracket100 component. Coincident4 is created.

20. **Insert** the Wheel1 part as illustrated.

21. **Insert** a Concentric mate between the outside cylindrical face of Axle40 and the inside front cylindrical face of the Wheel1 component. Concentric6 is created.

22. **Insert** a Coincident mate between the Front Plane of Axle40 and the Front Plane of Wheel1. Coincident5 is created.

23. **Mirror** the components. Mirror the Bracket100 and the two Pin-4 components about the Front Plane.

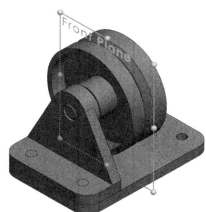

🔅 Click **Insert, Mirror Components** from the Menu bar menu or click the **Mirror Components** tool from the Linear Component Pattern Consolidated toolbar.

Create the coordinate location for the assembly.

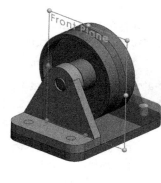

24. Click the **Coordinate System** tool from the Reference Geometry Consolidated toolbar. The Coordinate System PropertyManager is displayed.

25. **Select** the right bottom midpoint as the Origin location.

26. **Select** the bottom right edge as the X axis direction reference as illustrated.

27. Click **OK** from the Coordinate System PropertyManager. Coordinate System1 is displayed.

28. **Locate** the Center of mass based on the location of the illustrated coordinate system. Select Coordinate System1.

- X: = 0.00 millimeters

- Y: = 37.14 millimeters

- Z: = -50.00 millimeters

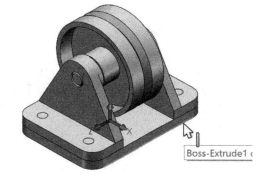

If you don't find an option within 1% of your answer on the exam re-check your assembly.

29. **Save** the part and name it Assembly Modeling 10-3.

30. **Close** the model.

```
Mass = 3797.32 grams

Volume = 474665.19 cubic millimeters

Surface area = 130119.83  square millimeters

Center of mass: ( millimeters )
    X = 0.00
    Y = 37.14
    Z = -50.00
```

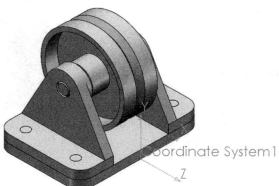

Summary

Assembly Creation and Modification is one of the five categories on the CSWA exam. In the last two chapters, a Basic, Intermediate or Advanced model was the focus. The Assembly Creation and Modification (Bottom-up) category addresses an assembly with numerous components.

There are four questions on the CSWA exam in the Assembly Creation and Modification category: (2) different assemblies - (4) questions - (2) multiple choice \ (2) single answers - 30 points each.

You are required to download the needed components from a provided zip file and insert them correctly to create the assembly as illustrated. You are then requested to modify the assembly and answer fill in the blank format questions.

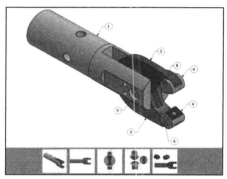

The **Official Guide to Certified SOLIDWORKS Associate Exams**: **CSWA, CSWA-SD, CSWSA-FEA and CSWA-AM** book is written to assist the SOLIDWORKS user to pass the associate level exams. Information is provided to aid the user to pass the Certified Associate - Mechanical Design exam, SOLIDWORKS Certified Associate - Sustainable Design exam, and SOLIDWORKS Certified Associate - Simulation (FEA) exam. The primary goal of the book is not only to help you pass the exams but also to ensure that you understand and comprehend the concepts and implementation details of the process.

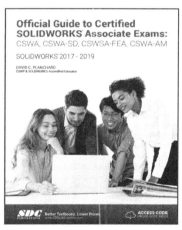

Click on the additional views during the CSWA exam to better understand the assembly/component. Read each question carefully. Identify the dimensions, center of mass and units. Apply needed material.

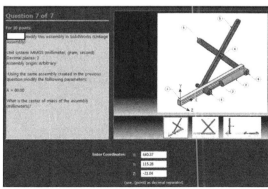

Questions

1. Build this ANSI MMGS assembly from the provided information in SOLIDWORKS.

Calculate the overall mass and volume of the assembly.

Locate the Center of mass using the illustrated coordinate system.

The assembly contains the following: one Base100 component, one Yoke component, and one AdjustingPin component.

- Base100, (Item 1): Material 1060 Alloy. The distance between the front face of the Base100 component and the front face of the Yoke = 60mm.
- Yoke, (Item 2): Material 1060 Alloy. The Yoke fits inside the left and right square channels of the Base100 component (no clearance). The top face of the Yoke contains a Ø12mm through all hole.
- AdjustingPin, (Item 3): Material 1060 alloy. The bottom face of the AdjustingPin head is located 40mm from the top face of the Yoke component. The AdjustingPin component contains an Ø5mm Through All hole.

The Coordinate system is located in the lower left corner of the Base100 component. The X axis points to the right.

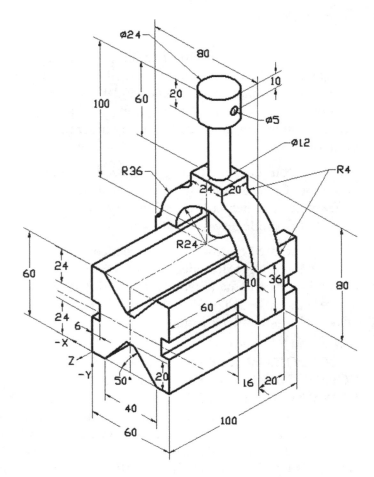

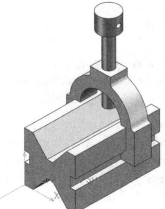

2. Build the assembly from the provided information in SOLIDWORKS. Calculate the overall mass and volume of the assembly. Locate the Center of mass using the illustrated coordinate system. The assembly contains the following: three MachinedBracket components and two Pin-5 components. Apply the MMGS unit system.

Insert the Base component, float the component, then mate the first component with respect to the assembly reference planes.

- MachinedBracket, (Item 1): Material 6061 Alloy. The MachineBracket component contains two Ø10mm through all holes. Each MachinedBracket component is mated with two Angle mates. The Angle mate = 45deg. The top edge of the notch is located 20mm from the top edge of the MachinedBracket.
- Pin-5, (Item 2): Material Titanium. The Pin-5 component is 5mms in length and equal in diameter. The Pin-5 component is mated Concentric to the MachinedBracket (no clearance). The end faces of the Pin-5 component is Coincident with the outer faces of the MachinedBracket. There is a 1mm gap between the Machined Bracket components.

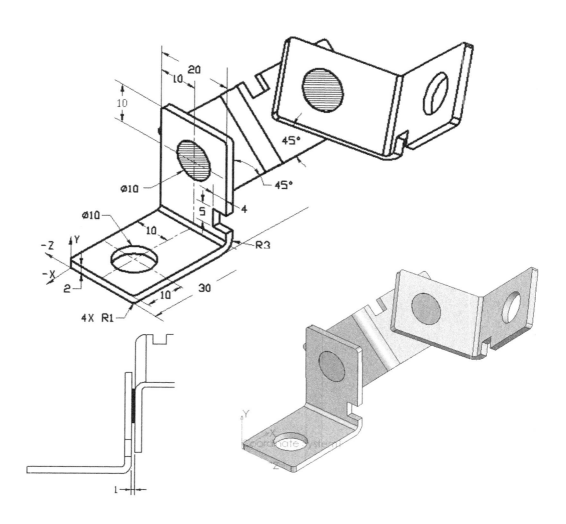

3. Build the assembly from the provided information in SOLIDWORKS. Calculate the overall mass and volume of the assembly. Locate the Center of mass using the illustrated coordinate system. The illustrated assembly contains the following components: three Machined-Bracket components and two Pin-6 components. Apply the MMGS unit system.

 Insert the Base component, float the component, then mate the first component with respect to the assembly reference planes.

- Machined-Bracket, (Item 1): Material 6061 Alloy. The Machine-Bracket component contains two Ø10mm through all holes. Each Machined-Bracket component is mated with two Angle mates. The Angle mate = 45deg. The top edge of the notch is located 20mm from the top edge of the MachinedBracket.
- Pin-6, (Item 2): Material Titanium. The Pin-6 component is 5mms in length and equal in diameter. The Pin-5 component is mated Concentric to the Machined-Bracket (no clearance). The end faces of the Pin-6 component is Coincident with the outer faces of the Machined-Bracket. There is a 1mm gap between the Machined-Bracket components.

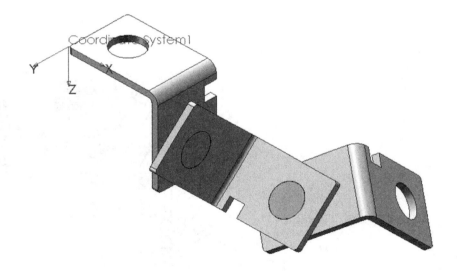

Sample screen shots from an older CSWA exam for an assembly. Click on the additional views to understand the assembly and provided information. Read each question carefully. Understand the dimensions, center of mass and units. Apply needed materials.

Zoom in on the part if needed.

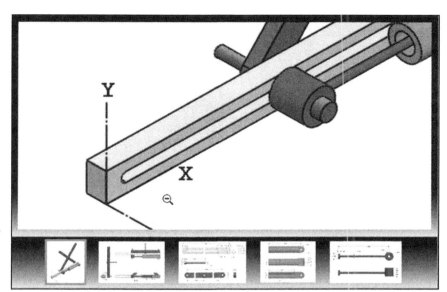

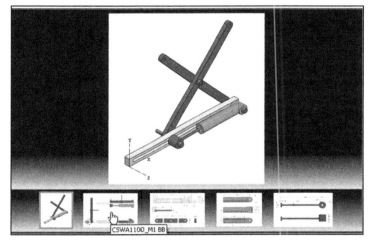

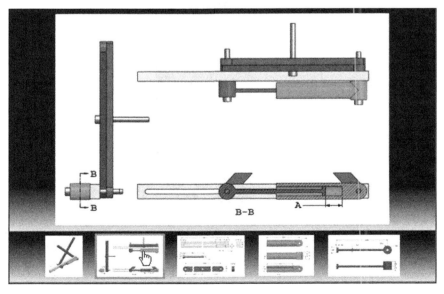

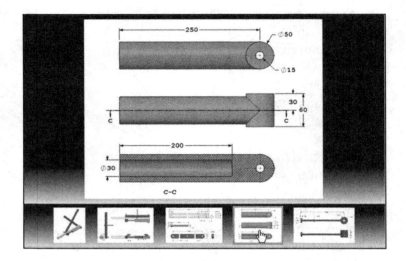

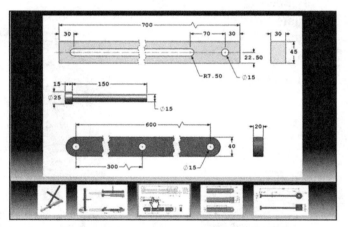

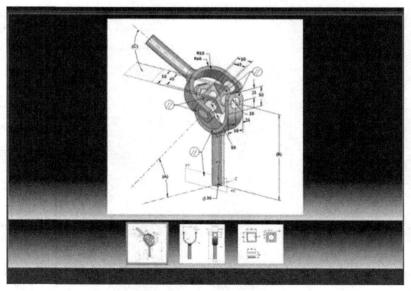

Screen shots from an exam

Zoom in on the part if needed.

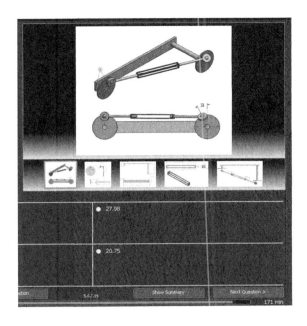

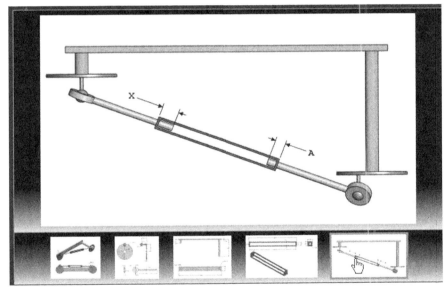

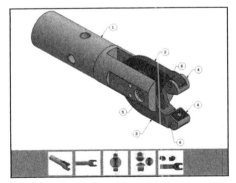

Screen shots from an exam

Notes:

Chapter 11

Additive Manufacturing - 3D Printing

Below are the desired outcomes and usage competencies based on the completion of Chapter 11.

Desired Outcomes:	Usage Competencies:
• Knowledge of Additive Manufacturing.	• Examine the advantages and disadvantages of Additive Manufacturing.
• Comprehend various 3D printer technologies and terminologies.	• Discuss 3D printer technology: Fused Filament Fabrication (FFF), STereoLithography (SLA), and Selective Laser Sintering (SLS).
• Familiarity with filament materials: PLA (Polylactic acid), FPLA (Flexible Polylactic acid), ABS (Acrylonitrile butadiene styrene), PVA (Polyvinyl alcohol), Nylon 618, and Nylon 645.	• Select the correct filament material.
	• Create an STL (*.stl) file, Additive Manufacturing file (*.amf), and a 3D Manufacturing Format file (*.3mf).
	• Choose optimum build orientation and slicer parameters.
• Understand the following file types: STL (*.stl), Additive Manufacturing (*.amf), and 3D Manufacturing format (*.3mf).	• Discuss 3D printer terminology: Raft, Skirt, Brim, Support, Touching, Build plate, Hot end, etc.
	• Address fit tolerance for interlocking parts.
• Awareness of slicer and 3D printer parameters.	• Define 3D Printing tips.
	• Print directly from SOLIDWORKS.
• Identify 3D Printer Add-ins.	

Notes:

Chapter 11 - Additive Manufacturing - 3D Printing

Chapter Objective

Provide a basic understanding between Additive vs. Subtractive manufacturing. Discuss Fused Filament Fabrication (FFF), STereoLithography (SLA), and Selective Laser Sintering (SLS) printer technology. Select suitable filament material. Comprehend 3D printer terminology. Knowledge of preparing, saving, and printing a model on a Fused Filament Fabrication 3D printer. Information on the Certified SOLIDWORKS Associate Additive Manufacturing (CSWA-AM) exam.

On the completion of this chapter, you will be able to:

- Discuss Additive vs Subtractive manufacturing.

- Review 3D printer technology: Fused Filament Fabrication (FFF), STereoLithography (SLA), and Selective Laser Sintering (SLS).

- Select the correct filament material:
  o PLA (Polylactic acid)
  o FPLA (Flexible Polylactic acid)
  o ABS (Acrylonitrile butadiene styrene)
  o PVA (Polyvinyl alcohol)
  o Nylon 618
  o Nylon 645

- Create an STL (*.stl) file, an Additive Manufacturing (*.amf) file and a 3D Manufacturing format (*.3mf) file.

- Prepare G-code.

- Comprehend general 3D printer terminology.

- Understand optimum build orientation.

- Enter slicer parameters:
  o   Raft, brim, skirt, layer height, percent infill, infill pattern, wall thickness, fan speed, print speed, bed temperature, and extruder (hot end) temperature.

- Address fit tolerance for interlocking parts.

- Define general 3D Printing tips.

- Print directly from SOLIDWORKS.

- Knowledge of the Certified SOLIDWORKS Associate Additive Manufacturing exam.

Additive vs. Subtractive Manufacturing

In April, 2012, *The Economist* published an article on 3D printing. In the article they stated that this was the "beginning of a third industrial revolution, offering the potential to revolutionize how the world makes just about everything."

Avi Reichental, who was President and CEO, 3D Systems stated, "With 3D printing, complexity is free. The printer doesn't care if it makes the most rudimentary shape or the most complex shape, and that is completely turning design and manufacturing on its head as we know it."

Over the past five years, companies are now using 3D printing to evaluate more concepts in less time to improve decisions early in product development. As the design process moves forward, technical decisions are iteratively tested at every step to guide decisions big and small, to achieve improved performance, lower manufacturing costs, delivering higher quality and more successful product introductions. In pre-production, 3D printing is enabling faster first article production to support marketing and sales functions, and early adopter customers. And in final production processes, 3D printing is enabling higher productivity, increased flexibility, reduced warehouse and other logistics costs, economical customization, improved quality, reduced product weight, and greater efficiency in a growing number of industries.

Technology for 3D printing continues to advance in three key areas: **printers** and **printing methods**, **design software**, and **materials** used in printing.

Already, 3D printing is being used in the medical industry to help save lives and in some space exploration efforts. But how will 3D printing affect the average, middle-class person in the future? Low cost 3D printers are addressing this consumer market.

Additive manufacturing is the process of joining materials to create an object from a 3D model, usually adding layer over layer.

Subtractive manufacturing relies upon the removal of material to create something. The blacksmith hammered away at heated metal to create a product. Today, a Computer Numerical Control CNC machine cuts and drills and otherwise removes material from a larger initial block to create a product.

☼ Additive manufacturing, sometimes known as *rapid prototyping* can be slower than Subtractive manufacturing. Both take skill in creating the G-code and understanding the machine limitations.

☼ Fused Filament Fabrication (FFF) and Fused Deposition Modeling (FDM) are used interchangeably in the book.

3D Printer Technology

Fused Filament Fabrication (FFF)

Fused filament fabrication (FFF) is a relatively new method of Additive manufacturing (also known as FDM) technology used for building three-dimensional prototypes or models layer by layer with a range of thermoplastics.

FFF technology is the most widely used form of 3D desktop printing at the consumer level, fueled by students, hobbyists, and office professionals.

The technology uses a continuous filament of a thermoplastic material. The thermoplastic filament is provided in a spool (open filament area) or in a refillable auto loading cartridge. Thermoplastic filament comes in two standard diameters: 1.75mm and 3mm (true size of 2.85mm).

Ultimaker 3 uses Near Field Communication technology with their 3mm filament. NFC technology informs the printer of the filament type and color. NFC spools tend to be more expensive than generic filament and you need to manually feed the filament correctly through the Bowden-tubes into the extruder (hot end). The spools are located in the back of the printer. This increases the footprint of the printer.

Ultimaker 3 3D printer

Sindoh 3DWOX printers uses a replacement filament spool (1.75mm) inside their cartridge with a smart chip. The benefit is the automatic loading and unloading feature along with informing the printer of the filament type and properties. Replacement spools are available. The Sindoh 3DWOX 1 printer provides the ability to either use their filament or open source filament.

Sindoh 3DWOX
replacement spool and
cartridge

The thermoplastic material is heated in the nozzle (hot end) (160C - 250ºC) to form liquid, which solidifies immediately when it's deposited onto the build plate or platform. The nozzle travels at a controlled rate and moves in the X and Y direction. The build plate moves in the Z direction. This creates mechanical adhesion (not chemical).

Fused Filament Fabrication (FFF) technology can be described as the inverse process of a computer numerical cutting (CNC) machine. This technology only uses the amount of material required for the part, as opposed to a CNC machine which requires significant amounts of scrap material.

Sindoh 3DWOX
DP200 printer

Fused Filament Fabrication printers are available with multiple-head extruders. The most common usage for multiple-head extruders is to print in different colors or a different material for support or increase bed adhesion (raft).

Sindoh 3DWOX 2X

A few advantages of Fused Filament Fabrication (FFF) Additive manufacturing:

- Lower cost (different entry levels) into the manufacturing environment.

- Lowers the barriers (space, power, safety, and training) to traditional subtractive manufacturing.

- Reduce part count in an assembly from traditional subtractive manufacturing (complex parts vs. assemblies).

- Numerous thermoplastic filament types and colors (open and closed source) are available and affordable.

- Remote monitoring ability with a camera and LED lighting.

- Print more than one thermoplastic filament type and color during the print cycle (multiple-head extruders).

- Change or load new filament material during a print cycle.

- Customize percent infill and infill pattern type in the print.

- Only uses the amount of material required for the part. One exception, when using a raft and or support during the print.

- No post-curing process required.

- Reduce prototyping time.

- Faster development cycle.

- Quicker customer feedback.

- Quicker product customization and configuration.

- Parallel verticals: develop and prototype at the same time.

- Most work on either the Window or Mac OSX platform.

- Open source slicing engines. To name a few, (Slic3r, Skeinforge, Netfabb, KISSkice, and Cura).

- Open source filament.

A slicer takes a 3D model, most often in STereoLithography (*.stl) file format, and translates the model into individual layers. It then generates the machine code that the 3D printer uses. You can also use Additive Manufacturing file (*.amf) or a 3D Manufacturing Format (*.3mf) file if supported.

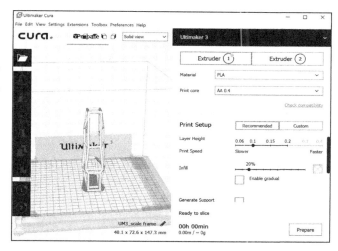

Ultimaker Cura slicer - Version: 3.4.1

Slicer software allows the user to calibrate 3D printer settings: filament type, part orientation, extruder speed, extruder temperature, bed temperature, cooling fan rate, raft type, support type, percent infill, infill pattern type, etc.

A few disadvantages of Fused Filament Fabrication (FFF) Additive manufacturing:

- Slow build rates. Many printers lay down material at a speed of one to five cubic inches per hour. Depending on the part needed, other manufacturing processes may be significantly faster.

- May require post-processing. The surface finish and dimensional accuracy may be lower quality than other manufacturing methods.

- Poor mechanical properties. Layering and multiple interfaces can cause defects in the product.

- Frequent calibration is required. Without frequent calibration, prints may not be the correct dimensions, they may not stick to the build plate, and a variety of other not-so-wanted effects can occur.

- Limited by the accuracy of the stepper motors, extruder nozzle diameter, user calibration as well as print speed.

- Print time increases linearly as part tolerances become tighter. In general, FFF print tolerances range from 0.05mm to 0.5mm.

- To print something, you require a CAD model. You either need to know how to design using CAD software (SOLIDWORKS) or download a CAD model (native format or .stl) from a website (3D ContentCentral, Thingiverse, etc.).

StereoLithography (SLA)

Stereolithography (SLA) was the world's first 3D printing technology. It was introduced in 1988 by 3D Systems, Inc., based on work by inventor Charles Hull.

SLA is one of the most popular resin-based 3D printing technologies for professionals today.

SLA technology is a form of 3D printing in which a computer-controlled low-power, highly focused ultraviolet (UV) laser is used to create layers (layer by layer) from a liquid polymer (photopolymerization) that hardens on contact.

When a layer is completed, a leveling blade is moved across the surface to smooth it before depositing the next layer. The platform moves by a distance equal to the layer thickness, and a subsequent layer is formed on top of the previously completed layers.

Markforged X7
SLS printer

The process of tracing and smoothing is repeated until the part is complete. Once complete, the part is removed from the resin tank and drained of any excess polymer.

The part is in a "green" state. This green state differs from the completely cured state in one very important way: there are still polymerizable groups on the surface that subsequent layers can covalently bond to.

Through the application of heat and light, the strength and stability of printed parts improves beyond their original "green" state. However, each resin behaves slightly differently when post-cured, and requires different amounts of time and temperature to arrive at the material's optimum properties.

After the post-cure, you need to address the post-process. Post-processing is the removal of the supports and any needed polishing or sanding.

SLA prints are watertight and fully dense. No Infill is required.

Resolution of SLA technology varies from 0.05mm to 0.15mm with the industry average tolerance around 0.1mm. On average, this is significantly more precise than FFF technology and is the preferred rapid prototyping solution when extremely tight tolerances are required.

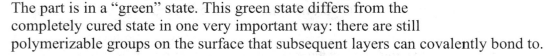 As the ultraviolet (UV) laser traces the layer, the liquid polymer solidifies and the excess areas are left as liquid in the tank.

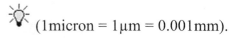 (1micron = 1μm = 0.001mm).

One example of a popular SLA 3D desktop printer is the Formlabs Form 2. Formlabs was founded in September 2011 by three MIT Media Lab students.

The build area is 45mm × 145mm × 175mm with a layer thickness (Axis Resolution) of 25, 50, and 100 microns.

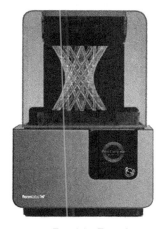

Formlabs Form 2
SLA 3D printer

🔅 Desktop area increases significantly if you include their Form Wash and Form Cure products.

The resin (liquid polymer) is provided in a cartridge. There are different cartridges for various colors and materials. Not all resin cartridges support the 25, 50, and 100 micron resolutions.

Once the print is finished, remove it from the resin tank. Resin tanks are consumables and require replacement. Expect to replace a standard resin tank after 1,000-3,000 layers of printing (1-1.5 liters) of resin.

Formlabs Form 2
resin cartridges

Wash the print with isopropyl alcohol (IPA). You should wait approximately 30 minutes for the IPA to fully evaporate after washing.

To ensure proper washing, Formlabs recommends their Form Wash. Form Wash automatically cleans uncured liquid resin printed part surfaces.

IPA dissolves uncured resin. The part is covered in uncured resin when it's removed from the resin tank. Use IPA in a well-ventilated area. Always wear protective gloves and eyewear. Find a safe way to dispose of the used IPA.

Formlabs
Form Wash

After the IPA wash and dry, perform a post-cure. At a basic level, exposure to sun light triggers the formation of additional chemical bonds within a printed part, making the material stronger and stiffer.

To ensure proper post-curing, Formlabs recommends their post-curing unit, Form Cure. Form Cure precisely combines temperature and 405 nm light to post-cure parts for peak material performance.

🔅 You cannot use a camera or webcam inside the build area. Formlabs Dashboard feature provides the ability to keep track of what layer the printer is on, resin consumptions, and various other stats.

Formlabs
Form Cure

Formlabs uses their PreForm software. PreForm provides the ability to select the One-Click Print option, to automatically orient, support and layout the part.

Formlabs Form 2 printer works with the Window or Mac OSX platform.

💡 Clean the resin tank after a print fails. Remove any cured resin on the elastic layer, discard print failures, and filter out debris. Clean any contamination on the clear acrylic tank window.

A few advantages of Stereolithography (SLA) Additive manufacturing:

- Final parts are stronger than using FFF technology.

- Parts are watertight and fully dense. No infill is required.

- Higher resolution than FFF technology.

A few disadvantages of Stereolithography (SLA) Additive manufacturing:

- Higher cost (different entry levels) into the manufacturing environment than FFF.

- Material costs are significantly higher than FFF due to the proprietary nature and limited availability of the photopolymers.

- Significantly slower fabrication speed than FFF.

- Suitable for low volume production runs of small, precise parts.

- Print only a single material type (color) at a time.

- Requires an isopropyl alcohol (IPA) wash.

- Ability to safely dispose the used isopropyl alcohol (IPA).

- Requires a post-cure.

- Parts are sensitive to long exposure to UV light.

- May need drain hole in the part to remove excess liquid polymer.

Selective Laser Sintering (SLS)

Selective Laser Sintering (SLS) technology is the most common additive manufacturing technology for industrial powder base applications.

SLS fuses particles together layer by layer through a high energy pulse laser. Similar to SLA, this process starts with a tank full of bulk material but is in a powder form vs. liquid. As the print continues, the bed lowers itself for each new layer as done in the SLS process.

3D Systems
ProX DMP 300

Both plastics and metals can be fused in this manner, creating much stronger and more durable prototypes.

Although the quality of the powders is dependent on the supplier's proprietary processes, the base materials used are typically more abundant than photopolymers, and therefore cheaper. However, there are additional costs in energy used for fabricating with this method which may reverse any savings realized in the material cost.

 Speed and resolution of SLS printers typically match that of SLA, with industry averages at around 0.1mm tolerances. Only suitable for low volume production runs of small, precise parts.

SLS is the preferred rapid-prototyping method of metals and exotic materials.

Select the Correct Filament Material for FFF

There are many materials that are being explored for 3D printing; however, the two most dominant plastics for FFF technology are PLA (Polylactic acid) and ABS (Acrylonitrile-Butadiene-Styrene).

Both PLA and ABS are known as thermoplastics; that is, they become soft and moldable when heated and return to a solid when cooled.

There are three key printing stages for both thermoplastics.

1. Cold to warm: The thermoplastic starts in a hard state. It stays this way until heated to its glass transition temperature.

2. Warm to hot: The thermoplastic is now in a viscous state. It stays this way until heated to its melting temperature.

3. Hot to melting: The thermoplastic is now in a liquid state.

For PLA, the glass transition temperature is approximately 60°C. The melting temperature is approximately 155°C with a printing temperature range of 190C - 220°C.

For ABS, the glass transition temperature is approximately 100°C with a printing temperature range between 210C - 250°C. This is why a heated print bed is needed for ABS and is optional for PLA. The print bed must be kept well below the glass transition temperature and well above the melting temperature.

PLA and ABS are hygroscopic, meaning it attracts and absorbs moisture from the air. The effects of attracting water may result in one or more of the following problems: increased brittleness, diameter augmentation (potential problems with Bowden-tube printers), or filament bubbling once reaching the extruder (hot-end).

It is recommended that you store used filament in an airtight plastic bag (container) with a few silica gel packs. Place the bag in a dry, dark, controlled environment.

Silica gel pack

PLA

PLA (Polylactic Acid) is a biodegradable thermoplastic, made from renewable resources like corn starch or sugarcane. Relevant information:

- **Strength**: High | Flexibility: Low | Durability: Medium

- **Difficulty to use**: Low

- **Print temperature**: 190C - 220°C

- **Print bed temperature**: 20C - 40°C (not required)

- **Shrinkage/warping**: Minimal

- **Soluble**: No

- **Hot Head**: Standard Polytetrafluoroethylene, PTFE (Teflon)

- **Filament size**: 1.75mm and 2.85mm

- **Food safety**: Refer to manufacturer guidelines

PLA has a lower printing temperature (20C - 30°C) than ABS, and it doesn't warp as easily. PLA does not require a heated bed. Most non-heated beds are made from glass or metal.

 A removable flexible bed makes it ideal to retrieve printed parts.

PLA is normally used for its nice finish, easy and fast printing characteristics and for the large amounts of colors and varieties available.

Avoid using PLA if the print is exposed to temperatures of 60°C or higher or might be bent, twisted or dropped repeatedly.

Outside of 3D printing, PLA is typically used in medical implants, food packaging, and disposable tableware.

Darker colors and glow in the dark materials, often require higher extruder temperatures (5C - 10°C).

PLA in general is more forgiving to temperature fluctuations and moisture than ABS and Nylon during a build cycle.

Flex/Soft PLA - Common flexible filaments are polyester-based (non-toxic). Recommended print temperature range is 220C - 250°C. It is highly recommended to drastically lower your printing speed to around 10-20mm/s. To take advantage of the filament's properties, print it with 10% infill or less. Most flexible filament adheres well to a heated bed.

PLA - Storage

PLA is mildly hygroscopic, meaning it attracts and absorbs moisture from the air.

PLA responds somewhat differently to moisture than ABS. Over time, it can become very brittle. In addition to bubbles or spurting at the nozzle (hot end), you may see discoloration and a reduction in 3D printed part properties.

Store the filament in an airtight plastic bag (container) with a few silica gel packs. Place the bag (container) in a dry, dark, temperature controlled environment. As an extra precaution, filament manufacturers often recommend using up rolls as soon as possible.

PLA can be dried using something as simple as a food dehydrator. It is important to note that this can alter the crystallinity ratio in the PLA and will lead to changes in extrusion temperature and other extrusion characteristics.

PLA - Part Accuracy

Compared to ABS, PLA demonstrates much less part warping. PLA is less sensitive to changes in temperature than ABS.

PLA undergoes more of a phase-change when heated and becomes much more liquid. If actively cooled, sharper details can be seen on printed corners without the risk of cracking or warping. The increased flow can also lead to stronger binding between layers, improving the strength of the printed part.

In a small enclosed space, it is recommended to have your printer enclosed with a HEPA filtration filter.

ABS

ABS (Acrylonitrile-Butadiene-Styrene) is an oil-based thermoplastic, commonly found in (DWV) pipe systems, automotive trim, bike helmets, and toys (LEGO). Relevant information:

- **Strength**: High | Flexibility: Medium | Durability: High
- **Difficulty to use**: Medium
- **Print temperature**: 210C - 250°C
- **Print bed temperature**: 80C - 110°C (required)
- **Shrinkage/warping**: Medium
- **Soluble**: In esters, ketones, and acetone
- **Hot Head**: Standard Polytetrafluoroethylene, PTFE (Teflon)
- **Filament size**: 1.75mm and 2.85mm
- **Food safety**: Not food safe

ABS boast slightly higher strength, flexibility, and durability. ABS is more sensitive to changes in temperature than PLA, which can result in cracking and warping if the print cools too quickly.

A heated bed plate is required for ABS. Most heated beds are made from glass or metal.

It is recommended to have a ventilated printing area when using ABS material (oil-based thermoplastic). It is also recommended to have the printer enclosed with a HEPA filtration filter.

ABS is better suited for items that are frequently handled, dropped, or heated. It can be used for mechanical parts, especially if they are subjected to stress or must interlock with other parts.

Sindoh 3DWOX 1, 2X HEPA filter is built into the printer.

For high temperature applications, ABS (glass transition temperature of 105°C) is more suitable than PLA (glass transition temperature of 60°C). PLA can rapidly lose its structural integrity as it approaches 60°C.

ABS - Storage

ABS is mildly hygroscopic. Diameter augmentation (potential problems with Bowden-tube printers) can be an issue.

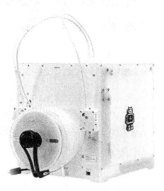

Store the filament in an airtight plastic bag (container) with a few silica gel packs. Place the bag in a dry, dark, temperature controlled environment. As an extra precaution, filament manufacturers often recommend using up rolls as soon as possible. ABS can be easily dried using a source of hot (preferably dry) air such as a food dehydrator.

Ultimaker 3 using Bowden-tubes

ABS - Part Accuracy

For most, the single greatest hurdle for accurate parts is good bed (platform) adhesion. Start with a clean, level, heated bed. Check the bed and extruder (hot end) are set to the correct temperature.

Eliminate all build area drafts (open windows, air conditioning vents, etc.). Use dry filament. Wet filament during printing prevents good layer adhesion and greatly weakens the part.

When printing on a glass plate, a (Polyvinyl Acetate) PVA based glue stick applied to the bed helps with bed adhesion. Elmer's or Scotch permanent glue sticks are inexpensive and easily found. Remember, less is more when applying the glue stick to the build plate.

You may need to add a raft (a horizontal latticework of filament located underneath the part) to the build. Over time, a heated metal bed can warp. Check for flatness.

ABS provides a more matte appearance than PLA, but it can become very shiny after acetone vapor smoothing.

Nylon

Nylon (618, 645) is a popular family of synthetic polymers used in many industrial applications. Compared to most other filaments, it ranks as the number one contender when together considering strength, flexibility, and durability.

Relevant information for Nylon 618:

- **Strength**: High | Flexibility: High | Durability: High

- **Difficulty to use**: Medium

- **Print temperature**: 240C - 255°C

- **Print bed temperature**: 50C - 60°C (required)

- **Shrinkage/warping**: Medium

- **Soluble**: No

- **Hot Head**: Metal

- **Filament size**: 1.75mm and 2.85mm

- **Food safety**: Refer to manufacturer guidelines

Relevant information for Nylon 645:

- **Strength**: High | Flexibility: High | Durability: High

- **Difficulty to use**: Medium

- **Print temperature**: 255C - 265°C

- **Print bed temperature**: 85C - 95°C (required)

- **Shrinkage/warping**: Considerable

- **Soluble**: No

- **Hot Head**: Metal

- **Filament size**: 1.75mm and 2.85mm

- **Food safety**: Refer to manufacturer guidelines

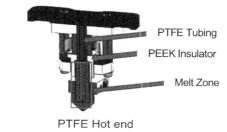

PTFE Hot end

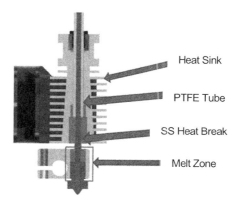

All metal Hot end

Nylon filament requires temperatures above 240°C to extrude. Most low-end 3D printers come standard with hot ends that use Polyether ether Ketone (PEEK) and Polytetrafluoroethylene (Teflon). Both PEEK and PTFE begin to break down above 240°C and will burn and emit noxious fumes. You should only use an all metal hot end.

☼ PLA and ABS are less likely to get stuck in the inner wall PTFE (Teflon) than an all metal hot end.

When should I use Nylon? Taking advantage of nylon's strength, flexibility, and durability use this 3D printer filament to create tools, functional prototypes, or mechanical parts (like hinges, buckles, or gears). Dry nylon prints buttery smooth and has a glossy finish.

Nylon - Storage

Nylon is very hygroscopic, more so than PLA or ABS. Nylon can absorb more than 10% of its weight in water in less than 24 hours. Successful 3D printing with nylon requires dry filament. When you print with nylon that isn't dry, the water in the filament explodes causing air bubbles during printing that prevents good layer adhesion and greatly weakens the part. It also ruins the surface finish. To dry nylon, place it in an oven at 50C - 60°C for 6-8 hours. After drying, store in an airtight container (Vacuum bag), preferably with dry silica gel packets. Place the container in a dry, dark, temperature controlled environment.

Nylon - Part Accuracy

Compared to ABS and PLA, Nylon and ABS warp approximately the same. PLA demonstrates much less part warping. A heated plate (50C - 90°C) is required for Nylon. When printing on a glass plate, a (Polyvinyl Acetate) PVA based glue stick applied to the bed is the best method of bed adhesion. Elmer's or Scotch permanent glue sticks are inexpensive and easily found. Remember, less is more when applying the glue stick to the plate. You will also have to clean it up after your build.

PVA (Polyvinyl Alcohol)

PVA (Polyvinyl Alcohol) filament is a water-soluble synthetic polymer. PVA filament dissolves in water. Many multi-head extruder users find PVA to be a useful support material because of its dissolvable properties. In general, PVA filament is used in conjunction with PLA not ABS. PVA adheres well to PLA and not ABS. Moreover, the extrusion temperature difference between PVA and ABS can be problematic.

PVA extrusion temperatures range between 160C - 190°C. A heated bed is recommended. Bed temperatures range between 40C - 50°C.

PVA is highly hygroscopic and is costly. PVA should be stored in an airtight box or container and may need to be dried before use.

Submerge the finished part in a bath of cold circulating water, until the PVA support structure is completely dissolved. This can be time consuming and messy.

Do not expose PVA filament to temperatures higher than 200°C for an extended period of time. An irreversible degradation of the material will occur, known as pyrolysis. It will jam the extruder nozzle (hot end). Unlike PLA and ABS, you cannot remove the jam by increasing the temperature. Clearing the jam in the nozzle will often require it to be re-drilled or replaced altogether.

A few filament companies provide breakaway material to replace the high cost of PVA. Breakaway is a support material used with multi head 3D printers. It is quick to remove and does not need further post-processing.

STereoLithography (*.stl) file

STereoLithography (*.stl) is a file format native to the Stereolithography CAD software created by 3D Systems. STL has several after-the-fact backronyms such as "Standard Triangle Language" and "Standard Tessellation Language".

An STL file describes only the surface geometry of a three-dimensional object without any representation of color, texture, or other common CAD model attributes. The STL format specifies both ASCII and Binary representations.

Binary files are more common, since they are more compact. An STL file describes a raw unstructured triangulated (point cloud) surface by the unit normal and vertices (ordered by the right-hand rule) of the triangles using a three-dimensional Cartesian coordinate system.

STL Save options allow you to control the number and size of the triangles by setting the various parameters in the CAD software.

Save an STL (*.stl) file in SOLIDWORKS

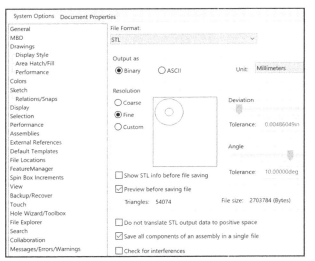

To save a SOLIDWORKS model as a STL file, click **File**, **Save As** from the Main menu or **Save As** from the Main menu toolbar. The Save As dialog is displayed. Select **STL(*.stl)** as the Save as type.

Click the **Options** button. The dialog box is displayed. View your options and the three types of Resolution: **Coarse**, **Fine** and **Custom**. The resolution options provide the ability to control the number and size of the triangles by setting various parameters in the CAD software. The Custom setting provides the ability to control the deviation and angle for the triangles. Click **OK**. View the generate point cloud of the part. Click **Yes**. The STL file is now ready to be imported into your 3D printer software.

In SOLIDWORKS for a smoother STL file, change the Resolution to Custom. Change the deviation to 0.0005in (0.01mm). Change the angle to 5. Smaller deviations and angles produce a smoother file but increase the file size and print time.

Additive Manufacturing (*.amf) file

Additive Manufacturing (*.amf) is a file format that includes the materials that have been applied to the parts or bodies in the 3D model.

Additive Manufacturing (*.amf) is an open standard for describing objects for additive manufacturing processes such as 3D printing. The official ISO/ASTM 52915:2013 standard is an XML-based format designed to allow any computer-aided design software to describe the shape and composition of any 3D object to be fabricated on any 3D printer. Unlike its predecessor STL format, AMF has native support for color, materials, lattices, and constellation (groups). Therefore, it requires less post-processing to define data such as the position of your model relative to the selected 3D printer, orientation, color, materials, etc.

Save an Additive Manufacturing (*.amf) file in SOLIDWORKS

To save a SOLIDWORKS model as an Additive Manufacturing (*.amf) file, Click **File**, **Save As** from the Main menu or **Save A**s from the Main menu toolbar. The Save As dialog is displayed. Select **Additive Manufacturing (*.amf)** as the Save as type.

Click the **Options** button. The dialog box is displayed. View your options. For most parts, utilize the default setting.

Close the dialog box.

Click **OK**. View the generate point cloud of the part.

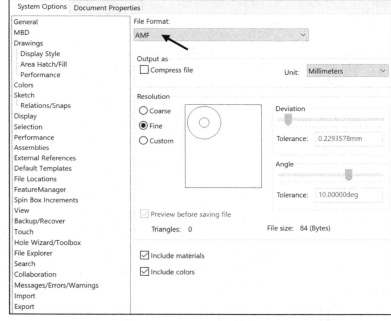

Click **Yes**. The file is now ready to be imported into your 3D printer software or printed directly from SOLIDWORKS.

3D Manufacturing Format (*.3mf) file

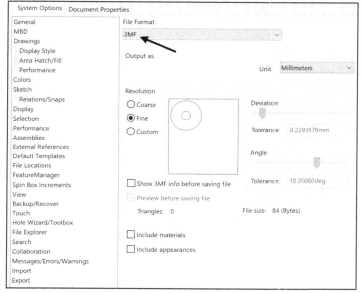

3D Manufacturing Format (*.3mf) is a file format developed and published by the 3MF Consortium. This format became natively supported in all Windows operating systems since Windows 8.1. 3MF has since garnered considerable support from large companies such as HP, 3D Systems, Stratasys, GE, Siemens, Autodesk and Dassault Systems although it is unknown how many actively use this file format. 3D Manufacturing Format file has similar native support for color, materials, lattices, and constellation (groups).

Save a 3D Manufacturing Format (*.3mf) file in SOLIDWORKS

To save a SOLIDWORKS model as a 3D Manufacturing Format (*.3mf) file, Click **File**, **Save As** from the Main menu or **Save As** from the Main menu toolbar. The Save As dialog is displayed. Select **3D Manufacturing Format (*.3mf)** as the Save as type.

Click the **Options** button. The dialog box is displayed. View your options. For most parts, utilize the default setting.

Close the dialog box. Click **OK**. View the generate point cloud of the part. Click **Yes**. The file is now ready to be imported into your 3D printer software or print directly from SOLIDWORKS.

The include materials and include appearances option is not selected by default.

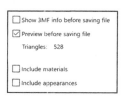

In SOLIDWORKS for a smoother STL file, change the Resolution to Custom. Change the deviation to 0.0005in (0.01mm). Change the angle to 5. Smaller deviations and angles produce a smoother file but increase the file size and print time.

What is a Slicer? How does a Slicer Work?

Creating the model is only the first step of 3D printing. You can 3D print directly from SOLIDWORKS. In addition, you can print directly from slicer software, if the slicer software supports this feature.

The slicer is software which converts a variety of file formats. The Additive Manufacturing file formats are: STL (*.stl), Additive Manufacturing (*.amf), and 3D Manufacturing format (*.3mf) file. The slicer turns these file formats into printing instructions (G-code) for your 3D printer.

Slicing is the process of turning the 3D model into a toolpath for the 3D printer. Most people call it slicing because the first thing the slicing engine does is cut the 3D model into thin horizontal layers. Open source slicing engines include Slic3r, Skeinforge, Netfabb, KISSkice, Cura, etc. to name a few.

The G-code file contains the instructions based on settings you choose and calculates how much material the printer will need and how long it will take to print.

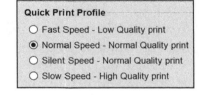

Slicer Parameters

Proper slicer parameters can mean the difference between a successful print, and a failed print. That's why it's important to know how slicers work and how various settings will affect the final print. There are open and close source slicers.

Layer Height

Layer height controls the resolution of the print. The setting specifies the height of each filament layer.

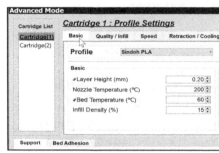

The higher values produce faster prints in lower resolution. Lower values produce slower prints in higher resolution.

Sindoh 3DWOX slicer - Version: 1.4.2102.0

The default value for (PLA) using the Sindoh 3DWOX 2X - Version: 1.4.2102.0 is .20mm. There are four default settings: Fast, Normal, Silent, and Slow. Settings range from .05mm - .4mm. Sindoh uses a .4mm nozzle with 1.75mm filament.

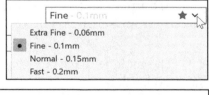

The default value for (PLA) using in the Ultimaker 3 Cura slicer - Version: 3.4.1 is .1mm. There are four default settings: Fast, Normal, Fine, and Extra Fine. Settings range from .06mm - .2mm.

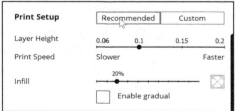

Ultimaker Cura slicer - Version: 3.4.1

Shell (Wall) Thickness

Wall thickness refers to the distance between one surface of the model and the opposite sheer surface. Set the wall thickness of the outside shell in the horizontal direction. Use in combination with the nozzle size to define the number of perimeter lines and the thickness of these perimeter lines.

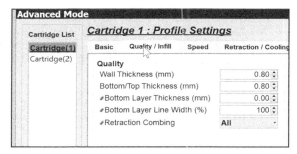

Sindoh 3DWOX slicer - Version: 1.4.2102.0

The default value for (PLA) using the Sindoh 3DWOX 2X - Version: 1.4.2102.0 is .80mm. The minimum is .40mm.

The default value for (PLA) using the Ultimaker 3 Cura - Version: 3.4.1 is 1mm.

Shell		
Wall Thickness	1	mm
Wall Line Count	3	
Top/Bottom Thickness	1	mm
Top Thickness	1	mm
Top Layers	10	
Bottom Thickness	1	mm
Bottom Layers	10	

Ultimaker Cura slicer - Version: 3.4.1

Infill Density/Overlap

Infill is the internal structure of your object, which can be as sparse or as substantial as you would like it to be. A higher percentage will result in a more solid object, so 100% (not recommended) infill will make your object completely solid, while 0% infill will give you something completely hollow.

The higher the infill percentage, the more material and longer the print time. It will also increase weight and material cost.

When using any infill percentage, a pattern is used to create a strong and durable structure inside the print. A few standard patterns are Rectilinear, Honeycomb, Circular, Tri-hexagon, Cubic, Octet, and Triangular. In general, use a 10% - 15% infill with a maximum infill of 60%. 100% infill is not recommended. Part warping can be a concern.

Shells are the outer layers of a print which make the walls of an object, prior to the various infill levels being printed within. The number of shells affects stability and translucency of the model.

View the different infill and number of shells between the two models.

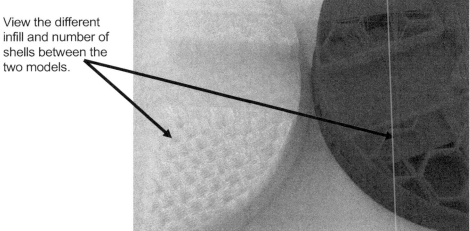

The default value for (PLA) using the Sindoh 3DWOX 2X - Version: 1.4.2102.0 is 15%.

The default value for (PLA) using the Ultimaker 3 Cura - Version: 3.4.1 is 20%.

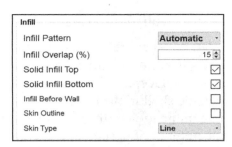

Sindoh 3DWOX slicer - Version: 1.4.2102.0

Strength corresponds to the maximum stress the print can take before breaking.

Print Speed

Print speed refers to the speed at which the extruder travels while it lays down filament.

Print speed affects the following areas: Infill speed, Wall speed, Top/Bottom speed, Initial layer travel speed, Raft print speed, and Maximum travel resolution.

The default value for (PLA) using the Sindoh 3DWOX 2X - Version: 1.4.2102.0 is 40mm/s.

The default value for (PLA) using the Ultimaker 3 Cura - Version: 3.4.1 is 70mm/s.

Supports

Generate structures to support parts (features) of the model which have overhangs to the build plate.

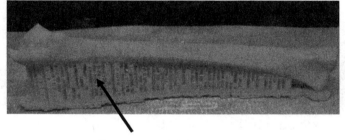

Without these structures, such parts would collapse during the print process.

In order to create an overhang at any angle less than vertical, your

Support material

printer offsets each successive layer. The lower the angle gets to horizontal, or 90°, the more each successive layer is offset.

General 45 degree rule. If your model has overhangs greater than 45 degrees, you need support material. If the part has numerous holes, sharp edges, long run (bridge), or thin bodies, support material may also be required.

Support Types

Touching Buildplate

The Touching Buildplate option provides supports only where the part touches the build plate. This reduces build time, clean up and support material.

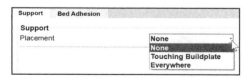

Sindoh 3DWOX slicer - Version: 1.4.2102.0

Use the touching Buildplate option when you have overhangs and tricky angles toward the bottom of a design, but do not wish to plug up holes, hollow spaces, or arches in the rest of the design.

Everywhere

The Everywhere option provides support material everywhere on the part, not only where the part touches the build plate.

Bed (Platform) Adhesion

Bed adhesion is one of the most important elements for getting a good 3D print. Set various options to ensure good bed adhesion and to prevent part warping. If needed, use blue painter's tape, hair spray or a glue stick.

Bed Adhesion Type- Raft

A Raft is a horizontal latticework of filament located underneath the part. A Raft is used to help the part stick to the build plate (heated or non-heated).

Rafts are also used to help stabilize thin tall parts with small build plate footprints.

When the print is complete, remove the part from the build plate. Peel the raft away from the part. If needed, use a scraper or spatula.

Bed Adhesion Type- Skirt

A Skirt is a layer of filament that surrounds the part with a 3mm - 4mm offset. The layer does not connect the part directly to the build plate. The Skirt primes the extruder and establishes a smooth flow of filament. In some slicers, the skirt is added automatically when you select the None option, for bed adhesion type.

Bed Adhesion Type- Brim

A Brim is basically like a Skirt for the part. A Brim has a zero offset from the part. It is a layer of filament laid down around the base of the part to increase its surface area. A Brim, however, does not extend underneath the part, which is the key difference between a brim and a raft.

Touching Build plate

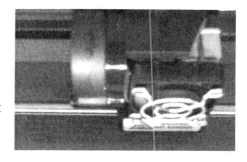

Part with a Raft on the Build plate

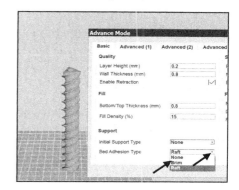

Part with a Skirt on the Build plate

Part Orientation

Insert the file into your printer's slicer. The model is displayed in the build plate area.

Depending on your printer's slicer software, you may or may not receive a message indicating the object is too large for the current build plate.

If the object is too large, you will need to scale it down or redesign it into separate parts.

Use caution when scaling critical features if you require fasteners or a minimum wall thickness.

You should always center the part and have it lay flat on the build plate. Bed adhesion is one of the most important elements for getting a good 3D print.

If you are printing more than one part, space them evenly on the plate or position them for a single build.

In SOLIDWORKS, lay the parts out in an assembly. Save the assembly as a part. Save the part file as an STL (*.stl), or Additive Manufacturing (*.amf), or 3D Manufacturing format (*.3mf).

Consideration should be used when printing an assembly. If the print takes 20 hours, and a failure happens after 19 hours, you just wasted a lot of time versus printing each part individually.

Example 1: Part Orientation

Part orientation is very important on build strength and the amount of raft and support material required for the build. Incorrect part orientations can lead to warping, curling, and delamination.

If maximizing strength is an issue, select the part orientation on the build plate so that the "grain" of the print is oriented to maximize the strength of the part.

Example 1: First Orientation - Vertical

In the first orientation (vertical), due to the number of holes and slots, additional support material is required (with minimum raft material) to print the model.

Removing the material in these geometrics can be very time consuming.

Example 1: Second Orientation - Horizontal

In the second orientation (horizontal), additional raft material is used, and the support material is reduced.

The raft material can be easily removed with a pair of needle-nose pliers and no support material clean-up is required for the holes and slots. Note: In some cases, raft material is not needed.

Example 2: Part Orientation

The lens part is orientated in a vertical position with the large face flat on the build plate. This reduces the required support material and ensures proper contact (maximum surface area) with the build plate.

Optimize Print Direction

Some slicers (Sindoh 3DWOX desktop) provide the ability to run an optimization print direction analysis of the part under their Advanced Mode. The areas in evaluations are **Thin Region**, **Area of Overhang Surface** and the **Amount of needed support** material. This can be very useful when you are unsure of the part orientation.

Suppress mates in an assembly to have the model lay flat on the build plate. If the model does not lay flat, you will require a raft and additional supports. This will increase build time and material cost.

First orientation - vertical

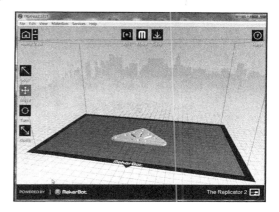

Second orientation - horizontal

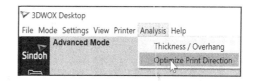

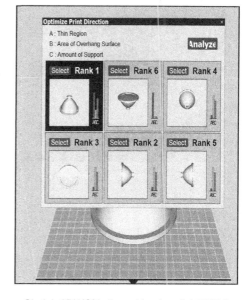

Sindoh 3DWOX slicer - Version: 1.4.2102.0

The needed support material is created mainly internal to the part to print the CBORE feature. Note: There is some outside support material on the top section of the part.

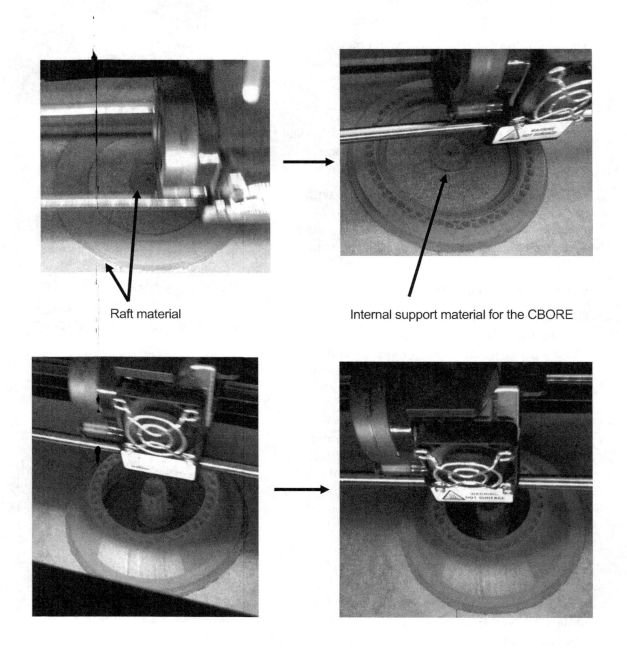

Raft material　　　　　　　　　Internal support material for the CBORE

Proper part orientation for thin parts will make the removal of the raft easier.

Remove the Model from the Build Plate

Non-heated Build Plate

Most Non-heated build plates are made from glass or metal. If needed, use blue painter's tape, glue stick or hair spray to assure good model adhesion. After the build, remove the plate from the printer.

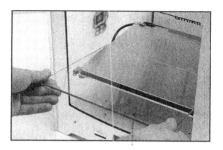

Ultimaker 3 glass build plate

Utilize a flat edge tool (thin steel spatula). Gently work under the part, and lift the part directly away from yourself. Clean the build plate. Return the plate. Re-level the plate after every build.

The Sindoh 3DWOX DP201 printer (non-heated) has a flexible magnet removable plate that does not require blue painter's tape, glue stick or hair spray to assure good model adhesion. This eliminates the need for scrapers or any sharp tools to remove the print. After the build, remove the plate and bend.

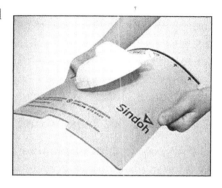

Sindoh 3DWOX DP201 flexible magnet build plate

🔆 Bed adhesion is one of the most important elements for getting a good 3D print.

Heated Build Plate

Most heated build plates are made from glass or metal. If needed, use a glue stick or hair spray to assure good model adhesion. If you have a heated build plate your temperatures can range between 30C - 90ºC depending on the material, so be careful. After the build, remove the plate from the printer. Utilize a flat edge tool (thin steel spatula). Gently work under the part, and lift the part directly away from yourself. Clean the build plate. Return the plate. Re-level the plate after every build.

Prusa i3 MK3 flexible metal magnet build plate

A few manufacturers (Sindoh 3DWOX 1 and 2X and Prusa i3 MK3) provide a flexible heated magnet metal build plate.

This eliminates the need for scrapers or any sharp tools most of the time. After the build, remove the flexible magnet plate and bend.

🔆 Most heated metal bed plates are coated. Over time, the coating wears and the plate needs to be replaced.

Sindoh 3DWOX 1 and 2X flexible metal magnet build plate

Know your Printer's Limitations

Overall part size can be an issue. Most affordable 3D printers typically are small enough to fit on your desktop. Typical build volumes range between 200 x 200 x 185 mm and 228 x 200 x 300 mm. The Sindoh 3DWOX 2X has one of the largest build volumes for an affordable desktop FFF 3D printer.

Sindoh 3DWOX 2X

There are features that are too small to be printed on a desktop 3D printer. An important, but often overlooked variable in what the printer can achieve is thread width. Thread width is determined by the diameter of the extruder nozzle. Most printers have a 0.4mm nozzle. A circle created by the printer is approximately two thread widths deep: 0.8mm thick with a 0.4mm nozzle to 1mm thick for a 0.5mm nozzle. A good rule of thumb is "The smallest feature you can create is double the thread width".

Tolerance for Interlocking Parts

For objects with multiple interlocking parts, design for a tolerance fit. Getting tolerances correct can be difficult using FFF technology.

In general, use the below suggested guidelines.

- Use ±0.1mm (±.004 in.) tolerance for a tight fit (press fit parts, connectors).

- Use ±0.2mm (±.008 in.) tolerance for a print in place (hinge).

- Use ±0.3mm (±.012 in.) tolerance for loose fit (pin in hole).

Test the fit yourself with the particular model to determine the right tolerance for the items you are creating and material you are using.

Tolerance may vary depending on filament type, manufacturer, color, humidity, build plate flatness, bed temperature, extruder temperature, etc.

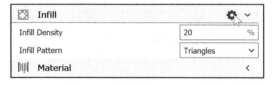

Ultimaker Cura slicer - Version: 3.4.1

General Printing Tips

Reduce Infill (Density/Overlap)

Infill is a settable variable in most Slicers. The amount of infill can affect the top layers, bottom layers, infill line distance and infill overlap.

Reduced infill can have a negative effect on part strength. There are always trade-offs.

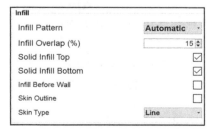

Sindoh 3DWOX slicer - Version: 1.4.2102.0

The material inside the part (infill) exerts a force on the entire printed part as it cools. More material increases cost of the part and build time.

Parts with a lower percentage of infill should have a lower internal force between layers and can reduce the chance of curling, cracking, and layer delamination along with a low build cost and time.

Sindoh 3DWOX - Standard enclosed build area

Control Build Area Temperature

For a consistent quality build, control the build area and environment temperature. Eliminate all drafts and control air flow that may cause a temperature gradient within the build area.

Changes in temperature during a build cycle can cause curling, cracking and layer delamination, especially on long thin parts. From 1000s of hours in 3D printing experience, having a top cover and sides along with a consistent room temperature provides the best and repeatable builds.

FlashForge 3D Printer Creator Pro

Some printer companies like Ultimaker and FlashForge, sell additional enclosure kits (doors, sides, and tops). This helps regulate the temperature and reduce drafts for improved print quality.

The kits also lower the sound of the printer and provide a more secure print area.

When troubleshooting issues with your printer, it is always best to know if the nozzle and heated bed are achieving the desired temperatures. A thermocouple and a thermometer come in handy. I prefer a Type-K Thermocouple connected to a multi-meter. A non-contact IR or Laser-based thermometer also works well.

One important thing to remember is that IR and Laser units are not 100% accurate when it comes to shiny reflective surfaces. A Type-K thermocouple can be taped to the nozzle or heated bed using Kapton tape, for a very accurate temperature measurement.

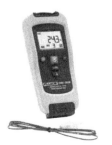

Cover/Door kit for Ultimaker 3

Add Pads

Sometimes, when you are printing a large flat object, you may view warping at the corners or extremities. One way to address this is to create small pads to your part during the modeling process. Create the model for the print. Think before you print and know your printer limitations. The pads can be any size and shape, but generally, diameter 10mm cylinders that are 1-2 layers thick work well. After the part is printed, remove them.

Makerbot Image

Unique Shape or a Large Part

If you need to make parts larger than your build area or create parts that have intricate projections, here are a few suggestions:

- Fuse smaller sections together using acetone (if using ABS). Glue if using PLA.

- Design smaller parts to be attached together (without hardware).

- Design smaller parts to be screwed together (with hardware).

Safe Zone Rule

Parts may have a safe zone. The safe zone is called "self-supporting" and no support material is required to build the part.

The safe zone can range between 30° to 150°. If the part's features are below 30° or greater than 150°, it should have support material during the build cycle. This is only a rule. Are there other factors to consider? Yes. They are layer thickness, extrusion speed, material type, length of the overhang along with the general model design of features.

Design your part for your printer. Use various modeling techniques (ribs, fillets, pads, etc.) during the design process to eliminate or to minimize the need for supports and clean up.

First Layer Not Sticking

One of the toughest aspects of 3D printing is to get your prints to stick to the build surface or bed platform. Investigate the following:

- Clean the bed. Remove any residue of tape, glue, hair spray, etc.

- Apply new blue painter's tape (non-heated), hair spray or glue to the build plate if needed.

- Level the bed (build platform). Perform an automatic (Assisted Bed Leveling) or manual leveling.

- Check extruder (hot end) temperature. Different filaments require different hot end temperatures.

- Check heated build plate temperature. Different filaments require different bed temperatures.

- Control the build area temperature around the printer. Eliminate all drafts.

- Layer height. Min layer height = 1/4 nozzle diameter. Max layer height = 1/2 nozzle diameter. Layer height too low, might cause the filament to be pushed back into the nozzle (plugging). If layer height is too high, the layers won't stick to the build plate.

Level Build Platform

An unleveled build platform will cause many headaches during a print. You can quickly check the platform by performing the business card test: use a single business card to judge the height of the extruder nozzle over the build platform. Achieve a consistent slight resistance when you position the business card between the tip of the extruder and the bed platform for all leveling positions.

Sindoh 3DWOX Printers

Most 3D printers have an automatic Assisted Bed Leveling feature as illustrated.

Minimize Internal Support

Design the part for the printer. Use various modeling techniques in SOLIDWORKS (ribs, fillets, pads, etc.) during the design process to eliminate or to minimize the need for support and final part clean up.

Design a Water-Tight Mesh

A water-tight mesh is achieved by having closed edges creating a solid volume. If you were to fill your geometry with water, would you see a leak? You may have to clean up any internal geometry that could have been left behind accidentally from Booleans.

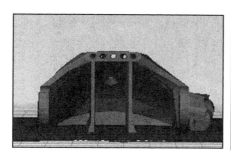
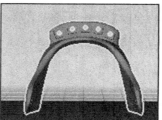
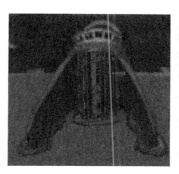

Clearance

If you are creating separate or interlocking parts, make sure there is a large enough distance between tight areas. 3D printing production makes moving parts without assembly a possibility. Take advantage of this strength by creating enough clearance that the model's pieces do not fuse together or trap support material inside.

In General (FFF Printers)

- Keep your software and firmware up to date.

- Think before you print. Design the model for your printer.

- Understand the printer's limitations. Adjust one thing at a time between prints and keep notes about the settings and effect on the print. Label test prints and take photographs.

- Control the build area and environment temperature. Eliminate all drafts and control air flow that may cause a temperature gradient within the build area.

- Level and clean the build plate before a build.

- Select the correct filament (material) for the application. Materials are still an area of active exploration.

- Set the correct extruder (hot end) and bed plate temperature.

- Most low-end 3D printers use an extruder (hot end) with Polyether ether Ketone (PEEK) or Polytetrafluoroethylene (Teflon). Both PEEK and PTFE begin to breakdown above 240°C and will burn and emit noxious fumes. Use an all metal hot end above 240°C.

- Select the correct part orientation.

- Suppress mates in an assembly to have the model lay flat on the build plate.

- Most parts can be printed successfully with 15 - 20% infill.

- Select the correct settings for your Slicer. If in doubt, use the factory default settings.

- Control the filament storage environment (temperature, humidity, etc.).

- General 45 degree rule. If your model doesn't have any overhangs greater than 45 degrees, you should not need support material. If the part has numerous holes, sharp edges, long run (bridge), or thin bodies, support material may be needed.

- If in doubt, create your first build with a raft and support.

- If needed, orientate the printed part on the bed plate for maximum strength (lines perpendicular to the force being applied).

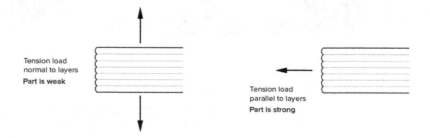

Print Directly from SOLIDWORKS

Download and install the printer drivers.

Download and install the SOLIDWORKS slicer Add-in. In this case, the SOLIDWORKS 3DWOX.

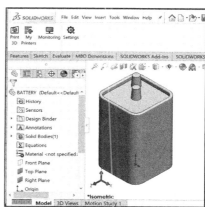

Screen shots and procedure will vary depending on the 3D printer manufacturer's slicer Add-in.

When the 3DWOX Add-in installation is complete, a 3DWOX tab is displayed in the CommandManager.

Open a SOLIDWORKS model.

Click the 3DWOX tab. View your options:

- **Print 3D**.

- **My Printers**.

- **Monitoring**.

- **Settings**.

The **Print 3D** button provides the ability to access the slicer within SOLIDWORKS. The SOLIDWORKS part model is automatically converted into an STL file and appears on the selected (Settings button) printer bed. The Sindoh 3DWOX slicer is displayed.

The **My Printers** button provides the ability to manage your printers (printer name, IP and availability) over a network.

The **Monitoring** button provides the ability to connect to the network and view your printing real time with an internal camera.

The **Settings** button displays the printers that you are connected to.

Certified SOLIDWORKS Associate Additive Manufacturing (CSWA-AM)

The Certified SOLIDWORKS Associate Additive Manufacturing (CSWA-AM) exam indicates a foundation in and apprentice knowledge of today's 3D printing technology and market.

The CSWA-AM exam is meant to be taken after the completion of the 10-part learning path located on MySOLIDWORKS.com.

The learning content is free, but the creation of a MySOLIDWORKS account is needed in order to access the content.

Lessons	Status	Languages
Introduction To Additive Manufacturing	Completed	ENG
Machine Types	Completed	ENG
Materials	Completed	ENG
Model Preparation	Completed	ENG
File Export Settings	Completed	ENG
Machine Preparation	Completed	ENG
Printing the Part	Completed	ENG
Post Printing	Completed	ENG
Part Finishing	Completed	ENG
Software Options	Completed	ENG

The lessons cover: **Introduction to Additive Manufacturing** (7 minutes), **Machine Types** (8 minutes), **Materials** (7 minutes), **Model Preparation** (9 minutes), **File Export Settings** (8 minutes), **Machine Preparation** (7 minutes), **Printing the Part** (6 minutes), **Post Printing** (7 minutes), **Part Finishing** (9 minutes) and **Software Options** (9 minutes).

After each lesson, there is a short online quiz covering the topic area. The lessons are focused on two types of 3D printer technology; Fused Filament Fabrication (FFF) and STereoLithography (SLA). There are a few questions on Selective Laser Sintering (SLS) technology and available software-based printing aids.

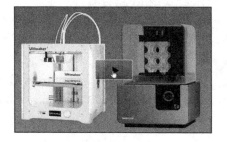

The exam covers the Ultimaker 3 (FFF) and the FormLabs Form 2 (SLA) machine as examples.

Fused Filament Fabrication (FFF) and Fused Deposition Modeling (FDM) are used interchangeably in the certification exam.

The CSWA-AM (Additive Manufacturing) exam covers numerous areas: material types, printing technologies, machine types and processes, part design and orientation for 3D printing, printer preparation, post printing finishing STereoLithography (SLA), Slicer software features and functionality and available software-based printing aids.

The exam is 50 questions. Each question is worth 2 points. You are allowed to answer the questions in any order. Total exam time is 60 minutes. You need a minimum passing score of 80 or higher. The exam is out of 100 points.

Summary

Stereolithography (SLA) was the world's first 3D printing technology. It was introduced in 1988 by 3D Systems, Inc., based on work by inventor Charles Hull.

SLA is one of the most popular resin-based 3D printing technologies for professionals today.

Selective Laser Sintering (SLS) technology is the most common additive manufacturing technology for industrial powder base applications. SLS fuses particles together layer by layer through a high energy pulse laser. Similar to SLA, this process starts with a tank full of bulk material but is in a powder form vs. liquid.

Sindoh 3DWOX 1

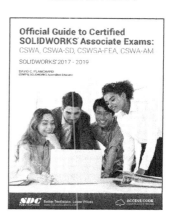

Fused filament fabrication (FFF) is an additive manufacturing technology used for building three-dimensional prototypes or models layer by layer with a range of thermoplastics.

FFF technology is the most widely used form of 3D desktop printing at the consumer level, fueled by students, hobbyists and office professionals.

The two most dominant plastics for FFF technology are PLA (Polylactic acid) and ABS (Acrylonitrile-Butadiene-Styrene).

With 1000s of hours using multiple low cost FFF 3D printers, we have found that learning about additive print technology is a great experience. Students face a multitude of obstacles with their first 3D prints. Understanding what went wrong and knowing the capabilities of the 3D printer produces positive results.

Never focus too much on one single issue. These machines are complex, and trouble often arises from multiple reasons. A slipping filament may not only be caused by a bad gear drive, but also by an obstructed nozzle, a wrong feed value, a too low (or too high) temperature or a combination of all these.

Always store filament in an airtight plastic bag (container) with a few silica gel packs. Place the bag in a dry, dark, temperature controlled environment. As an extra precaution, filament manufacturers often recommend using up rolls as soon as possible.

It's best to monitor the first 3 - 5 layers when starting a print. Never assume you will have perfect bed adhesion. If there is a bed adhesion issue, check to see if the bed is level. Check the bed and extruder (hot end) temperatures and then either apply blue painter's tape (non-heated), a glue stick or hair spray. You can also add a raft to the part.

Always clean and level the build plate before every print.

As printed parts cool (PLA, ABS and Nylon), various areas of the object cool at different rates. Depending on the model being printed and the filament material, this effect can lead to warping, curling and or layer delamination.

Acetone is often used in post processing to smooth ABS, also giving the part a glossy finish. ABS can be sanded and is often machined (for example, drilled) after printing. PLA can also be sanded and machined, however greater care is required.

For high temperature applications, ABS (glass transition temperature of 105°C) is more suitable than PLA (glass transition temperature of 60°C). PLA can rapidly lose its structural integrity as it approaches 55C - 60°C.

Design the part for your printer. Try to use the 45 degree rule. If your model doesn't have any overhangs greater than 45 degrees, you should not need support. There are exceptions to the 45 degree rule. The most common ones are straight overhangs, and fully suspended islands.

Use various CAD modeling techniques (ribs, fillets, pads, etc.) during the design process to eliminate or to minimize the need for supports and clean up.

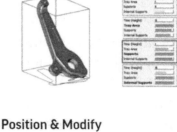

Native Data Transfer

Maintain Design Integrity

Position & Modify

Ensure Printable Geometry

From a structural situation, select the correct % infill, infill pattern and orientate the part on the build plate to print the layers perpendicular in relation to the movement of the build platform.

Design Supports

Ensure Quality Prints with Minimal Supports

Optimize Structure

Minimize Weight & Material Usage and Apply Surface Textures

3DXpert is a SOLIDWORKS Add-in. 3DXpert for SOLIDWORKS provides an extensive toolset to analyze, prepare and optimize your design for additive manufacturing. It also provides the ability to print an assembly file as a single part.

Key tools are:

- **Native Data Transfer**.

- **Position & Modification**.

- **Optimize Structure**.

- **Arrange Build Plate**.

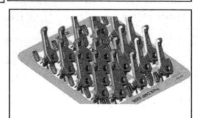

Arrange Build Plate

Best Utilization of Tray Area and Printer Time

Appendix

SOLIDWORKS Keyboard Shortcuts

Below are some of the pre-defined keyboard shortcuts in SOLIDWORKS:

Action:	Key Combination:
Model Views	
Rotate the model horizontally or vertically	**Arrow** keys
Rotate the model horizontally or vertically 90 degrees	**Shift** + **Arrow** keys
Rotate the model clockwise or counterclockwise	**Alt** + left of right **Arrow** keys
Pan the model	**Ctrl** + **Arrow** keys
Magnifying glass	**g**
Zoom in	**Shift + z**
Zoom out	**z**
Zoom to fit	**f**
Previous view	**Ctrl + Shift + z**
View Orientation	
View Orientation menu	**Spacebar**
Front view	**Ctrl + 1**
Back view	**Ctrl + 2**
Left view	**Ctrl + 3**
Right view	**Ctrl + 4**
Top view	**Ctrl + 5**
Bottom view	**Ctrl + 6**
Isometric view	**Ctrl + 7**
NormalTo view	**Ctrl + 8**
Selection Filters	
Filter edges	**e**
Filter vertices	**v**
Filter faces	**x**
Toggle Selection Filter toolbar	**F5**
Toggle selection filters on/off	**F6**
File menu items	
New SOLIDWORKS document	**Ctrl + n**
Open document	**Ctrl + o**
Open From Web Folder	**Ctrl + w**
Make Drawing from Part	**Ctrl + d**
Make Assembly from Part	**Ctrl + a**
Save	**Ctrl +s**
Print	**Ctrl + p**
Additional items	
Access online help inside of PropertyManager or dialog box	**F1**
Rename an item in the FeatureManager design tree	**F2**

Action:	Key Combination:
Rebuild the model	**Ctrl + b**
Force rebuild - Rebuild the model and all its features	**Ctrl + q**
Redraw the screen	**Ctrl + r**
Cycle between open SOLIDWORKS document	**Ctrl + Tab**
Line to arc/arc to line in the Sketch	**a**
Undo	**Ctrl + z**
Redo	**Ctrl + y**
Cut	**Ctrl + x**
Copy	**Ctrl + c**
Paste	**Ctrl + v**
Delete	**Delete**
Next window	**Ctrl + F6**
Close window	**Ctrl + F4**
View previous tools	**s**
Selects all text inside an Annotations text box	**Ctrl + a**

In a sketch, the **Esc** key un-selects geometry items currently selected in the Properties box and Add Relations box.

In the model, the **Esc** key closes the PropertyManager and cancels the selections.

Use the **g** key to activate the Magnifying glass tool. Use the Magnifying glass tool to inspect a model and make selections without changing the overall view.

Use the **s** key to view/access previous command tools in the Graphics window.

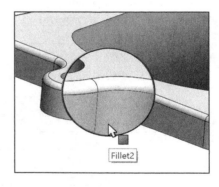

Fillet2

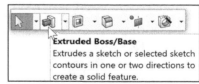

Extruded Boss/Base
Extrudes a sketch or selected sketch contours in one or two directions to create a solid feature.

Modeling - Best Practices

Best practices are simply ways of bringing about better results in easier, more reliable ways. The Modeling - Best Practice list is a set of rules helpful for new users and users who are trying to experiment with the limits of the software.

These rules are not inflexible, but conservative starting places; they are concepts that you can default to but that can be broken if you have good reason. The following is a list of suggested best practices:

- Create a folder structure (parts, drawings, assemblies, simulations, etc.). Organize into project or file folders.

- Construct sound document templates. The document template provides the foundation that all models are built on. This is especially important if working with other SOLIDWORKS users on the same project; it will ensure consistency across the project.

- Generate unique part filenames. SOLIDWORKS assemblies and drawings may pick up incorrect references if you use parts with identical names.

- Apply Custom Properties. Custom Properties is a great way to enter text-based information into the SOLIDWORKS parts. Users can view this information from outside the file by using applications such as Windows Explorer, SOLIDWORKS Explorer, and Product Data Management (PDM) applications.

- Understand part orientation. When you create a new part or assembly, the three default Planes (Front, Right and Top) are aligned with specific views. The plane you select for the Base sketch determines the orientation.

- Learn to sketch using automatic relations.

- Limit your usage of the Fixed constraint.

- Add geometric relations, then dimensions in a 2D sketch. This keeps the part from having too many unnecessary dimensions. This also helps to show the design intent of the model. Dimension what geometry you intend to modify or adjust.

- Fully define all sketches in the model. However, there are times when this is not practical, generally when using the Spline tool to create a freeform shape.

- When possible, make relations to sketches or stable reference geometry; such as the Origin or standard planes, instead of edges or faces. Sketches are far more stable than faces, edges, or model vertices, which change their internal ID at the slightest change and may disappear entirely with fillets, chamfers, split lines, and so on.

- Do not dimension to edges created by fillets or other cosmetic or temporary features.

- Apply names to sketches, features, dimensions, and mates that help to make their function clear.

- When possible, use feature fillets and feature patterns rather than sketch fillets and sketch patterns.

- Apply the Shell feature before the Fillet feature, and the inside corners remain perpendicular.

- Apply cosmetic fillets and chamfers last in the modeling procedure.

- Combine fillets into as few fillet features as possible. This enables you to control fillets that need to be controlled separately; such as fillets to be removed and simplified configurations.

- Create a simplified configuration when building very complex parts or working with large assemblies.

- Use symmetry during the modeling process. Utilize feature patterns and mirroring when possible. Think End Conditions.

- Use global variables and equations to control commonly applied dimensions (design intent).

- Add comments to equations to document your design intent. Place a single quote (') at the end of the equation, then enter the comment. Anything after the single quote is ignored when the equation is evaluated.

- Avoid redundant mates. Although SOLIDWORKS allows some redundant mates (all except distance and angle), these mates take longer to solve and make the mating scheme harder to understand and diagnose if problems occur.

- Fix modeling errors in the part or assembly when they occur. Errors cause rebuild time to increase, and if you wait until additional errors exist, troubleshooting will be more difficult.

- Create a Library of Standardized notes and parts.

- Utilize the Rollback bar. Troubleshoot feature and sketch errors from the top of the design tree.

- Determine the static and dynamic behavior of mates in each sub-assembly before creating the top level assembly.

- Plan the assembly and sub-assemblies in an assembly layout diagram. Group components together to form smaller sub-assemblies.

- When you create an assembly document, the base component should be fixed, fully defined or mated to an axis about the assembly origin.

- In an assembly, group fasteners into a folder at the bottom of the FeatureManager. Suppress fasteners and their assembly patterns to save rebuild time and file size.

- When comparing mass, volume and other properties with assembly visualization, utilize similar units.

- Use limit mates sparingly because they take longer to solve and whenever possible, mate all components to one or two fixed components or references. Long chains of components take longer to solve and are more prone to mate errors.

Helpful On-line Information

The SOLIDWORKS URL:
http://www.SOLIDWORKS.com
contains information on Local
Resellers, Solution Partners,
Certifications, SOLIDWORKS users
groups and more.

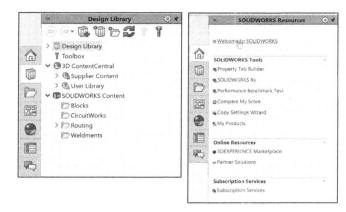

Access 3D ContentCentral using the
Task Pane to obtain engineering
electronic catalog model and part
information.

Use the SOLIDWORKS Resources tab in the Task Pane to obtain access to Customer
Portals, Discussion Forums, User Groups, Manufacturers, Solution Partners, Labs and
more.

Helpful on-line SOLIDWORKS information is available from the following URLs:

- http://www.swugn.org/

List of all SOLIDWORKS User groups.

- https://www.solidworks.com/sw/education/certification-programs-cad-students.htm

The SOLIDWORKS Academic
Certification Programs.

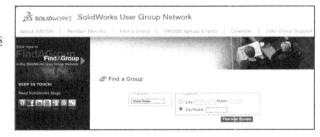

- http://www.solidworks.com/sw/indus
  tries/education/engineering-
  education-software.htm

The SOLIDWORKS Education
Program.

- https://solidworks.virtualtester.com/#
  home_button

The SOLIDWORKS Certification Center - Virtual tester site.

*On-line tutorials are for educational purposes only. Tutorials are copyrighted by their respective owners.

SOLIDWORKS Document Types

SOLIDWORKS has three main document file types: Part, Assembly and Drawing, but there are many additional supporting types that you may want to know. Below is a brief list of these supporting file types:

Design Documents	**Description**
.sldprt	SOLIDWORKS Part document
.slddrw	SOLIDWORKS Drawing document
.sldasm	SOLIDWORKS Assembly document

Templates and Formats	**Description**
.asmdot	Assembly Template
.asmprp	Assembly Template Custom Properties tab
.drwdot	Drawing Template
.drwprp	Drawing Template Custom Properties tab
.prtdot	Part Template
.prtprp	Part Template Custom Properties tab
.sldtbt	General Table Template
.slddrt	Drawing Sheet Template
.sldbombt	Bill of Materials Template (Table-based)
.sldholtbt	Hole Table Template
.sldrevbt	Revision Table Template
.sldwldbt	Weldment Cutlist Template
.xls	Bill of Materials Template (Excel-based)

Library Files	**Description**
.sldlfp	Library Part file
.sldblk	Blocks

Other	**Description**
.sldstd	Drafting standard
.sldmat	Material Database
.sldclr	Color Palette File
.xls	Sheet metal gauge table

CSWA Section: Answer key

Chapter 7

1. Identify the illustrated Drawing view.

- The correct answer is B: Alternative Position View.

2. Identify the illustrated Drawing view.

- The correct answer is B: Break View.

3. Identify the illustrated Drawing view.

- The correct answer is D: Aligned Section View.

4. Identify the view procedure. To create the following view, you need to insert a:

- The correct answer is B: Closed Profile: Spline.

5. Identify the view procedure. To create the following view, you need to insert a:

- The correct answer is B: Closed Spline.

6. Identify the illustrated view type.

- The correct answer is A: Crop View.

7. To create View B from Drawing View A insert which View Type?

- The correct answer is Aligned Section View.

8. To create View B it is necessary to sketch a closed spline on View A and insert which View type?

- The correct answer is Broken out Section View.

9. To create View B it is necessary to sketch a closed spline on View A and insert which View type?

- The correct answer is Horizontal Break View.

Chapter 8

1. Part 1. Calculate the overall mass of the part, volume, and locate the Center of mass with the provided illustrated information.

- Overall mass of the part = 1280.91 grams

- Volume of the part = 474411.54 cubic millimeters

- Center of Mass Location: X = 0.00 millimeters, Y = -29.17 millimeters, Z = 3.18 millimeters

2. Part 2. Calculate the overall mass of the part, volume, and locate the Center of mass with the provided information.

- Overall mass of the part = 248.04 grams

- Volume of the part = 91868.29 cubic millimeters

- Center of Mass Location: X = -51.88 millimeters, Y = 24.70 millimeters, Z = 29.47 millimeters

3. Part 3. Calculate the overall mass of the part, volume, and locate the Center of mass with the provided information.

- Overall mass of the part = 0.45 pounds, Volume of the part = 4.60 cubic inches and the Center of Mass Location: X = 0.17 inches, Y = 0.39 inches, Z = 0.00 inches

4. Part 4. Calculate the overall mass of the part, volume, and locate the Center of mass with the provided information.

- Overall mass of the part = 0.28 pounds, Volume of the part = 2.86 cubic inches and the Center of Mass Location: X = 0.70 inches, Y = 0.06 inches, Z = 0.00 inches

```
Mass properties of Homework 8-1
  Configuration: Default
  Coordinate system: -- default --

Density = 0.00 grams per cubic millimeter

Mass = 1280.91 grams

Volume = 474411.54 cubic millimeters

Surface area = 86851.60 square millimeters

Center of mass: ( millimeters )
  X = 0.00
  Y = -29.17
  Z = 3.18
```

```
Mass properties of Homework 8-2
  Configuration: Default
  Coordinate system: -- default --

Density = 0.00 grams per cubic millimeter

Mass = 248.04 grams

Volume = 91868.29 cubic millimeters

Surface area = 29140.64  square millimeters

Center of mass: ( millimeters )
  X = -51.88
  Y = 24.70
  Z = 29.47
```

```
Mass properties of Homework 8-3
  Configuration: Default
  Coordinate system: -- default --

Density = 0.10 pounds per cubic inch

Mass = 0.45 pounds

Volume = 4.60 cubic inches

Surface area = 32.66 square inches

Center of mass: ( inches )
  X = 0.17
  Y = 0.39
  Z = 0.00
```

```
Mass properties of Homework 8-4
  Configuration: Default
  Coordinate system: -- default --

Density = 0.10 pounds per cubic inch

Mass = 0.28 pounds

Volume = 2.86 cubic inches

Surface area = 22.54 square inches

Center of mass: ( inches )
  X = 0.70
  Y = 0.06
  Z = 0.00
```

5. Part 5. Calculate the overall mass and volume of the part with the provided information.

- Overall mass of the part = 888.48grams. Overall volume of the part = 111059.43 cubic millimeters.

6. Part 5A Modify. Calculate the Center of mass with a new Coordinate System.

- X= -80.39 millimeters, Y = 15.93 millimeters, Z = -22.65 millimeters.

7. Part 5B Modify. Modify the model. Calculate the mass and volume of the model.

- Mass = 309.75 grams. Volume = 114721.22 cubic millimeters

8. Part 6. Calculate the overall mass and volume of the part with the provided information.

- Overall mass of the part = 509.92 grams. Overall volume of the part = 188860.93 cubic millimeters.

9. Part 6 A modify. Calculate the overall mass and volume of the part with the provided information.

- Overall mass of the part = 1465.70 grams. Overall volume of the part = 187910.60 cubic millimeters.

10. Part 6 B modify. Create a new coordinate system. Enter the Center of Mass location.

- X = 31.29 millimeters
- Y = 16.45 millimeters
- Z = -48.67 millimeters

Density = 0.01 grams per cubic millimeter

Mass = 888.48 grams

Volume = 111059.43 cubic millimeters

Surface area = 25814.97 square millimeters

Center of mass: (millimeters)
   X = -15.39
   Y = 15.93
   Z = -2.65

Mass = 888.48 grams

Volume = 111059.43 cubic millimeters

Surface area = 25814.97 square millimeters

Center of mass: (millimeters)
   X = -80.39
   Y = 15.93
   Z = -22.65

Mass = 309.75 grams

Volume = 114721.22 cubic millimeters

Surface area = 27094.41 square millimeters

Mass = 509.92 grams

Volume = 188860.93 cubic millimeters

Surface area = 32545.06 square millimeters

Center of mass: (millimeters)
   X = 31.39
   Y = 16.55
   Z = 1.37

Configuration: modified
   Coordinate system: -- default --

Density = 0.01 grams per cubic millimeter

Mass = 1465.70 grams

Volume = 187910.60 cubic millimeters

Surface area = 32373.16 square millimeters

Center of mass: (millimeters)
   X = 31.29
   Y = 16.45
   Z = 1.33

Configuration: modified
   Coordinate system: Coordinate System1

Density = 0.01 grams per cubic millimeter

Mass = 1465.70 grams

Volume = 187910.60 cubic millimeters

Surface area = 32373.16 square millimeters

Center of mass: (millimeters)
   X = 31.29
   Y = 16.45
   Z = -48.67

11. Part 7. Calculate the overall mass and volume of the part with the provided information.

- Mass = 4079.32 grams

- Volume = 522989.22 cubic millimeter

```
Mass properties of Homework 8-7
    Configuration: Default
    Coordinate system: -- default --

Density = 0.01 grams per cubic millimeter

Mass = 4079.32 grams

Volume = 522989.22 cubic millimeters

Surface area = 92824.57 square millimeters

Center of mass: ( millimeters )
    X = 35.91
    Y = 0.00
    Z = 0.70
```

Chapter 9

1. Calculate the overall mass of the part, volume, and locate the Center of mass with the provided information.

- Overall mass of the part = 1.99 pounds

- Volume of the part = 6.47 cubic inches

- Center of Mass Location: X = 0.00 inches, Y = 0.00 inches, Z = 1.49 inches

```
Mass properties of Homework Problem 9-1
    Configuration: Default
    Coordinate system: -- default --

Density = 0.31 pounds per cubic inch

Mass = 1.99 pounds

Volume = 6.47 cubic inches

Surface area = 48.10 square inches

Center of mass: ( inches )
    X = 0.00
    Y = 0.00
    Z = 1.49
```

2. Calculate the overall mass of the part, volume, and locate the Center of mass with the provided information.

- Overall mass of the part = 279.00 grams

- Volume of the part = 103333.73 cubic millimeters

- Center of Mass Location: X = 0.00 millimeters, Y = 0.00 millimeters, Z = 21.75 millimeters

```
Mass = 279.00 grams

Volume = 103333.73 cubic millimeters

Surface area = 29853.94 square millimeters

Center of mass: ( millimeters )
    X = 0.00
    Y = 0.00
    Z = 21.75
```

3. Calculate the overall mass of the part, volume and locate the Center of mass with the provided information.

- Overall mass of the part = 2040.57 grams

- Volume of the part = 755765.04 cubic millimeters

- Center of Mass Location: X = -0.71 millimeters, Y = 16.66 millimeters, Z = -9.31 millimeters

```
Mass = 2040.57 grams

Volume = 755765.04 cubic millimeters

Surface area = 75833.18 square millimeters

Center of mass: ( millimeters )
    X = -0.71
    Y = 16.66
    Z = -9.31
```

4. Calculate the overall mass of the part, volume and locate the Center of mass with the provided information. Create Coordinate System1 to locate the Center of mass for the model.

-   Overall mass of the part = 2040.57 grams

-   Volume of the part = 755765.04 cubic millimeters

-   Center of Mass Location: X = 49.29 millimeters, Y = 16.66 millimeters, Z = -109.31 millimeters

```
Mass = 2040.57 grams

Volume = 755765.04 cubic millimeters

Surface area = 75833.18  square millimeters

Center of mass: ( millimeters )
    X = 49.29
    Y = 16.66
    Z = -109.31
```

5. Calculate the overall mass of the part, volume and locate the Center of mass with the provided information.

-   Overall mass of the part = 37021.48 grams

-   Volume of the part = 13711657.53 cubic millimeters

-   Center of Mass Location: X = 0.00 millimeters, Y = 0.11 millimeters, Z = 0.00 millimeters

```
Mass = 37021.48 grams

Volume = 13711657.53 cubic millimeters

Surface area = 539553.61  square millimeters

Center of mass: ( millimeters )
    X = 0.00
    Y = 0.11
    Z = 0.00
```

6. Create a new Coordinate system with the provided information. Center of Mass location:

-   X = 0.00 millimeters

-   Y = 0.11 millimeters

-   Z = 0.00 millimeters

```
Mass = 37021.48 grams

Volume = 13711657.53 cubic millimeters

Surface area = 539553.61  square millimeters

Center of mass: ( millimeters )
    X = 0.00
    Y = 0.11
    Z = 0.00
```

7. Build the provided model. Calculate the overall mass and volume of the part.

-   Mass = 3442.00 grams

-   Volume = 447012.75 cubic millimeters

```
Mass properties of Homework Part 9-7
    Configuration: Default
    Coordinate system: -- default --

Density = 0.01 grams per cubic millimeter

Mass = 3442.00 grams

Volume = 447012.75 cubic millimeters

Surface area = 67768.84  square millimeters

Center of mass: ( millimeters )
    X = 0.00
    Y = 50.00
    Z = 0.00
```

Notes:

GLOSSARY

Alphabet of Lines: Each line on a technical drawing has a definite meaning and is drawn in a certain way. The line conventions recommended by the American National Standards Institute (ANSI) are presented in this text.

Alternate Position View: A drawing view superimposed in phantom lines on the original view. Utilized to show range of motion of an assembly.

Anchor Point: The origin of the Bill of Material in a sheet format.

Annotation: An annotation is a text note or a symbol that adds specific information and design intent to a part, assembly, or drawing. Annotations in a drawing include specific note, hole callout, surface finish symbol, datum feature symbol, datum target, geometric tolerance symbol, weld symbol, balloon and stacked balloon, center mark, centerline marks, area hatch and block.

ANSI: American National Standards Institute.

Area Hatch: Apply a crosshatch pattern or solid fill to a model face, to a closed sketch profile, or to a region bounded by a combination of model edges and sketch entities. Area hatch can be applied only in drawings.

ASME: American Society of Mechanical Engineering, publisher of ASME Y14 Engineering Drawing and Documentation Practices that controls drawing, dimensioning and tolerancing.

Assembly: An assembly is a document in which parts, features and other assemblies (sub-assemblies) are put together. A part in an assembly is called a component. Adding a component to an assembly creates a link between the assembly and the component. When SOLIDWORKS opens the assembly, it finds the component file to show it in the assembly. Changes in the component are automatically reflected in the assembly. The filename extension for a SOLIDWORKS assembly file name is *.sldasm.

Attachment Point: An attachment point is the end of a leader that attaches to an edge, vertex, or face in a drawing sheet.

AutoDimension: The Autodimension tool provides the ability to insert reference dimensions into drawing views such as baseline, chain, and ordinate dimensions.

Auxiliary View: An Auxiliary View is similar to a Projected View, but it is unfolded normal to a reference edge in an existing view.

AWS: American Welding Society, publisher of AWS A2.4, Standard Location of Elements of a Welding Symbol.

Axonometric Projection: A type of parallel projection, more specifically a type of orthographic projection, used to create a pictorial drawing of an object, where the object is rotated along one or more of its axes relative to the plane of projection.

Balloon: A balloon labels the parts in the assembly and relates them to item numbers on the bill of materials (BOM) added in the drawing. The balloon item number corresponds to the order in the Feature Tree. The order controls the initial BOM Item Number.

Baseline Dimensions: Dimensions referenced from the same edge or vertex in a drawing view.

Bill of Materials: A table inserted into a drawing to keep a record of the parts and materials used in an assembly.

Block: A symbol in the drawing that combines geometry into a single entity.

BOM: Abbreviation for Bill of Materials.

Broken-out Section: A broken-out section exposes inner details of a drawing view by removing material from a closed profile. In an assembly, the Broken-out Section displays multiple components.

CAD: The use of computer technology for the design of objects, real or virtual. CAD often involves more than just shapes.

Cartesian Coordinate System: Specifies each point uniquely in a plane by a pair of numerical coordinates, which are the signed distances from the point to two fixed perpendicular directed lines, measured in the same unit of length. Each reference line is called a coordinate axis or just axis of the system, and the point where they meet is its origin.

Cell: Area to enter a value in an EXCEL spreadsheet, identified by a Row and Column.

Center Mark: A cross that marks the center of a circle or arc.

Centerline: An axis of symmetry in a sketch or drawing displayed in a phantom font.

CommandManager: The CommandManager is a Context-sensitive toolbar that dynamically updates based on the toolbar you want to access. By default, it has toolbars embedded in it based on the document type. When you click a tab below the Command Manager, it updates to display that toolbar. For example, if you click the Sketch tab, the Sketch toolbar is displayed.

Component: A part or sub-assembly within an assembly.

ConfigurationManager: The ConfigurationManager is located on the left side of the SOLIDWORKS window and provides the means to create, select and view multiple configurations of parts and assemblies in an active document. You can split the

ConfigurationManager and either display two ConfigurationManager instances, or combine the ConfigurationManager with the FeatureManager design tree, PropertyManager or third party applications that use the panel.

Configurations: Variations of a part or assembly that control dimensions, display and state of a model.

Coordinate System: SOLIDWORKS uses a coordinate system with origins. A part document contains an original origin. Whenever you select a plane or face and open a sketch, an origin is created in alignment with the plane or face. An origin can be used as an anchor for the sketch entities, and it helps orient perspective of the axes. A three-dimensional reference triad orients you to the X, Y, and Z directions in part and assembly documents.

Copy and Paste: Utilize copy/paste to copy views from one sheet to another sheet in a drawing or between different drawings.

Cosmetic Thread: An annotation that represents threads.

Crosshatch: A pattern (or fill) applied to drawing views such as section views and broken-out sections.

Cursor Feedback: The system feedback symbol indicates what you are selecting or what the system is expecting you to select. As you move the mouse pointer across your model, system feedback is provided.

Datum Feature: An annotation that represents the primary, secondary and other reference planes of a model utilized in manufacturing.

Depth: The horizontal (front to back) distance between two features in frontal planes. Depth is often identified in the shop as the thickness of a part or feature.

Design Table: An Excel spreadsheet that is used to create multiple configurations in a part or assembly document.

Detail View: A portion of a larger view, usually at a larger scale than the original view. Create a detail view in a drawing to display a portion of a view, usually at an enlarged scale. This detail may be of an orthographic view, a non-planar (isometric) view, a section view, a crop view, an exploded assembly view or another detail view.

Detailing: Detailing refers to the SOLIDWORKS module used to insert, add and modify dimensions and notes in an engineering drawing.

Dimension Line: A line that references dimension text to extension lines indicating the feature being measured.

Dimension Tolerance: Controls the dimension tolerance values and the display of non-integer dimensions. The tolerance types are *None, Basic, Bilateral, Limit, Symmetric, MIN, MAX, Fit, Fit with tolerance* or *Fit (tolerance only)*.

Dimension: A value indicating the size of the 2D sketch entity or 3D feature. Dimensions in a SOLIDWORKS drawing are associated with the model, and changes in the model are reflected in the drawing, if you DO NOT USE DimXpert.

Dimensioning Standard - Metric: - ASME standards for the use of metric dimensioning required all the dimensions to be expressed in millimeters (mm). The (mm) is not needed on each dimension, but it is used when a dimension is used in a notation. No trailing zeroes are used. The Metric or International System of Units (S.I.) unit system in drafting is also known as the Millimeter, Gram Second (MMGS) unit system.

Dimensioning Standard - U.S: - ASME standard for U.S. dimensioning uses the decimal inch value. When the decimal inch system is used, a zero is not used to the left of the decimal point for values less than one inch, and trailing zeroes are used. The U.S. unit system is also known as the Inch, Pound, Second (IPS) unit system.

DimXpert for Parts: A set of tools that applies dimensions and tolerances to parts according to the requirements of the ASME Y.14.41-2009 standard.

DimXpertManager: The DimXpertManager lists the tolerance features defined by DimXpert for a part. It also displays DimXpert tools that you use to insert dimensions and tolerances into a part. You can import these dimensions and tolerances into drawings. DimXpert is not associative.

Document: In SOLIDWORKS, each part, assembly, and drawing is referred to as a document, and each document is displayed in a separate window.

Drawing Sheet: A page in a drawing document.

Drawing Template: A document that is the foundation of a new drawing. The drawing template contains document properties and user-defined parameters such as sheet format. The extension for the drawing template filename is .DRWDOT.

Drawing: A 2D representation of a 3D part or assembly. The extension for a SOLIDWORKS drawing file name is .SLDDRW. Drawing refers to the SOLIDWORKS module used to insert, add, and modify views in an engineering drawing.

Edit Sheet Format: The drawing sheet contains two modes. Utilize the Edit Sheet Format command to add or modify notes and Title block information. Edit in the Edit Sheet Format mode.

Edit Sheet: The drawing sheet contains two modes. Utilize the Edit Sheet command to insert views and dimensions.

eDrawing: A compressed document that does not require the referenced part or assembly. eDrawings are animated to display multiple views in a drawing.

Empty View: An Empty View creates a blank view not tied to a part or assembly document.

Engineering Graphics: Translates ideas from design layouts, specifications, rough sketches, and calculations of engineers & architects into working drawings, maps, plans and illustrations which are used in making products.

Equation: Creates a mathematical relation between sketch dimensions, using dimension names as variables, or between feature parameters, such as the depth of an extruded feature or the instance count in a pattern.

Exploded view: A configuration in an assembly that displays its components separated from one another.

Export: The process to save a SOLIDWORKS document in another format for use in other CAD/CAM, rapid prototyping, web or graphics software applications.

Extension Line: The line extending from the profile line indicating the point from which a dimension is measured.

Extruded Cut Feature: Projects a sketch perpendicular to a Sketch plane to remove material from a part.

Face: A selectable area (planar or otherwise) of a model or surface with boundaries that help define the shape of the model or surface. For example, a rectangular solid has six faces.

Family Cell: A named empty cell in a Design Table that indicates the start of the evaluated parameters and configuration names. Locate Comments in a Design Table to the left or above the Family Cell.

Fasteners: Includes Bolts and nuts (threaded), Set screws (threaded), Washers, Keys, and Pins to name a few. Fasteners are not a permanent means of assembly such as welding or adhesives.

Feature: Features are geometry building blocks. Features add or remove material. Features are created from 2D or 3D sketched profiles or from edges and faces of existing geometry.

FeatureManager: The FeatureManager design tree located on the left side of the SOLIDWORKS window provides an outline view of the active part, assembly, or drawing. This makes it easy to see how the model or assembly was constructed or to examine the various sheets and views in a drawing. The FeatureManager and the Graphics window are dynamically linked. You can select features, sketches, drawing views and construction geometry in either pane.

First Angle Projection: In First Angle Projection the Top view is looking at the bottom of the part. First Angle Projection is used in Europe and most of the world. However, America and Australia use a method known as Third Angle Projection.

Fully defined: A sketch where all lines and curves in the sketch, and their positions, are described by dimensions or relations, or both, and cannot be moved. Fully defined sketch entities are shown in black.

Foreshortened radius: Helpful when the centerpoint of a radius is outside of the drawing or interferes with another drawing view: Broken Leader.

Foreshortening: The way things appear to get smaller in both height and depth as they recede into the distance.

French curve: A template made out of plastic, metal or wood composed of many different curves. It is used in manual drafting to draw smooth curves of varying radii.

Fully Defined: A sketch where all lines and curves in the sketch, and their positions, are described by dimensions or relations, or both, and cannot be moved. Fully defined sketch entities are displayed in black.

Geometric Tolerance: A set of standard symbols that specify the geometric characteristics and dimensional requirements of a feature.

Glass Box method: A traditional method of placing an object in an *imaginary glass box* to view the six principle views.

Global Coordinate System: Directional input refers by default to the Global coordinate system (X-, Y- and Z-), which is based on Plane1 with its origin located at the origin of the part or assembly.

Graphics Window: The area in the SOLIDWORKS window where the part, assembly, or drawing is displayed.

Grid: A system of fixed horizontal and vertical divisions.

Handle: An arrow, square or circle that you drag to adjust the size or position of an entity such as a view or dimension.

Heads-up View Toolbar: A transparent toolbar located at the top of the Graphic window.

Height: The vertical distance between two or more lines or surfaces (features) which are in horizontal planes.

Hidden Lines Removed (HLR): A view mode. All edges of the model that are not visible from the current view angle are removed from the display.

Hidden Lines Visible (HLV): A view mode. All edges of the model that are not visible from the current view angle are shown gray or dashed.

Hole Callouts: Hole callouts are available in drawings. If you modify a hole dimension in the model, the callout updates automatically in the drawing if you did not use DimXpert.

Hole Table: A table in a drawing document that displays the positions of selected holes from a specified origin datum. The tool labels each hole with a tag. The tag corresponds to a row in the table.

Import: The ability to open files from other software applications into a SOLIDWORKS document. The A-size sheet format was created as an AutoCAD file and imported into SOLIDWORKS.

Isometric Projection: A form of graphical projection, more specifically, a form of axonometric projection. It is a method of visually representing three-dimensional objects in two dimensions, in which the three coordinate axes appear equally foreshortened and the angles between any two of them are 120º.

Layers: Simplifies a drawing by combining dimensions, annotations, geometry and components. Properties such as display, line style and thickness are assigned to a named layer.

Leader: A solid line created from an annotation to the referenced feature.

Line Format: A series of tools that controls Line Thickness, Line Style, Color, Layer and other properties.

Local (Reference) Coordinate System: Coordinate system other than the Global coordinate system. You can specify restraints and loads in any desired direction.

Lock Sheet Focus: Adds sketch entities and annotations to the selected sheet. Double-click the sheet to activate Lock Sheet Focus. To unlock a sheet, right-click and select Unlock Sheet Focus or double click inside the sheet boundary.

Lock View Position: Secures the view at its current position in the sheet. Right-click in the drawing view to Lock View Position. To unlock a view position, right-click and select Unlock View Position.

Mass Properties: The physical properties of a model based upon geometry and material.

Menus: Menus provide access to the commands that the SOLIDWORKS software offers. Menus are Context-sensitive and can be customized through a dialog box.

Model Item: Provides the ability to insert dimensions, annotations, and reference geometry from a model document (part or assembly) into a drawing.

Model View: A specific view of a part or assembly. Standard named views are listed in the view orientation dialog box such as isometric or front. Named views can be user-defined names for a specific view.

Model: 3D solid geometry in a part or assembly document. If a part or assembly document contains multiple configurations, each configuration is a separate model.

Motion Studies: Graphical simulations of motion and visual properties with assembly models. Analogous to a configuration, they do not actually change the original assembly model or its properties. They display the model as it changes based on simulation elements you add.

Mouse Buttons: The left, middle, and right mouse buttons have distinct meanings in SOLIDWORKS. Use the middle mouse button to rotate and Zoom in/out on the part or assembly document.

Oblique Projection: A simple type of graphical projection used for producing pictorial, two-dimensional images of three-dimensional objects.

OLE (Object Linking and Embedding): A Windows file format. A company logo or EXCEL spreadsheet placed inside a SOLIDWORKS document are examples of OLE files.

Ordinate Dimensions: Chain of dimensions referenced from a zero ordinate in a drawing or sketch.

Origin: The model origin is displayed in blue and represents the (0,0,0) coordinate of the model. When a sketch is active, a sketch origin is displayed in red and represents the (0,0,0) coordinate of the sketch. Dimensions and relations can be added to the model origin but not to a sketch origin.

Orthographic Projection: A means of representing a three-dimensional object in two dimensions. It is a form of parallel projection, where the view direction is orthogonal to the projection plane, resulting in every plane of the scene appearing in affine transformation on the viewing surface.

Parametric Note: A Note annotation that links text to a feature dimension or property value.

Parent View: A Parent view is an existing view on which other views are dependent.

Part Dimension: Used in creating a part, they are sometimes called construction dimensions.

Part: A 3D object that consists of one or more features. A part inserted into an assembly is called a component. Insert part views, feature dimensions and annotations into 2D drawing. The extension for a SOLIDWORKS part filename is .SLDPRT.

Perspective Projection: The two most characteristic features of perspective are that objects are drawn smaller as their distance from the observer increases and foreshortened, the size of an object's dimensions along the line of sight are relatively shorter than dimensions across the line of sight.

Plane: To create a sketch, choose a plane. Planes are flat and infinite. Planes are represented on the screen with visible edges.

Precedence of Line Types: When obtaining orthographic views, it is common for one type of line to overlap another type. When this occurs, drawing conventions have established an order of precedence.

Precision: Controls the number of decimal places displayed in a dimension.

Projected View: Projected views are created for Orthogonal views using one of the following tools: Standard 3 View, Model View or the Projected View tool from the View Layout toolbar.

Properties: Variables shared between documents through linked notes.

PropertyManager: Most sketch, feature, and drawing tools in SOLIDWORKS open a PropertyManager located on the left side of the SOLIDWORKS window. The PropertyManager displays the properties of the entity or feature so you specify the properties without a dialog box covering the Graphics window.

RealView: Provides a simplified way to display models in a photo-realistic setting using a library of appearances and scenes. RealView requires graphics card support and is memory intensive.

Rebuild: A tool that updates (or regenerates) the document with any changes made since the last time the model was rebuilt. Rebuild is typically used after changing a model dimension.

Reference Dimension: Dimensions added to a drawing document are called Reference dimensions and are driven; you cannot edit the value of reference dimensions to modify the model. However, the values of reference dimensions change when the model dimensions change.

Relation: A relation is a geometric constraint between sketch entities or between a sketch entity and a plane, axis, edge or vertex.

Relative view: The Relative View defines an Orthographic view based on two orthogonal faces or places in the model.

Revision Table: The Revision Table lists the Engineering Change Orders (ECO), in a table form, issued over the life of the model and the drawing. The current Revision letter or number is placed in the Title block of the Drawing.

Right-Hand Rule: Is a common mnemonic for understanding notation conventions for vectors in 3 dimensions.

Rollback: Suppresses all items below the rollback bar.

Scale: A relative term meaning "size" in relationship to some system of measurement.

Section Line: A line or centerline sketched in a drawing view to create a section view.

Section Scope: Specifies the components to be left uncut when you create an assembly drawing section view.

Section View: You create a section view in a drawing by cutting the parent view with a cutting, or section line. The section view can be a straight cut section or an offset section defined by a stepped section line. The section line can also include concentric arcs. Create a Section View in a drawing by cutting the Parent view with a section line.

Sheet Format: A document that contains the following: page size and orientation, standard text, borders, logos, and Title block information. Customize the Sheet format to save time. The extension for the Sheet format filename is .SLDDRT.

Sheet Properties: Sheet Properties display properties of the selected sheet. Sheet Properties define the following: Name of the Sheet, Sheet Scale, Type of Projection (First angle or Third angle), Sheet Format, Sheet Size, View label, and Datum label.

Sheet: A page in a drawing document.

Silhouette Edge: A curve representing the extent of a cylindrical or curved face when viewed from the side.

Sketch: The name to describe a 2D profile is called a sketch. 2D sketches are created on flat faces and planes within the model. Typical geometry types are lines, arcs, corner rectangles, circles, polygons, and ellipses.

Spline: A sketched 2D or 3D curve defined by a set of control points.

Stacked Balloon: A group of balloons with only one leader. The balloons can be stacked vertically (up or down) or horizontally (left or right).

Standard views: The three orthographic projection views, Front, Top and Right positioned on the drawing according to First angle or Third angle projection.

Suppress: Removes an entity from the display and from any calculations in which it is involved. You can suppress features, assembly components, and so on. Suppressing an entity does not delete the entity; you can unsuppress the entity to restore it.

Surface Finish: An annotation that represents the texture of a part.

System Feedback: Feedback is provided by a symbol attached to the cursor arrow indicating your selection. As the cursor floats across the model, feedback is provided in the form of symbols riding next to the cursor.

System Options: System Options are stored in the registry of the computer. System Options are not part of the document. Changes to the System Options affect all current and future documents. There are hundreds of Systems Options.

Tangent Edge: The transition edge between rounded or filleted faces in hidden lines visible or hidden lines removed modes in drawings.

Task Pane: The Task Pane is displayed when you open the SOLIDWORKS software. It contains the following tabs: SOLIDWORKS Resources, Design Library, File Explorer, Search, View Palette, Document Recovery and RealView/PhotoWorks.

Templates: Templates are part, drawing and assembly documents that include user-defined parameters and are the basis for new documents.

Third Angle Projection: In Third angle projection the Top View is looking at the Top of the part. First Angle Projection is used in Europe and most of the world. America and Australia use the Third Angle Projection method.

Thread Class or Fit: Classes of fit are tolerance standards; they set a plus or minus figure that is applied to the pitch diameter of bolts or nuts. The classes of fit used with almost all bolts sized in inches are specified by the ANSI/ASME Unified Screw Thread standards (which differ from the previous American National standards).

Thread Lead: The distance advanced parallel to the axis when the screw is turned one revolution. For a single thread, lead is equal to the pitch; for a double thread, lead is twice the pitch.

Tolerance: The permissible range of variation in a dimension of an object. Tolerance may be specified as a factor or percentage of the nominal value, a maximum deviation from a nominal value, an explicit range of allowed values, specified by a note or published standard with this information, or implied by the numeric accuracy of the nominal value.

Toolbars: The toolbar menus provide shortcuts enabling you to access the most frequently used commands. Toolbars are Context-sensitive and can be customized through a dialog box.

T-Square: A technical drawing instrument, primarily a guide for drawing horizontal lines on a drafting table. It is used to guide the triangle that draws vertical lines. Its name comes from the general shape of the instrument where the horizontal member of the T slides on the side of the drafting table. Common lengths are 18", 24", 30", 36" and 42".

Under-defined: A sketch is under defined when there are not enough dimensions and relations to prevent entities from moving or changing size.

Units: Used in the measurement of physical quantities. Decimal inch dimensioning and Millimeter dimensioning are the two types of common units specified for engineering parts and drawings.

Vertex: A point at which two or more lines or edges intersect. Vertices can be selected for sketching, dimensioning, and many other operations.

View Palette: Use the View Palette, located in the Task Pane, to insert drawing views. It contains images of standard views, annotation views, section views, and flat patterns (sheet metal parts) of the selected model. You can drag views onto the drawing sheet to create a drawing view.

Weld Bead: An assembly feature that represents a weld between multiple parts.

Weld Finish: A weld symbol representing the parameters you specify.

Weld Symbol: An annotation in the part or drawing that represents the parameters of the weld.

Width: The horizontal distance between surfaces in profile planes. In the machine shop, the terms length and width are used interchangeably.

Zebra Stripes: Simulate the reflection of long strips of light on a very shiny surface. They allow you to see small changes in a surface that may be hard to see with a standard display.

Index